T0418687

.

Chemistry of 2-Oxoaldehydes and 2-Oxoacids

Chemistry of 2-Oxoaldehydes and 2-Oxoacids

Atul Kumar
Natural Product and Medicinal Chemistry Division, CSIR-Indian Institute of Integrative Medicine, Jammu, India

Javeed Rasool
Natural Product and Medicinal Chemistry Division, CSIR-Indian Institute of Integrative Medicine, Jammu, India

Qazi Naveed Ahmed
Natural Product and Medicinal Chemistry Division, CSIR-Indian Institute of Integrative Medicine, Jammu, India

ELSEVIER

Elsevier
Radarweg 29, PO Box 211, 1000 AE Amsterdam, Netherlands
The Boulevard, Langford Lane, Kidlington, Oxford OX5 1GB, United Kingdom
50 Hampshire Street, 5th Floor, Cambridge, MA 02139, United States

Notices
Knowledge and best practice in this field are constantly changing. As new research and experience broaden our understanding, changes in research methods, professional practices, or medical treatment may become necessary.

Practitioners and researchers must always rely on their own experience and knowledge in evaluating and using any information, methods, compounds, or experiments described herein. In using such information or methods they should be mindful of their own safety and the safety of others, including parties for whom they have a professional responsibility.

To the fullest extent of the law, neither the Publisher nor the authors, contributors, or editors, assume any liability for any injury and/or damage to persons or property as a matter of products liability, negligence or otherwise, or from any use or operation of any methods, products, instructions, or ideas contained in the material herein.

British Library Cataloguing-in-Publication Data
A catalogue record for this book is available from the British Library

Library of Congress Cataloging-in-Publication Data
A catalog record for this book is available from the Library of Congress

ISBN: 978-0-12-824285-8

For Information on all Elsevier publications
visit our website at https://www.elsevier.com/books-and-journals

Publisher: Susan Dennis
Editorial Project Manager: Hillary Carr
Production Project Manager: Bharatwaj Varatharajan
Cover Designer: Matthew Limbert

Typeset by MPS Limited, Chennai, India

Working together
to grow libraries in
developing countries

www.elsevier.com • www.bookaid.org

Contents

Chapter 1

Synthesis and Physical Properties of 2-Oxoaldehydes and 2-Oxoacids

1.1 Introduction of 2-Oxoaldehydes

Carbonyl functionality holds immense significance in organic chemistry and has been developed and employed in fundamental chemistry for more than a century. The ultimate chemistry of carbonyl functional group continues to pave its way in achieving never before objectives through its applications on various scientific platforms.[1] Basically, the carbonyl group is composed of a carbon double-bonded to an oxygen atom in a planer geometry. The carbonyl carbon bears sp^2 hybridization that makes it susceptible to enormous chemical changes. The carbonyl functional group is further subdivided mainly into ketone and aldehyde, depending on the substitution of its electronically deficient central carbon atom.

The common perspectives of both these functionalities have been particularly known on the basis of their fundamental nature. However, there is a lack of sense in understanding their phenomenal divergent chemical character when the two carbonyl groups come adjacent to each other. In this context, 2-oxoaldehydes **1** and 2-oxoacids **2** play a very crucial role among 1,2-dicarbonyl precursors, which could build our vision on their divergent chemistry (Fig. 1.1).[2]

In both of these scaffolds, aldehyde and ketone functionality co-exist at the adjacent positions and exhibit reactions in a number of unusual ways. Thus, the existence of more electron-withdrawing character in 2-oxoaldehydes renders them more reactive as compared with their simple benzaldehyde or acetaldehyde counterparts. 2-Oxoaldehydes easily captures water and are normally present in unstable hydrate form as compared to simple aldehydes (Fig. 1.2).[3]

In synthetic organic chemistry, aldehydes in general and 2-oxoaldehydes in particular, are among the key precursors widely launched for N−C, N−N, N−O, O−C bond formation strategies.[4] The significance of 2-oxoaldehydes also arises because of their synthetic utility in protecting groups, chiral auxiliaries, reaction intermediates, and catalysis.[5] Therefore, there is an open space for the development of new strategies to access novel heterocycles by

Chemistry of 2-Oxoaldehydes and 2-Oxoacids. DOI: https://doi.org/10.1016/B978-0-12-824285-8.00005-4

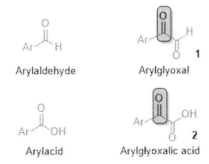

FIGURE 1.1   The general structure of 2-oxoaldehydes and 2-oxoacids.

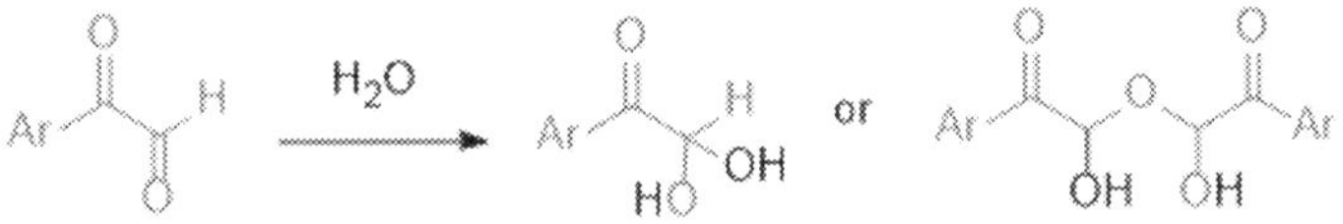

FIGURE 1.2   Hydrate forms of 2-oxoaldehydes.

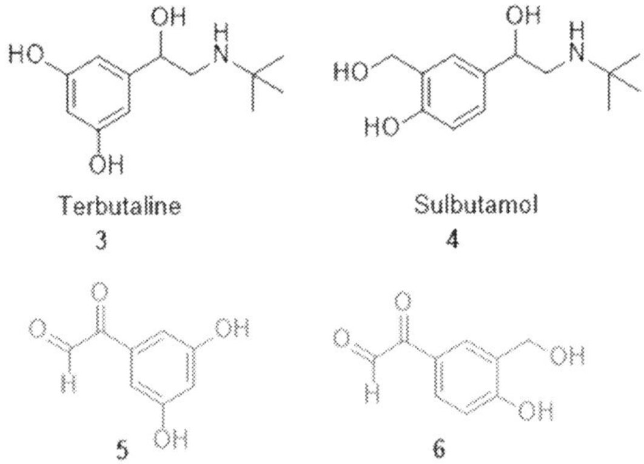

Terbutaline
3

Sulbutamol
4

5

6

FIGURE 1.3   Biologically active scaffolds.

exploiting 2-oxoaldehydes.[6] The aldehyde and ketone functional group of arylglyoxal can be converted into various other useful groups such as hydroxyl, carboxylic acid, and ester which are smart chemical entities in organic synthetic chemistry.[7] Along these lines, especially the co-existence of adjacent aldehyde and ketone functional groups in 2-oxoaldehyde (AG) makes it more demanding in synthetic and biological chemistry.[8]

In biology, phenylglyoxal (PG) as a reagent, is useful for the selective chemical modification of arginine residues in proteins.[9] PGs have a special role owing to the synthesis of bronchodilators such as Terbutaline **3** and Salbutamol **4** which are obtained from 3,5-dihydroxyphenylglyoxal **5** and 4-hydroxy-3 hydroxymethylphenylglyoxal **6** (Fig. 1.3).

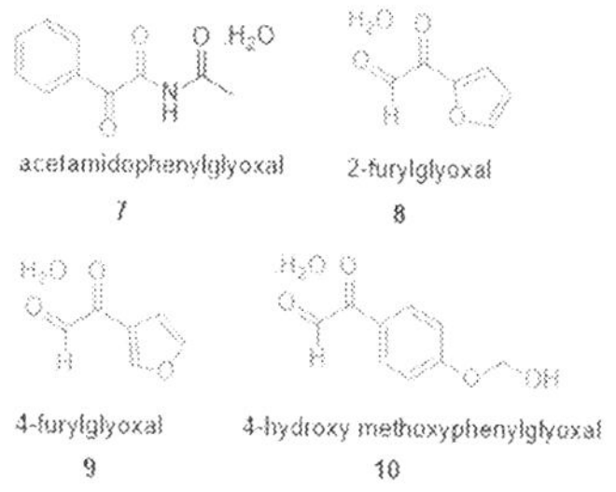

acetamidophenylglyoxal
7

2-furylglyoxal
8

4-furylglyoxal
9

4-hydroxy methoxyphenylglyoxal
10

FIGURE 1.4 Importance of AGs in medicinal chemistry.

Moreover, AG-hydrates, like acetamidophenylglyoxal hydrate **7**, 2- and 4-furylglyoxal hydrate **8** and **9**, 4-hydroxy methoxyphenylglyoxal hydrate **10** exhibit antiviral activities in embryonated egg including influenza (PR-8) and Newcastle disease (NJKD strain) viruses.[10] These examples illustrate the importance of AGs as synthetic tools in medicinal chemistry (Fig. 1.4).

Although, the initial work pertaining to the synthesis of heterocycles and their derivatives form 2-oxoaldehydes is classical and their developments have increased in recent years.[11] Therefore, keeping in view their dire significance, the aim is to cover various core aspects of chemistry related to 2-oxoaldehydes **1** and 2-oxoacids **2** in the form of a standard text. This text will highlight the reactions and applications related to oxoaldehydes and 2-oxoacids in the synthesis of various compounds and other aspects like medicine and biology along with some miscellaneous work on the ligand and material chemistry since last a few years. This work will also be articulated around their functional groups both participating separately or in combination as their reactivity, reaction pathways, and other domains are concerned. Different types of common and named reactions such as cyclo condensation, Ugi cyclo condensation, Ugi-Wittig, Wittig dehydrative cyclization, Aldol-Paal-Knorr cycloaddition, Pictet- Spengler, and so on will also be described throughout the course of this text.

This chapter particularly deals with the preparatory methods of 2-oxoaldehydes and 2-oxoacids from various substrates. Many of these methods are classical; however, there are some modern developments in the synthesis of 2-oxoaldehydes and 2-oxoacids that are duly discussed in this chapter as well as in the following chapters. In addition, some of the basic properties related to 2-oxoaldehydes and 2-oxoacids such as stability, planarity, and reactivity are also highlighted.

1.2   Methods for the Preparation of 2-Oxoaldehydes

Due to the broad significance of 2-oxoaldehydes in synthetic as well as in medicinal chemistry, various synthetic procedures are reported in the literature for the preparation of 2-oxoaldehyde or arylglyoxals (AGs).

In a first report, 2-oxoaldehydes were prepared from the sulfite derivative of the oxime **5** in a thermal decomposition process. The addition of excess water regenerates the corresponding aldehydes **7** from the sulfite salt **6** (Fig. 1.5).[11]

In addition, 2-oxoaldehydes were synthesized from aryl methyl ketones in the presence of SeO_2 as an oxidizing reagent in 1,4-dioxane-H_2O (9:1) under reflux conditions (Fig. 1.6).[12]

Apart from SeO_2, aryl methyl ketones are oxidized to arylglyoxals (AGs) in presence of selenious acid (H_2SeO_3) (Fig. 1.7). Similarly, treatment of aryl and heteroaryl methyl ketones with an excess amount of $(NH_4)_2S_2O_8$ and catalytic amounts of $(PhSe)_2$ in MeOH under reflux conditions afford 2-oxoaldehyde-acetals in 60−95% yields (Fig. 1.8).[13]

Phenacyl bromides **11** are also used as starting materials for the synthesis of various substituted 2-oxoaldehydes or arylglyoxals (AGs). With DMSO at room temperature with 48-95% conversion in 4h.[14] The reaction of **11** with $AgNO_3$ in acetonitrile yields nitrate esters that further react with NaOAc in DMSO to afford AGs at room temperature in 82−86% yields.[15] In a similar

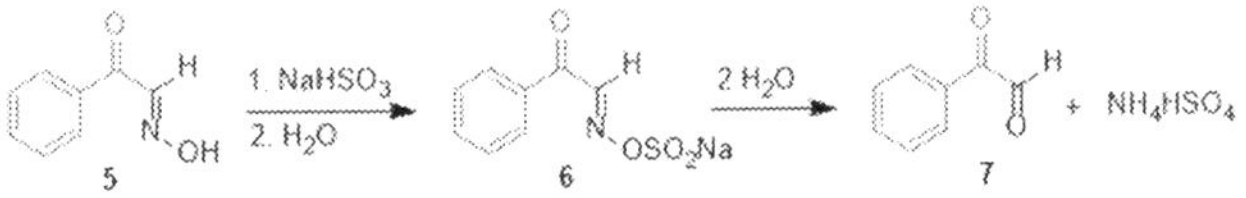

FIGURE 1.5 Synthesis of 2-oxoaldehyde from sulfite derivative.

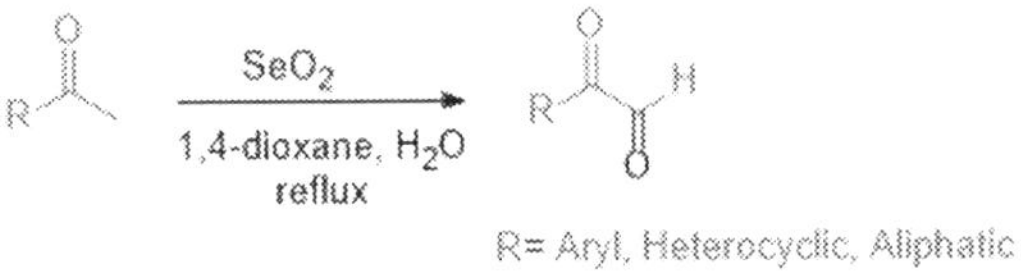

FIGURE 1.6 Synthesis of 2-oxoaldehyde from aryl ketones.

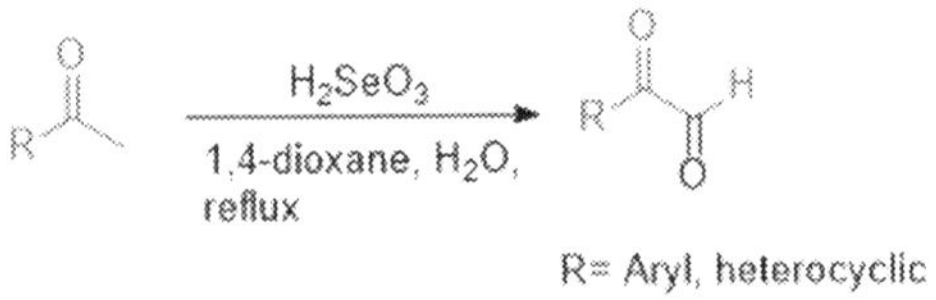

FIGURE 1.7 Synthesis of 2-oxoaldehyde from aryl and heteroaryl ketones.

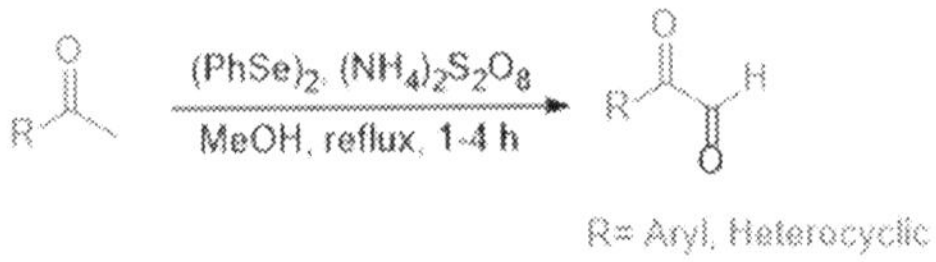

FIGURE 1.8 Synthesis of 2-oxoaldehyde from aryl ketones.

manner, 2-oxoaldehydes are formed in 55−90% yields, when **11** was refluxed with *N,N*-diethylhydroxylamine in MeOH for 2 hours (Fig. 1.9).[16]

The formation of 2-oxoaldehydes also takes place under mild conditions by the reaction of **11** and α-picoline N-oxide at 0°C, followed by the addition of Na_2CO_3 in water.[17]

Phenyl acetylenes **12** and **13** are also used for the synthesis of 2-oxoaldehydes in different optimized conditions. 2-oxoaldehyde synthesis is achieved by the reaction of phenyl acetylenes with $(HMPA)MoO(O_2)_2$, a metal-peroxide complex, with $Hg(OAc)_2$ at 0°C in DCE (Fig. 1.10).[18] In yet another strategy, oxidation of phenylacetylene with NBS in DMSO at room temperature easily affords 2-oxoaldehydes in 86−90% yield. (Fig. 1.11).[19]

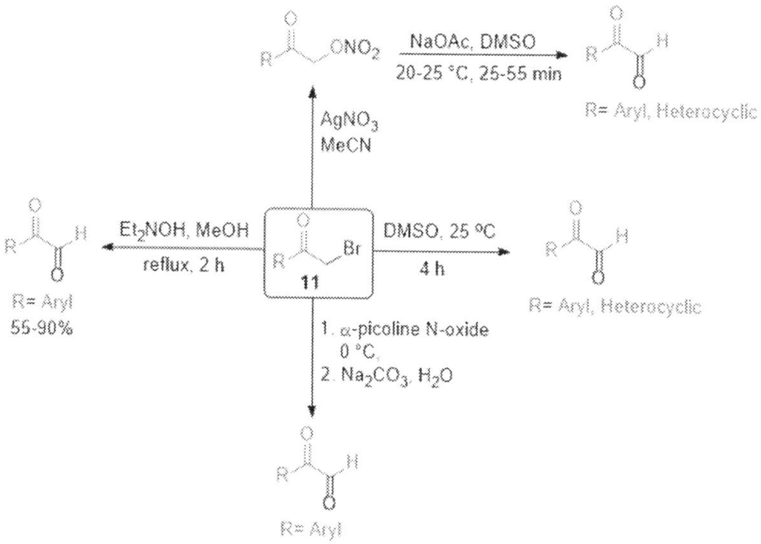

FIGURE 1.9   Synthesis of 2-oxoaldehyde from phenacyl bromide.

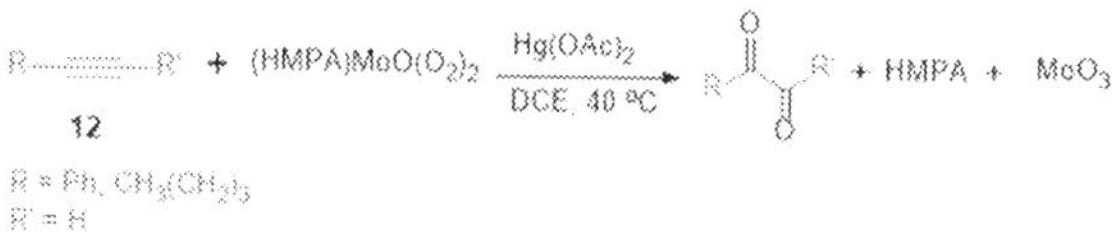

FIGURE 1.10   Synthesis of 2-oxoaldehyde from internal alkynes.

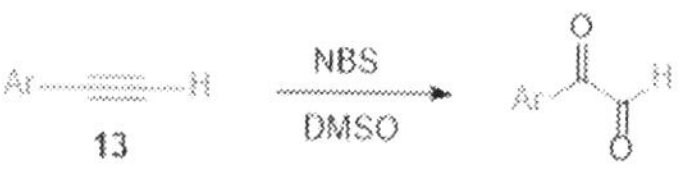

FIGURE 1.11   Synthesis of 2-oxoaldehyde from phenyl acetylenes.

Methyl benzoates **14** are common reagents and readily available precursors that react with $KCH_2S(O)CH_3$ in presence of K^tOBu, followed by the addition of aq. HCl to form intermediate **A** which undergoes oxidation with $Cu(OAc)_2$ to furnish phenyl glyoxal.(Fig. 1.12).[20]

The hydrated form of 2-oxoaldehydes **15** can be obtained through oxidation of α-diazo ketones **16** driven by DMDO in acetone. The hydrated form easily loses water molecule upon heating and regenerates anhydrous 2-oxoaldehyde. The products are obtained in pure form without further purification. This is yet another important oxidative process in which the synthesis of crucial heterocyclic derivatives of 2-oxoaldehydes like 2-furyl, 2-thienyl, 2-pyridyl, and 3-pyridyl-2-oxoaldehydes are obtained (Fig. 1.13).[21]

2-Oxoaldehydes are obtained in high yields when oxidation of acetophenone **17** takes place with aqueous HBr in DMSO. The same product is obtained when the reaction is carried out with SeO_2 and aqueous nitric acid in dioxane or ethanol. Selenium metal formed in the reaction is reconverted to SeO_2 by nitric acid. Selenium dioxide, thus, is used as a catalyst instead of a stoichiometric oxidizing agent in its redox cycle (Fig. 1.14).[22]

Acetal forms of 2-oxoaldehydes have also been utilized for the synthesis of different derivatives of 2-oxoaldehyde. For example, diethyl acetal form of p-N,N-dimethylaminophenylglyoxal **19** generates p-N,N-dimethylaminophenylglyoxal **20** by simple hydrolysis. In this case, **19** was previously

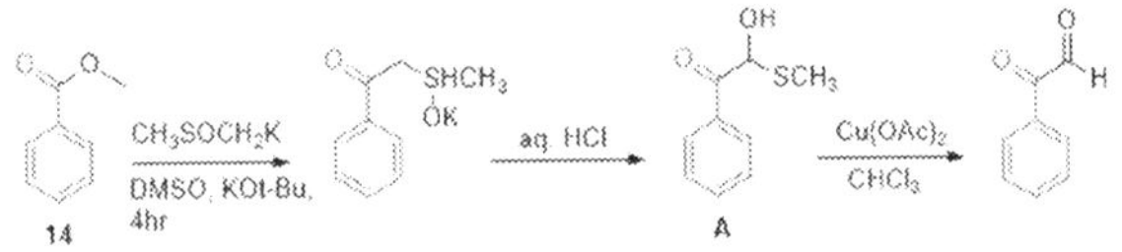

FIGURE 1.12 Synthesis of 2-oxoaldehyde from methyl benzoates.

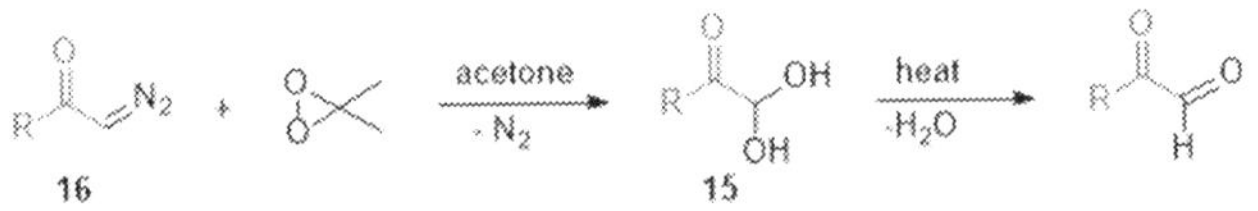

FIGURE 1.13 Synthesis of 2-oxoaldehyde from diazoketones.

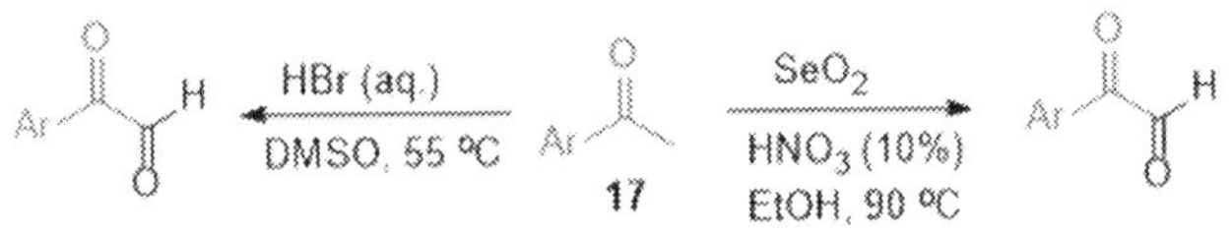

FIGURE 1.14 Synthesis of 2-oxoaldehyde from acetophenoe.

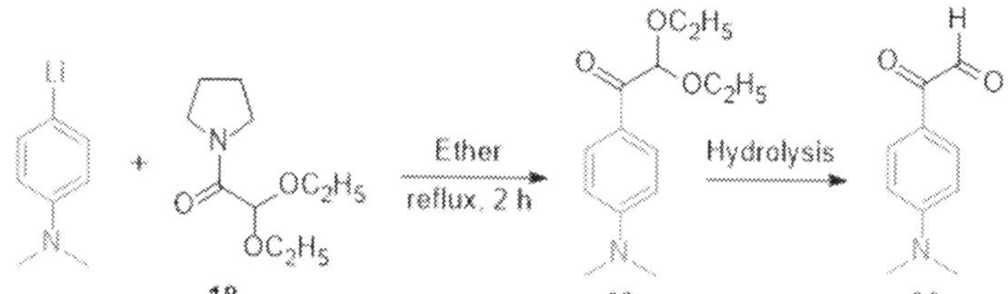

FIGURE 1.15   Synthesis of 2-oxoaldehyde from acetals.

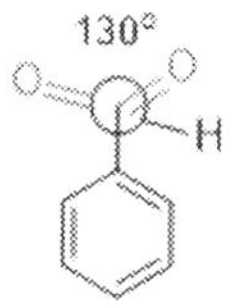

FIGURE 1.16   The torsional angle between two C=O groups.

obtained from the reaction between p-$(Me_2N)C_6H_4Li$ and diethoxy acetyl piperidine **18** (Fig. 1.15).[23]

1.3   Physical Properties and Thermal Stability of 2-Oxoaldehydes

Among the simplest arylglyoxals (AG), phenylglyoxal (PG) is a yellow liquid that polymerizes with the passage of time. Its crystalline hydrate form is colorless and unstable among simple aldehydes. However, upon heating, the polymer returns to its original yellow form. The arylglyoxal-hydrate may contain one or half molecule of water, however, upon heating hydrated form loses its fraction of water and regenerates anhydrous arylglyoxal.[3]

1.4   Reactivity of 2-oxoaldehydes

Gas-phase electron diffraction conformational studies reveal the two different carbonyl groups in PG are nonplanar in nature in which the torsional angle between two carbonyl groups OC−CO corresponds to 130 degrees while CO and phenyl ring lie in a coplanar fashion (Fig. 1.16).[24]

The two adjacent carbonyl motifs incorporated in arylglyoxal (AGs) are contrasted in their chemical nature owing to the more electron pulling capacity due to the presence of an additional carbonyl group as compared with that of the single carbonyl group.

The comparative reactivity of AG corresponding to the aldehyde fragment is more than the rest part of molecule and its simple aromatic benzaldehyde counterpart due to the existence of adjacent electron-withdrawing ketone functionality. Thus, due to the more reactivity of the aldehyde unit of AG, it reacts rapidly with different nucleophiles and can further cyclize with ketone moiety of the arylglyoxal to construct different aryl heterocycles by one or two carbon atoms, respectively.

From a pericyclic point of view, AGs and arylglyoxal-imines are key dienophiles in [2 + 2] and [4 + 2] cycloadditions via C=N and CO bonds as entry points, respectively. In addition, AG-hydrates work as nucleophiles through the OH group.

1.5 Introduction of 2-Oxoacids

The α-keto acid or 2-oxoacids **2** is distinguished by the presence of an additional keto moiety at the α-position of the carboxylic acid group **2′** (Fig. 1.17). Keeping in view their significance, numerous aryl, heteroaryl, alkyl, or H-based α-keto acids have been synthesized since last a few years.[25]

Waters and co-workers have made significant contributions in this field, in order to address their preparation and physical and chemical properties.[26] Many more detailed revisions were summarized by Cooper and co-workers[27] in order to access α-keto acids and their related properties along with some of their basic applications.

The development of α-keto acid or α-oxocaboxylic acid chemistry has shown a significant role in organic synthesis and medicinal chemistry such as cross-coupling reactions, acylating reagents, and so on. The role of α-keto acids in organic synthesis initially started on the basis of condensation and esterification reactions.[28] However, in order to construct new molecular scaffolds, the reactivity of carbonyl chemistry in α-keto acid was understood and exploited by targeting electron-deficient carbonyl unit by O-, N-, and S-based nucleophiles.[29]

The α-keto acids are incredibly important as acylating agents like monoacylation of pyridines catalyzed by silver through an acyl radical intermediate.[30] They are also important in the formation of C—C, C—N, and C—S bondbased heterocycles. In addition, the use of α-keto acids as acylating agents is a much better, more reliable, and greener approach than using carbonyl diimidazole (CDI),[31] anhydrides, acyl halides, and thioesters,[32] which produce waste and toxic substances in the reaction. While employing α-keto acids no additives or activating agents such as

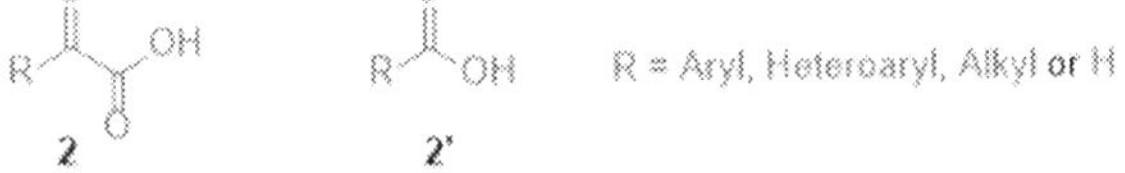

FIGURE 1.17 General structure of 2-oxoacids and acids.

DCC, DIC, HATU, PyBOP, and so on are required that generate adverse toxic materials (Fig. 1.18). In a similar manner, transition metal-catalyzed α-keto acids driven decarboxylative cross-coupling reactions through radical intermediates,[33] for example, Pd-catalyzed cross-coupling of α- keto acids in acylation reactions[34] that will be discussed in the coming chapters.

Acylating agent	Name	Waste material
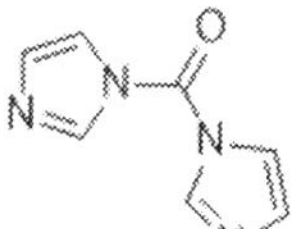	(CDI) Carbodiimidazole	Imidazole
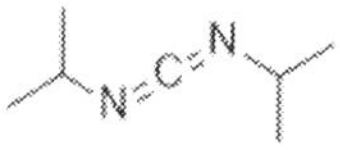	(DIC) Diisopropylcarbodiimide	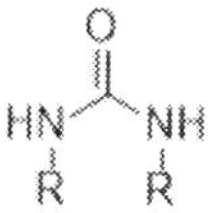
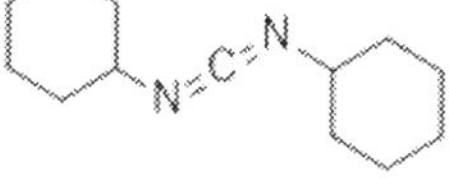	(DCC) Dicyclohexylcarbodiimide	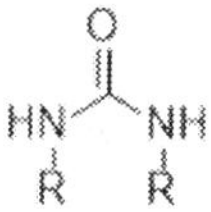
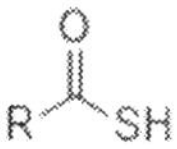	Thiocarboxylic acid	RSH
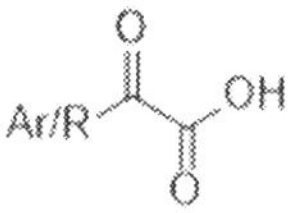	Glyoxalic acid	CO_2
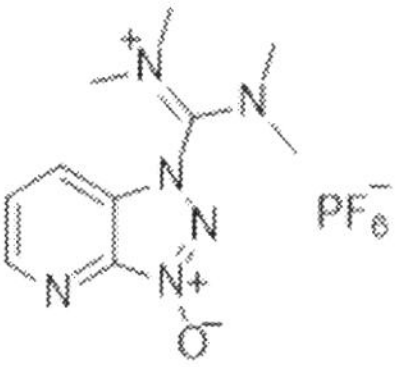	HATU Hexafluorophosphate Azabenzotriazole Tetramethyl Uronium	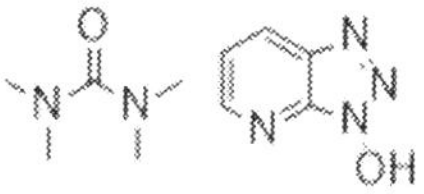
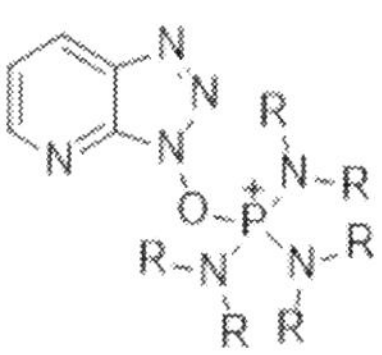	PyBOP (benzotriazolyloxy-tris [pyrrolidino]- phosphonium hexafluorophosphat)	PyBOP, R= $(CH_2)_2NH$ 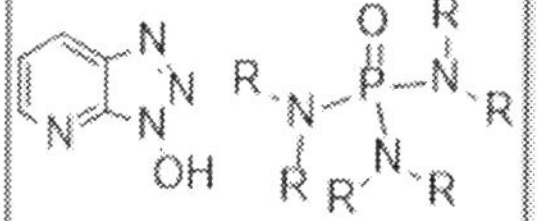

FIGURE 1.18   Reagents or additives in some acylation reactions.

In biological systems, α-keto acid is essential in energy delivery processes to the cells.[35] Along these lines, an anionic pyruvic acid **21** derivative is the key metabolite in a series of enzyme-catalyzed reactions that generate ATP in the Krebs pathway. **21** is basically involved in de facto acylation of acetyl CoA **22** in a decarboxylation process catalyzed by pyruvate dehydrogenase via CoA−SH and NAD$^+$. Oxaloacetate dianion **23** generated from **21** is another key intermediate in the Krebs cycle (Fig. 1.19).[36]

As mentioned earlier, α-ketoacid **2** exclusively being a different chemical entity as compared with carboxylic acid **2′**; both of them cannot be treated on the same platform because of their divergent reactivity. Chemists, therefore, have devoted substantial efforts to studying their preparation, chemical reactions, and physical properties. Thus the aim of the current and following chapters is to highlight the fundamental and divergent chemistry related to α-keto acids. In addition to the application in medicine, synthetic methods including cyclization reactions, acylations of aryl, and alkyl substrates are also elaborated in this chapter.

1.6 Methods for the Preparation of 2-Oxoacids

There are various classical and recent approaches for the synthesis of different α-keto acids by using a variety of precursors. Most of these reactions were carried out in the first half of the 20th century and the common reaction for the synthesis of α-keto acids in multigram scale are those reactions involving the oxidation of C−H and C−C bonds by using strong oxidizing reagents.

In 1891, $KMnO_4$ was first used by Claus and Neukranz for the transformation of acetophenones into α-ketoacids **2**.[37] This oxidation of acetophenone was carried in the presence of base (KOH) to form α-phenyl glyoxylic acid **2** (PGA) in 70% yields (Fig. 1.20).

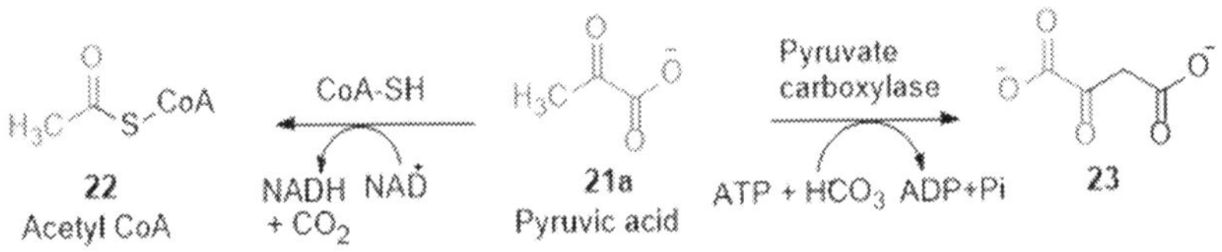

FIGURE 1.19 Decarboxylative acylation of CoA−SH catalyzed by pyruvate dehydrogenase.

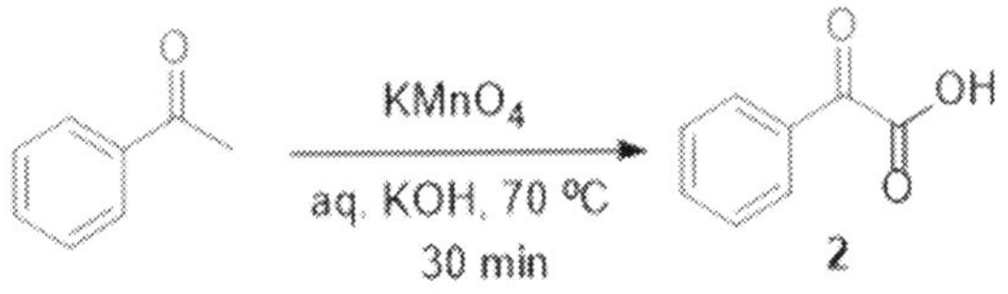

FIGURE 1.20 Synthesis of 2-ketoacids from aryl ketones.

The concept of using KMnO$_4$ as an oxidizing agent was further used by Hurd and coworkers in 1939 in aqueous NaOH to convert styrene **24** into α-phenyl glyoxylic acid **2** (PGA) in 55% yield (Fig. 1.21).[38]

Later on, Oakwood and Weisgerber reported a new protocol for the synthesis of α-phenyl glyoxylic acid by the hydrolysis of benzoyl cyanide **25** by using aqueous hydrochloric acid. Although this protocol is quite robust in which the formation of α-phenyl glyoxylic acid is completed in 5 days (Fig. 1.22).[39]

Although these protocols are widely used today for the synthesis of α-keto acids, they have the limitations of using strong oxidizing KMnO$_4$ in the presence of a strong base or use of strong acidic medium and longer reaction times. In 1960s, SeO$_2$ with pyridine was used as an alternative for KMnO$_4$ for the oxidation of Csp^3 − H bond in acetophenones **26** for the synthesis of different substituted α-keto acids **27** (Fig. 1.23).[40]

Recently, an ecofriendly approach was established by Crich and coworkers for the oxidation of ketones **28** to generate different aryl-substituted glyoxylic acids **29** by using iodoxybenzene (PhIO$_2$) as the oxidizing agent and fluorous seleninic acid as a catalyst that can be recovered and reused in new reactions. Different aryl-substituted glyoxylic acids were synthesized from this approach in 89 − 92% yields (Fig. 1.24).

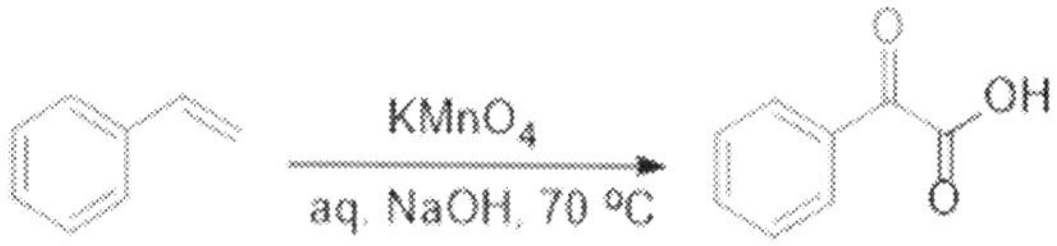

FIGURE 1.21   Synthesis of 2-ketoacids from styrene.

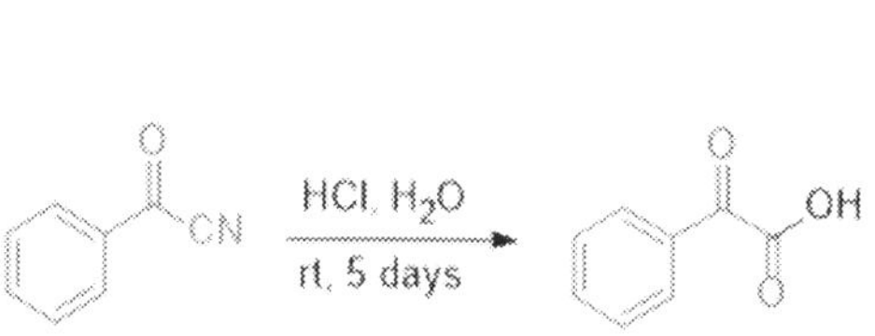

FIGURE 1.22   Synthesis of 2-ketoacids from benzoyl cyanides.

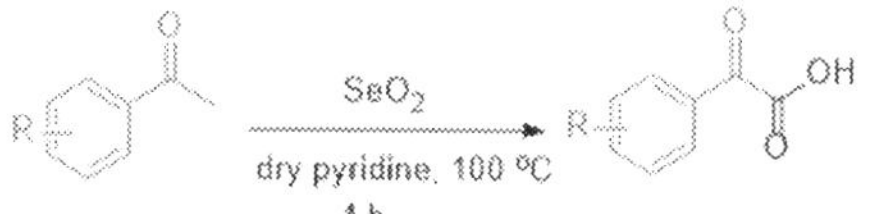

FIGURE 1.23   Synthesis of 2-ketoacids from oxidation of aryl ketones.

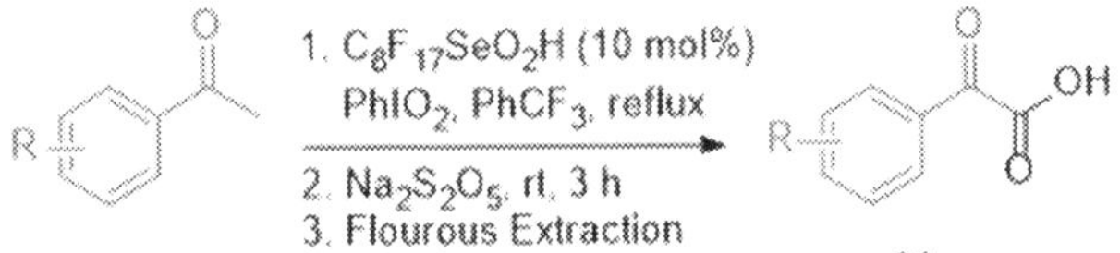

FIGURE 1.24   Synthesis of 2-ketoacids from iodo benzene and fluorous seleninic acid.

FIGURE 1.25   Structure of pyruvic acid and 2-oxobutyric acid.

1.7   Physical Properties of 2-Oxoacids

Glyoxylic acid **2**, pyruvic acid **21**, and 2-oxobutyric acid **30** are soluble in water. The long-chain and aryl-substituted acids are soluble in a variety of organic solvents, such as acetone, toluene, 1,4-dioxane, dichloromethane, NMP, dichloroethane, THF, DMF, DMA, acetonitrile, DMSO, and ethanol (Fig. 1.25). A mixture of H_2O and these organic solvents are also used in some examples in different ratios depending on the solubility of the reactants. It was observed by Waters and his group that some straight-chain and aryl derivatives of α-keto acids are decomposed to aldehydes, carboxylic acid, and oxalic acid in air or moisture at room temperature. Straight-chain, aryl- and heteroaryl α-keto acids are solid in nature. Aryl and heteroaryl α-keto acids are fairly stable at room temperature for up to six months. The physical appearance of straight-chain α-keto acids depends on their chain length. With an increase in chain length, their state varies from liquid to solid and show very similar behavior as that of fatty acids.[26]

1.8   Importance of 2-Oxoacids in Biology

The simplest and fundamental α-keto acid in the biological system is pyruvic acid **21**. Its conjugate base, pyruvate ion **21a** functions as a key intermediate in various metabolic pathways in living cells. Pyruvate **21a** is converted into carbohydrates, fatty acids, alanine, lactic acid (in animals), and ethanol (in plants). It supplies energy to living cells through Kreb's cycle in aerobic and produces lactic acid in anaerobic respiration.[41]

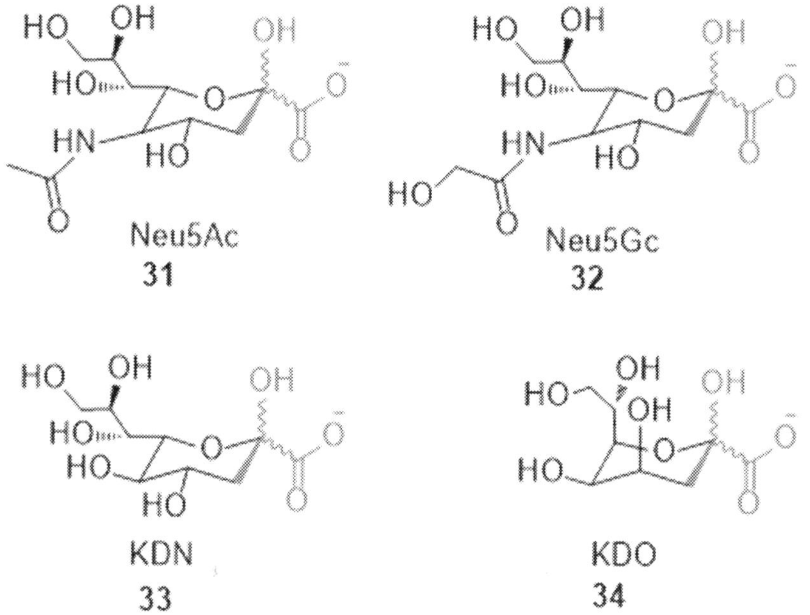

FIGURE 1.26   Structure of some sialic acids.

On the other hand, a naturally-occurring family of 9-carbon α-keto acids called sialic acids is present at the terminal sites of oligosaccharide chains on glycoproteins and glycolipids. They are the derivatives of Neu5Ac **31**, 5-glycolylneuraminic acid **32**(Neu5Gc), and deaminated neuraminic acid **33** (KDN) (Fig. 1.26). Among them, 3-deoxy-D-manno-octulosonic acid **34** (KDO) is an eight-carbon keto-deoxyacid present in some plant and prokaryotic cells. Sialic acid specifically binds to lectins and mediates cell-host attachment by microorganisms like protozoa, bacteria, viruses, and fungi.[42]

Pyruvate and mannosamines undergo a condensation reaction and create a new carbon-carbon bond to produce nine-carbon saccharides in a reversible aldol reaction catalyzed by sialic synthase or aldolase (Fig. 1.27).

2-Oxoacid dehydrogenase complexes are important metabolic checkpoints that allow sugar and amino acid metabolism. 2-oxoacid dehydrogenase based complexes like pyruvate dehydrogenase (PDHC) and 2-oxoglutarate (OGDHC) are multienzyme complexes that are important in central metabolism, macromolecular assembly, protein structure, and function. Among the three enzymes of the complexes involved, E1 is a 2-oxoacid dehydrogenase in which thiamin pyrophosphate is an essential cofactor.[43]

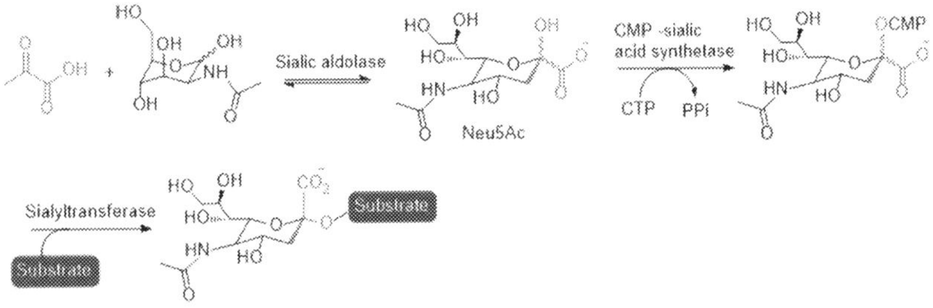

FIGURE 1.27 Biosynthesis of sialosides.

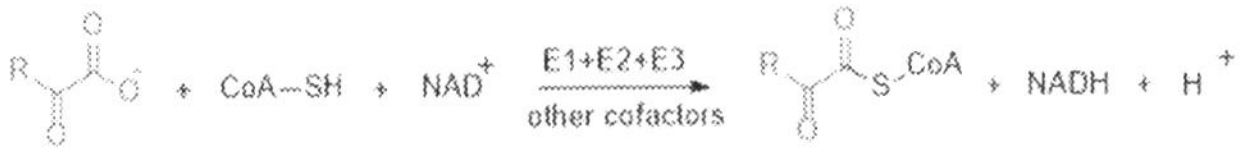

FIGURE 1.28 $R = CH_3$ pyruvate dehydrogenase, $(CH_2)COOH$ 2-oxoglutarate dehydrogenase, $CH(CH_3)_2$ or $CH_2CH(CH_3)_2$ or $CH(CH_3)CH_2CH_3$ branched-chain 2-oxoacid dehydrogenase.

   The branched-chain 2-oxoacid dehydrogenase complex (BCDHC), contribute to the degradation of branched-chain amino acids valine, leucine, and isoleucine after their transamination into corresponding 2-oxoacids. 2-oxoglutarate dehydrogenase complex (OGDHC) allows the metabolism of histidine, proline, glutamate, glutamine, and arginine, which belong to the 2-oxoglutarate family. 2-oxoacid dehydrogenase complexes also couples 2-oxoglutarate oxidative decarboxylation to CoA succinylation followed by NAD^+ reduction in an overall reaction of OGDHC and 2-oxoadipate dehydrogenase complex OADHC (Fig. 1.28). The formation of CoA thioester in a different type of complex enzymes (E1, E2, E3) based on pyruvate and 2-oxoglutarate are formed with the release of carbon dioxide and NADH by oxidative decarboxylation of 2-oxoacids.[44,45]

   In addition, phosphonate- and phosphinate-based synthetic analogs of 2-oxoacids are known for the inhibition of thiamine diphosphate (ThDP) dependent dehydrogenases, oxidases, and decarboxylases of 2-oxoacids. Moreover, they act as inhibitors of the cognate complexes in bacterial, plant, and animal cells (Fig. 1.29). They are known to form inhibitor−enzyme complexes that imitate transition states of their enzymatic reactions, which renders them more selective in vivo. The application of such molecules is capable of fixing key problems in medicine, metabolic engineering, and system biology.[46] These compounds are systematically synthesized and disintegrated naturally which has developed an interest in such analogs as ecofriendly herbicides.[47]

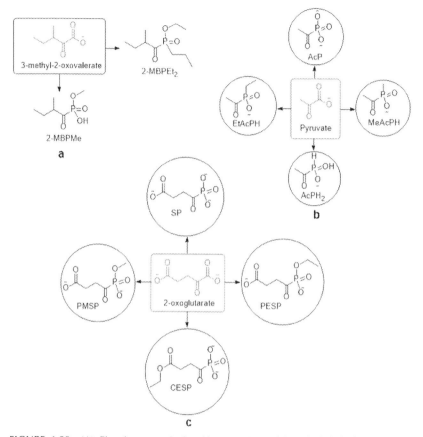

FIGURE 1.29 (A) Phosphonate and phosphinate analogs of branched-chain 2-oxo acid 3-methyl-2-oxovalerate, (B) pyruvate, and (C) 2-oxoglutarate.

References

1. (a) Vartanian, P. F. The Chemistry of Modern Petroleum Product Additives. *J. Chem. Educ.* **1991**, *68* (12), 1015–1020.
   (b) Urbansky, E. T. Carbinolamines and Geminal Diols in Aqueous Environmental Organic Chemistry. *J. Chem. Educ.* **2000**, *77* (12), 1644–1647.
   (c) Atterholt, C.; Butcher, D. J.; Bacon, J. R.; Kwochka, W. R.; Woosley, R. Implementation of an Environmental Focus in an Undergraduate Chemistry Curriculum by the Addition of Gas Chromatography-Mass Spectrometry. *J. Chem. Educ.* **2000**, *77* (12), 1550–1551.
   (c) Goldish, D. M. Let's Talk about the Organic Chemistry Course. *J. Chem. Educ.* **1988**, *65* (7), 603–604.
   (d) Westheimer, F. The Application of Physical Organic Chemistry to Biochemical Problems. *J. Chem. Educ.* **1986**, *63* (5), 409–413.

(e) Shulman, J. Chemistry in the Premedical Curriculum: Considering the Options. *J. Chem. Educ.* **2013,** *90* (7), 813−815.

(f) Mueller, W. J. A Selective Study Approach to Organic Chemistry for Health Science Majors. *J. Chem. Educ.* **1974,** *51* (10). 674 − 674.

(g) Gero, A. Some Remarks on the Role of Chemistry in Pre-medical Curriculum. *J. Chem. Educ.* **1956,** *33* (6), 278−281.

2. Penteado, F.; Lopes, E. F.; Alves, D.; Perin, G.; Jacob, R. G.; Lenardao, E. J. *Chem. Rev.* **2019,** *119,* 7113−7278.

3. Akbari, A. *Synlett* **2012,** *23,* 951.

4. (a) Akiyama, T.; Suzuki, T.; Mori, K. *Org. Lett* **2009,** *11,* 2445.

   (b) Konakahara, T.; Watanabe, A.; Maehara, K.; Nagata, M.;

   (c) Hojahmat, M. *Heterocycles* **1993,** *35,* 1171.

   (d) Zhang, C.; Moran, E. J.; Woiwode, T. F.; Short, K. M.; Mjalli, A. M. M. *Tetrahedron Lett.* **1996,** *37,* 751.

   (e) Oi, S.; Terada, E.; Ohuchi, K.; Kato, T.; Tachibana, Y.; Inoue, Y. *J. Org. Chem.* **1999,** *64,* 8660.

   (f) Luo, H.-K.; Khim, L. B.; Schumann, H.; Lim, C.; Jie, T. X.; Yang, H.-Y. *Adv. Synth. Catal* **2007,** *349,* 1781.

   (g) Luo, H.-K.; Woo, Y.-L.; Schumann, H.; Jacob, C.; van Meurs, M.; Yang, H.-Y.; Tan, Y.-T. *Adv. Synth. Catal* **2010,** *352,* 1356.

   (h) Becker, J. J.; Van Orden, L. J.; White, P. S.; Gagne, M. R. *Org. Lett.* **2002,** *4,* 727.

   (i) Tonoi, T.; Mikami, K. *Tetrahedron Lett.* **2005,** *46,* 6355.

   (j) Shchepin, V. V.; Korzun, A. E.; Nedugov, A. N.; Sazhneva, Y. K.; Shurov, S. N. *Russ. J. Org. Chem.* **2002,** *38,* 248.

5. Kumar, A.; Gannedi, V.; Rather, S. A.; Vishwakarma, R. A.; Ahmed, Q. N. *J. Org. Chem.* **2019,** *84,* 4131−4148.

6. (a) Ishihara, K.; Yano, T.; Fushimi, M. *J. Fluorine Chem.* **2008,** *129,* 994.

   (b) Schmitt, E.; Schiffers, I.; Bolm, C. *Tetrahedron Lett.* **2009,** *50,* 3185.

   (c) Blay, G.; Hernandez-Olmos, V.; Pedro, J. R. *Chem. Eur. J.* **2011,** *17,* 3768.

   (d) Kobayashi, S.; Araki, M.; Yasuda, M. *Tetrahedron Lett.* **1995,** *36,* 5773.

7. Bagher, E.-S.; Maryam, Z.; Ali, A. *Chem. Rev.* **2013,** *113* (5), 2958−3043.

8. (a) Moffett, R. B.; Tiffany, B. D.; Aspergren, B. D.; Heinzelman, R. V. *J. Am. Chem. Soc.* **1957,** *79,* 1687.

   (b) Takahashi, K. *J. Biol. Chem.* **1968,** *243,* 6171.

   Takahashi, K. *J. Biol. Chem.* **1968,** *243,* 6171.

   (c) Tiffany, B. D.; Wright, J. B.; Moffett, R. B.; Heinzelman, R. V.; Strube, R. E.; Aspergren, B. D.; Lincoln, E. H.; White, J. L. *J. Am. Chem. Soc.* **1957,** *79,* 1682.

9. Takahashi, K. *J. Biol. Chem.* **1968,** *243,* 6171.

10. (a) Tiffany, B. D.; Wright, J. B.; Moffett, R. B.; Heinzelman, R. V.; Strube, R. E.; Aspergren, B. D.; Lincoln, E. H.; White, J. L. *J. Am. Chem. Soc.* **1957,** *79,* 1682.

    (b) Moffett, R. B.; Tiffany, B. D.; Aspergren, B. D.; Heinzelman, R. V. *J. Am. Chem. Soc.* **1957,** *79,* 1687.

11. Von Pechmann, H. *Chem. Ber.* **1887,** *20,* 2904.

12. (a) Riley, H. L.; Morley, J. F.; Friend, N. A. C. *J. Chem. Soc.* **1932,** 1875.

    (b) Saldabol, N. O.; Popelis, J.; Slavinska, V. *Chem. Heterocycl. Compd.* **2002,** *38,* 783.

13. Tiecco, M.; Testaferri, L.; Tingoli, M.; Bartoli, D. *J. Org. Chem.* **1990,** *55,* 4523.

14. Kornblum, N.; Powers, J. W.; Anderson, G. J.; Jones, W. J.; Larson, H. O.; Levand, O.; Weaver, W. M. *J. Am. Chem. Soc.* **1957,** *79,* 6562.

15. Kornblum, N.; Frazier, H. W. *J. Am. Chem. Soc.* **1966**, *88*, 865.
16. Gunn, V. E.; Anselme, J.-P. *J. Org. Chem.* **1977**, *42*, 754.
17. Kato, T.; Goto, Y.; Yamamoto, Y. *Yakugaku Zasshi* **1964**, *84*, 287.
18. Ballistreri, F.; Failla, S.; Tomaselli, G. A.; Curci, R. *Tetrahedron Lett.* **1986**, *27*, 5139.
19. Wolfe, S.; Pilgrim, W. R.; Garrard, T. F.; Chamberlain, P. *Can. J. Chem.* **1971**, *49*, 1099.
20. Mikol, G. J.; Russell, G. A. *Org. Synth.* **1968**, *48*, 109 Org. Synth. Collect. Vol. 1973, 5, 937.
21. Ihmels, H.; Maggini, M.; Prato, M.; Scorrano, G. *Tetrahedron Lett.* **1991**, *32*, 6215.
22. Sharma, V. K.; Chandalia, S. B. *J. Chem. Technol. Biotechnol.* **1986**, *36*, 456.
23. Tiffany, B. D.; Wright, J. B.; Moffett, R. B.; Heinzelman, R. V.; Strube, R. E.; Aspergren, B. D.; Lincoln, E. H.; White, J. L. *J. Am. Chem. Soc.* **1957**, *79*, 1682.
24. Shen, Q.; Hagen, K. *J. Phys. Chem.* **1993**, *97*, 985.
25. (a) Earle, R. H.; Hurst, D. T., Jr; Viney, M. Synthesis and Hydrolysis of some Fused-ring β-Lactams. *J. Chem. Soc. C* **1969**, *0*, 2093–2098.

     (b) Draber, W.; Timmler, H.; Dickore, K.; Donner, W. Synthesé und Reaktionen von 3-Alkyl-4-amino-1,2,4-triazin-5-onen. *Justus Liebigs Ann. Chem.* **1976**, *1976*, 2206–2221.

     (c) Domagala, J. F. A Mild, Rapid, and Convenient Esterification of α-Keto Acids. *Tetrahedron Lett.* **1980**, *21*, 4997–5000.

     (d) Ibrahim, Y. A.; Eid, M. M.; Badawy, M. A.; Abdel-Hady, S. A. L. Reaction of 4-Aryl-1,2,4-triazines with Hydrazine. *J. Heterocycl. Chem.* **1981**, *18*, 953–956.

     (e) Styles, V. L.; Morrison, R. W., Jr. Pyrimido[4,5-c]pyridazines., Cyclizations with α-Keto Acids. *J. Org. Chem.* **1982**, *47*, 585–587.

     (f) Clerici, A.; Porta, O. A Novel Reaction Type Promoted by Aqueous Titanium Trichloride. Synthesis of Unsymmetrical 1,2-Diols. *J. Org. Chem.* **1982**, *47*, 2852–2856.

     (g) Cohen, M. J.; McNelis, E. Oxidative Decarboxylative of Propiolic Acids. *J. Org. Chem.* **1984**, *49*, 515–518.

     (h) Le-Bris, M.-T. Reaction de l' 'amino-2 nitro-5 phenol et du' diamino-2,5 phenol avec quelques acides et esters' α-cetoniques.' Synthese d' 'amino-7 benzoxazines-1,4 ones-2. *J. Heterocycl. Chem.* **1984**, *21*, 551–555.

     (i) Arndt, F.; Franke, W.; Klose, W.; Lorenz, J.; Schwarz, K. Synthesen von Thiazolo- und [1,3]Thiazino[1,2,4]triazinonen. *Justus Liebigs Ann. Chem.* **1984**, *1984*, 1302–1307.

     (j) Le-Bris, M.-T. Synthesis and Properties of Some 7- Dimethylamino-1,4-benzoxazin-2-ones. *J. Heterocycl. Chem.* **1985**, *22*, 1275–1280.
26. Waters, K. L. The α-Keto Acids. *Chem. Rev.* **1947**, *41*, 585–598.
27. Cooper, A. J. L.; Ginos, J. Z.; Meister, A. Synthesis and Properties of the α-Keto Acids. *Chem. Rev.* **1983**, *83*, 321–358.
28. (a) Corson, B. B.; Dodge, R. A.; Harris, S. A.; Hazen, R. K. Ethyl Benzoylformate. *Org. Synth.* **1928**, *8*, 68–72.

     (b) Baer, E. Oxidative Cleavage of α-Keto Acids and α-Keto Alcohols by Means of Lead Tetra-acetate. *J. Am. Chem. Soc.* **1940**, *62*, 1597–1606.
29. (a) Domagala, J. F. A Mild, Rapid, and Convenient Esterification of α-Keto Acids. *Tetrahedron Lett.* **1980**, *21*, 4997–5000.

     (b) Ibrahim, Y. A.; Eid, M. M.; Badawy, M. A.; Abdel-Hady, S. A. L. Reaction of 4-Aryl-1,2,4-triazines with Hydrazine. *J. Heterocycl. Chem.* **1981**, *18*, 953–956.

     (c) Styles, V. L.; Morrison, R. W., Jr. Pyrimido[4,5-c]pyridazines., Cyclizations with α-Keto Acids. *J. Org. Chem.* **1982**, *47*, 585–587.

     (d) Earle, R. H.; Hurst, D. T., Jr; Viney, M. Synthesis and Hydrolysis of some Fused-ring β-Lactams. *J. Chem. Soc. C* **1969**, *0*, 2093–2098.

30. Ke, Q.; Ferrara, E.; Radicchi, F.; Flammini, A. Defining and Identifying Sleeping Beauties in Science. *Proc. Natl. Acad. Sci. U. S. A.* **2015**, *112*, 7426–7431.

31. Carey, J. S.; Laffan, D.; Thomson, C.; Williams, M. T. Analysis of the Reactions Used for the Preparation of Drug Candidate Molecules. *Org. Biomol. Chem.* **2006**, *4*, 2337–2347.

32. Blanco-Canosa, J. B.; Dawson, P. E. An Efficient Fmoc-SPPS Approach for the Generation of Thioester Peptide Precursors for Use in Native Chemical Ligation. *Angew Chem.* **2008**, *120*, 6957–6961.

33. Guo, L.-N.; Wang, H.; Duan, X.-H. Recent Advances in Catalytic Decarboxylative Acylation Reactions via a Radical Process. *Org. Biomol. Chem.* **2016**, *14*, 7380–7390.

34. Miao, J.; Ge, H. Palladium-Catalyzed Decarboxylative CrossCoupling of α-Oxocarboxylic Acids and their Derivatives. *Synlett* **2014**, *25*, 911–919.

35. Kay, J.; Weitzman, P. D. J. *Biochem. Soc. Symp.*, Vol. 54. The Biochemical Society: London, 1987.

36. Voet, D.; Voet, J. G. *Biochemistry*, 3rd ed.; John Wiley & Sons, Inc: New York, 2004.

37. Claus, A.; Neukranz, W. Ueber die Oxidation der Gemischten Fettaromatischen Ketone Durch Kaliumpermauganat-Kali. *J. Prakt. Chem.* **1891**, *44*, 77–85.

38. Hurd, C. D.; McNamee, R. W.; Green, F. O. Benzoylformic Acid from Styrene. *J. Am. Chem. Soc.* **1939**, *61*, 2979–2980.

39. Oakwood, T. S.; Weisgerber, C. A. Benzoylformic Acid. *Org. Synth.* **1944**, *24*, 16.

40. Yang, N.; Zhang, H.; Yuan, G. KI-catalyzed Reactions of Aryl Hydrazines with Acids in the Presence of CO_2: Access to 1,3,4-oxadiazol-2(3H)-ones. *Org. Chem. Front.* **2019**, *6*, 532–536.

41. Sprenger, R. D.; Rouff, P. M.; Frazer, A. H. Critinin Studies. Hydroxy- and Methoxy-phenylglyoxylic Acids. *J. Am. Chem. Soc.* **1950**, *72*, 2874–2876.

42. Faculty of Bioengineering and Bioinformatics. Lomonosov Moscow State University, Moscow, Russia. 2A. N. Belozersky Institute of Physico-Chemical Biology, Lomonosov Moscow State University, Moscow, Russia.

43. Comb, D. G.; Roseman, S. *J. Am. Chem. Soc.* **1958**, *80*, 497–499 2. 6.07 Chemical Glycobiology Chad M. Whitman and Michelle R. Bond, Stanford University, Stanford, CA, USA, Jennifer J. Kohler, University of Texas Southwestern Medical Center, Dallas, TX, USA 2010 Elsevier Ltd. All rights reserved.

44 V., Bunik; A.V., Artiukhov. Synthetic analogues of 2-oxo acids discriminate metabolic contribution of the 2-oxoglutarate and 2-oxoadipate dehydrogenases in mammalian cells and tissues. *Sci. Rep.* **2020**, *10*, 1–22.

45 V., Bunik; A. V., Artiukhov. Thiamine induces long-term changes in amino acid profiles and activities of 2-oxoglutarate and 2-oxoadipate dehydrogenases in rat brain. *Biochemistry (Moscow) volume* **2017**, *82*, 723–736.

46 N., Lukashev; V. I., Bunik. Thiamin diphosphate-dependent enzymes: from enzymology to metabolic regulation, drug design and disease models. *FEBS* **2013**, *280*, 6412–6442.

47 V., Bunik'; V. I., Bunik. *Eur. J. Biochem.* **2003**, *270*, 1036–1042.

Chapter 2

Structure and Spectral Characteristics of 2-Oxoaldehydes and 2-Oxoacids

2.1 Reactivity comparison of 2-oxoaldehydes with aldehydes

A carbonyl group is prevalent in most organic compounds and is used in functional transformations in organic syntheses. Although many different kinds of carbonyl compounds like aldehydes, carboxylic acids, esters, anhydrides, and so on exist, nearly all carbonyl compounds adhere to some common basic principles during their reaction courses. The $C-O$ double bond is highly polarized due to the more electronegative oxygen atom. Thus, the carbon atom in the carbonyl group carries a partial positive charge and is a Lewis acid site that is vulnerable to nucleophilic attack with or without a catalyst or promoter. These interactions perhaps produce new bonding connections, including $C-C$, $N-C$, $P-C$, $O-C$, $S-C$, $X-C$, and so on. The overall fate of the reaction depends on the tetrahedral geometry of the intermediate. Condensation involves the elimination of carbonyl oxygen, which results in the formation of a double bond, while as the substitution involves elimination of a group from the tetrahedral intermediate, leading to regeneration of the carbonyl group. Fig. 2.1 summarizes some features of the carbonyl group of acetaldehydes.

On the other hand, the electrophilic nature of the carbonyl group can be increased or decreased depending upon various Lewis or Bronsted acid groups inherently attached to it or after the addition of some external reagents like amines or other activating groups. Imines and iminium ions are generated after aldehydes react with amines with the elimination of H_2O molecule and OH ion, respectively[1] (Fig. 2.2).

Since imines are less electrophilic as compared to aldehydes, some external activating reagents may be required in order to carry out or increase the their rate of reaction with nucleophiles. However, the electrophilic nature of imines increases due to electron-withdrawing groups like -phosphoryl (-PO (Ph)$_2$), -tert-butyloxycarbonyl (-CO$_2$C(CH$_3$)$_3$), and -tosyl (-OTs) (Fig. 2.3).

Comparative electrophilicity and nucleophilicity is a necessary condition that ensures the bonding between the reactants and reagents to form a

Chemistry of 2-Oxoaldehydes and 2-Oxoacids. DOI: https://doi.org/10.1016/B978-0-12-824285-8.00002-9

19

Bond Angle (°)		Bond Length (pm)	
H—C—C	118	C=O	122
C—C=O	121	C=C	150
H—C=O	121	O—C—H	109

FIGURE 2.1 Bond angles and bond lengths of acetaldehyde.

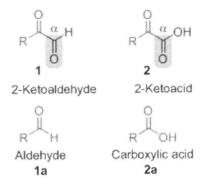

FIGURE 2.2 Structure of related carbonyl groups.

particular product. In contrast to the carbonyl group, the electrophilicity of iminium ion is shifted to a normalized state and brought at an intermediate level giving rise to new reactions that are not feasible otherwise.

In general, electron-donating groups tend to decrease the electrophilic nature of aldehydes. Environmental features like temperature, solvent, nitrogen, or argon atmospheres also control the reactivity of aldehydes.[2] Large numbers of reactions involving weak nucleophiles take place under acidic (Lewis acid, Bronsted acid) and basic (1°-amine, 2°-amine) conditions (Fig. 2.4).

In addition, nucleophiles generated at α-position of the carbonyl are stable and add to electrophiles after quickly passing through enol or enolate form. Another feature of aldehydes is their inverse behavior when the electrophilic carbon switches to nucleophilic carbon through a reversal in polarity or umpolung reactivity, which is a key factor behind the synthesis of 1,2- and 1,4-difunctionalized compounds. (Fig. 2.5).

Lewis acids or metals are employed in order to make aldehydes stronger electrophiles. Some of the acid-mediated reactions of aldehydes with different nucleophiles are given in Table 2.1.

Furthermore, aldehyde **1** reacts with enolizable carbonyl compound **3** in presence of a Bronsted or Lewis acid and other catalysts where it is transformed into more active electrophilic intermediate oxonium ion **3a** by protonation, which is then readily attacked by nucleophiles to form aldol **4** or

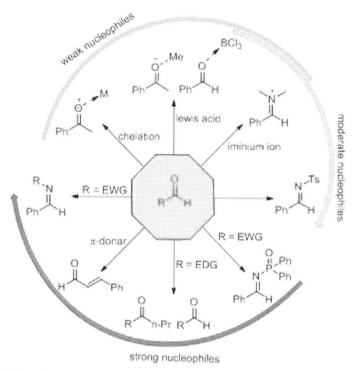

FIGURE 2.3 Representation of various aldehyde-based electrophiles and its relative strengths and interactions with different nucleophiles.

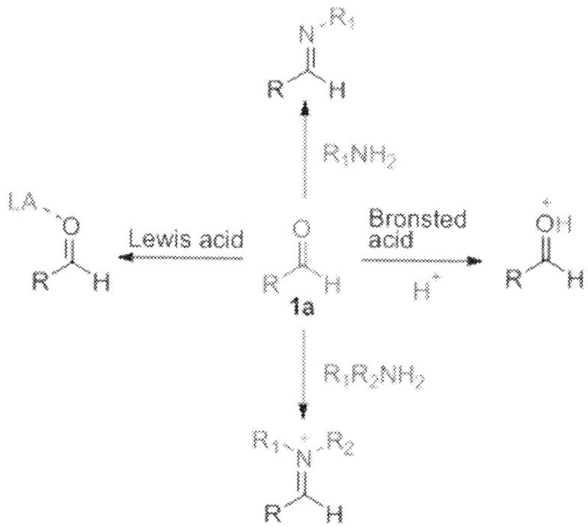

FIGURE 2.4 Generation of various electrophilic forms of aldehyde.

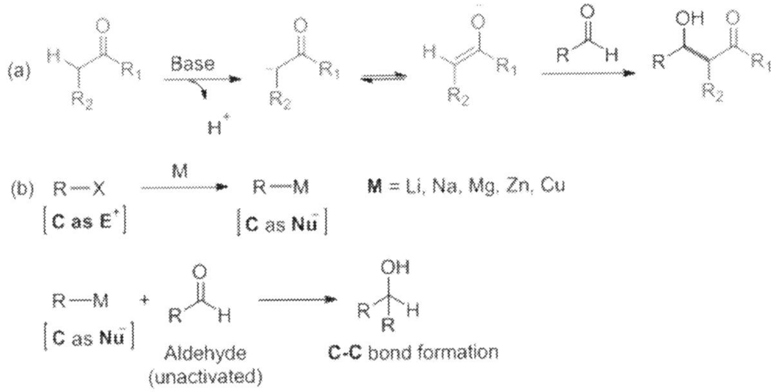

FIGURE 2.5 (A) Generation of carbon-nucleophiles under basic conditions and (B) generation of carbon-nucleophiles by umpolung reactivity.

unsaturated carbonyl compound **5**.[3] Multiproduct formation in aldol reaction is avoided due to preformed enolate by employing metal-enol or silyl-enol ether reagent. For example, Mukaiyama aldol reaction uses silyl-enol ether **6** and Lewis acid (TiCl$_4$) for C-C bond formation to give **4** and **5** (Fig. 2.6).[4]

In a Passerini reaction, carboxamides **10** are formed directly in a multi-component reaction between aldehydes **1a**, carboxylic acid **8** and isocyanides **9**. In this reaction, electrophilicity of aldehyde is enhanced due to protonation by acid which renders nucleophilic attack more favorable (Fig. 2.7).

Reductive coupling reaction of aldehydes **1a** takes place to give alcohols **13** in presence of halides **11** and chromium salts in which organochromium species **12** bearing more electrophilic carbon acts as an intermediate. This reaction is chromium-induced redox reaction bearing chemoselectivity toward aldehydes however, the use of excess the use of excess toxic chromium salt is an imminent problem (Fig. 2.8).[5]

Triphosphites and H-phosphonates **14** are important nucleophilic reagents in Pudovik reaction. They react with aldehydes **1a** to produce α-hydroxy phosphonates **15** through C-P bond formation promoted by acid (Fig. 2.9).[6]

Aldehydes are easily attacked by alkenes in Lewis or Bronsted acid media media and lead to the formation to the formation of new C-C bonds (Scheme 1.4). A mixture of products are usually obtained at lower temperatures and acetals **18** are formed when excess of electrophilic aldehydes are used, while as allylic alcohols **20** and diols **19** are formed at elevated temperatures. The hydroxycarbenium ion **17** is supposed to be intermediate, which reacts with the olefin to form carbenium ion (Fig. 2.10).[7]

Aromatic compounds have low nucleophilic character that makes their reaction with electrophilic aldehydes less feasible. Therefore, lewis acids are used to promote such processes, for example, the installation of a chloromethyl group on aromatic compounds **22** in Blanc chloromethylation

TABLE 2.1 Examples of acid promoted reactions in aldehydes.

S. No	Electrophile	Nucleophile	Reaction
1.			Passerini
2.			Nozaki-Hiyama-Kishi
3.			Pudovic
4.			Aldol
5.			Mukhayama Aldol
6.			Hosomi-Sakurai
7.			Prins
8.			Blanc
9.			Tishchenko

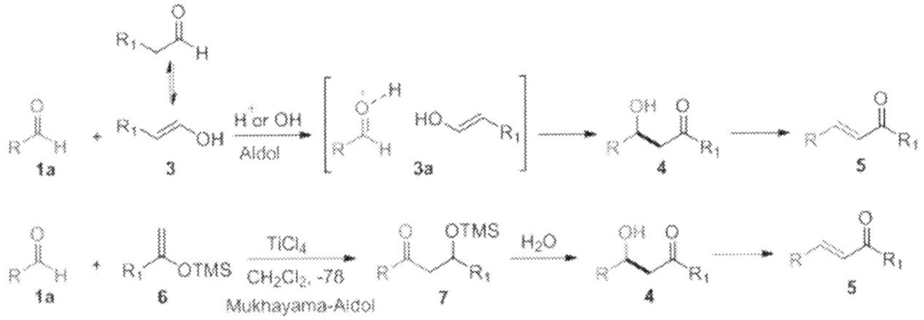

FIGURE 2.6 Reaction of aldehydes with enolizable carbonyl compounds.

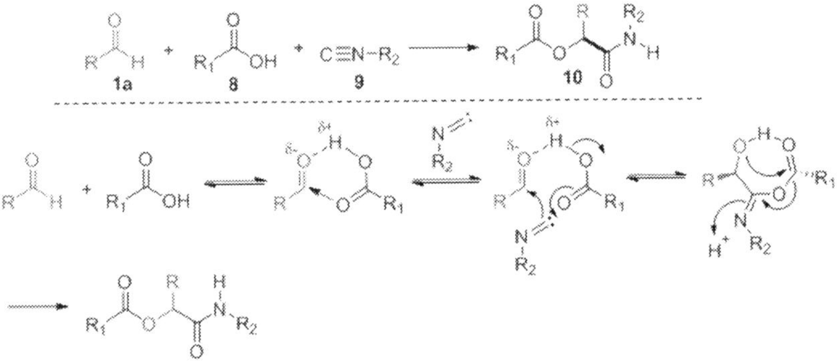

FIGURE 2.7 Multicomponent Passerini reaction.

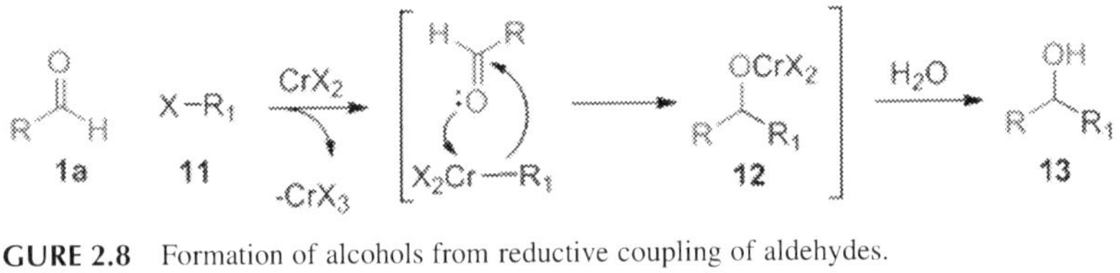

FIGURE 2.8 Formation of alcohols from reductive coupling of aldehydes.

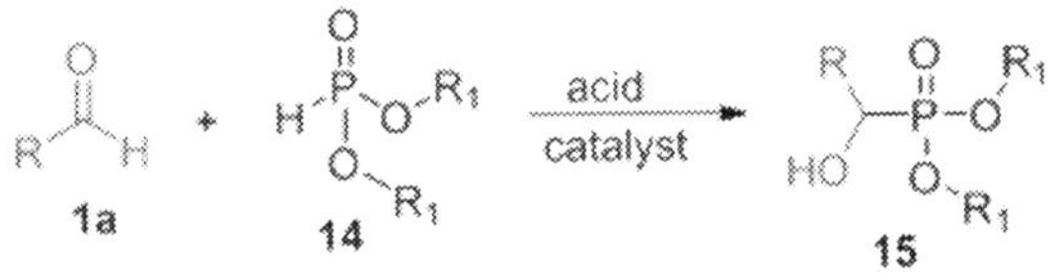

acid: TFA/TfOH, In/HCl, AlCl$_3$, Ti(O^{i}Pr)$_4$

catalyst: TMSCl, LiClO$_4$, Et$_2$O,NH$_4$VO$_3$, Guanidie/HCl, MoO$_2$Cl$_2$

FIGURE 2.9 Acid catalyzed Pudovik reaction.

involves the coupling of phenyl derivatives with aldehydes in presence of $ZnCl_2$ and $HCl_{(g)}$. The oxonium ion thus formed is more susceptible to electrophilic attack as the electrophilicity of aldehyde is augmented, thereby increasing the feasibility of the processes. On the other hand, electron rich aromatic compounds can react without any catalyst (Fig. 2.11).[8]

In a similar way, Tishchenko reaction is a disproportionation reaction that uses aluminum alkoxide as a lewis acid to coordinate with one mole of aldehyde, which in turn facilitates the addition of a second molecule of aldehyde to form a hemiacetal intermediate 23, which undergoes intramolecular 1, 3-hydride shift to from esters 24. In this reaction, Lewis acid improves the electrophilicity of aldehyde facilitating the nucleophilic attack by another aldehydes molecule through its oxygen atom (Fig. 2.12).[9]

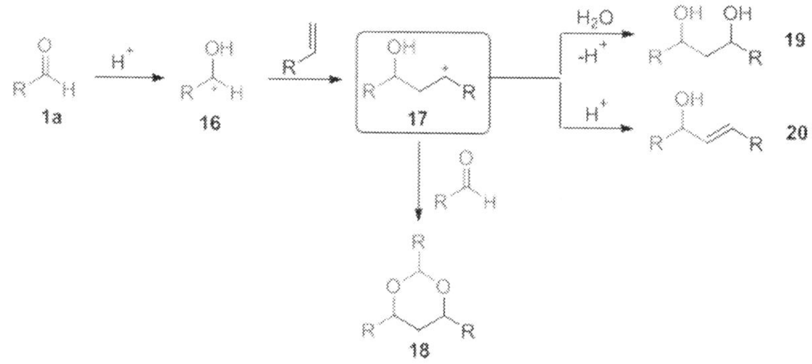

FIGURE 2.10   Formation of cyclic ethers and alcohols from aldehydes.

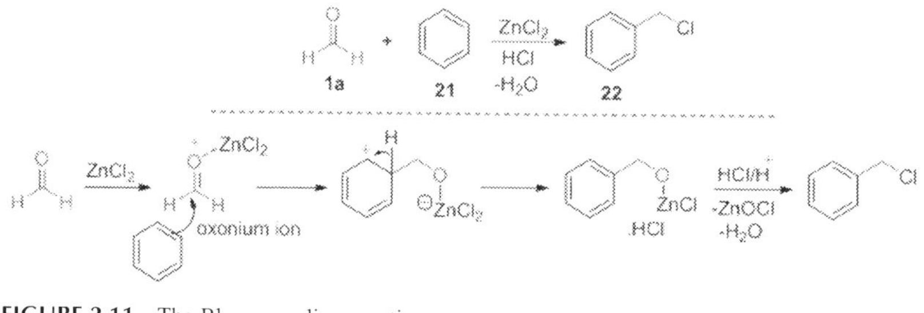

FIGURE 2.11   The Blanc coupling reaction.

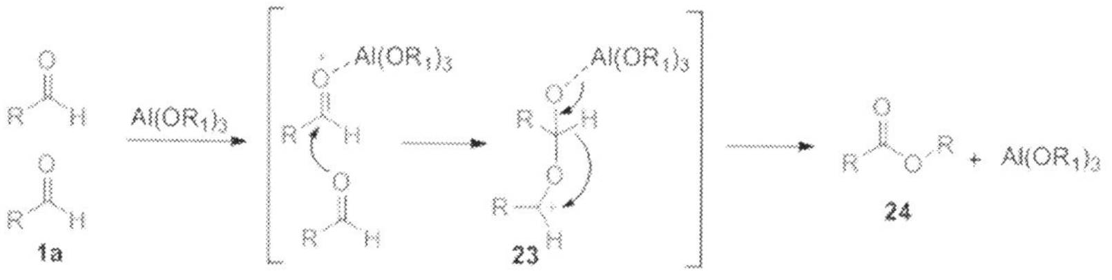

FIGURE 2.12   The Tishchenko reaction.

Aldehydes show diverse reactions with range of activated nucleophiles through the capture of acidic proton by using a weak or strong base. Some common base promoted reactions are given in Table 2.2.

Aldehydes that lack α-hydrogen undergo disproportionation reaction in strongly basic conditions to yield carboxylic acids **2a** and alcohol **25**. Nucleophilic addition of hydroxide anion to the carbonyl group of an aldehyde molecule occurs to give anion which allows the hydride shift to other aldehyde molecules. In contrast, 2-oxoaldehydes exclusively form α-hydroxycarboxylic acid **27** through intramolecular disproportionation reaction (Fig. 2.13).[10]

β-Nitro alcohols **30** are formed when nitroalkanes **28** react with aldehydes in presence of a base. This aldol condensation between nitroalkanes and carbonyl compounds is called Henry reaction or nitro-aldol reaction (Fig. 2.14). All the steps in the reaction are reversible and α-deprotonation occurs to form anion **29**. C-alkylation of **29** occurs with aldehydes to form diastereomeric β-nitro alkoxide **30**, which can also undergo disproportionation reaction to yield alcohols.

Base promoted aldol reaction between an aldehydes **1a** as an electrophile and an enolizable carbonyl nucleophile **3** occurs to produce aldols β-hydroxy carbonyl **4**, which may further loose water molecule to give α,β-unsaturated aldehydes **5** or ketones. Enolate formation takes place by deprotonation using a basewhich adds to the electrophilic carbonyl substrate (Equation 2.1).[3]

Wittig olefination is an important protocol in a C–C bond forming reactions with carbonyl compounds involving the use of pentavalent phosphorus or phosphoranes **31** to form alkene **32**. The key substrate phosphorus ylide obtained is nucleophilic enough and adds to the electrophilic carbon of carbonyl group. The cleavage of oxaphosphetane ring formed during the processes yields the final product with the elimination of triphenylphosphine oxide **33** as a byproduct (Equation 2.2).[11]

Aldehydes are easily converted into epoxides **35** through Corey-Chaykovsky reaction.[12] In this method, dimethyl sulfonium methylide (sulfur ylide) generated from deprotonation of trimethyl sulphonium halides **34**

TABLE 2.2 Examples of base promoted reactions in aldehydes.

S. No	Electrophile	Nucleophile	Reaction
1.			Cannizaro
2.			Henry
3.			Pudovic
4.			Aldol
5.			Wittig
6.			Sayferth-Gilbert
7.			Julia-Kocienski
8.			Corey-Chaykovski
9.			Corey-Fuchs
10.			Organometallic

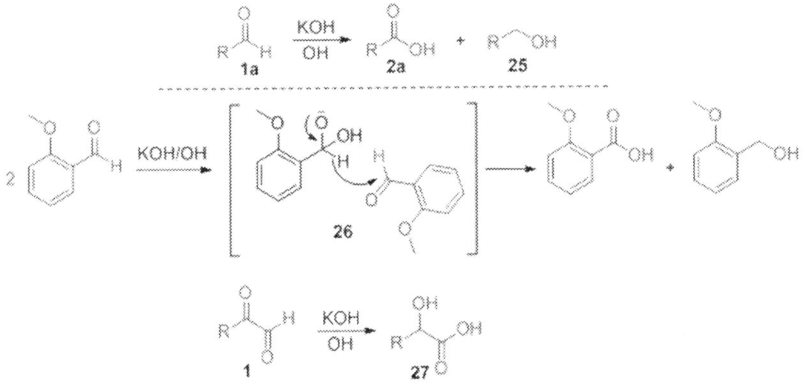
KOH
OH
1a
2a
25
KOH/OH
26
KOH
OH
1
27

FIGURE 2.13 Intramolecular disproportionation reaction of aldehydes in basic medium.

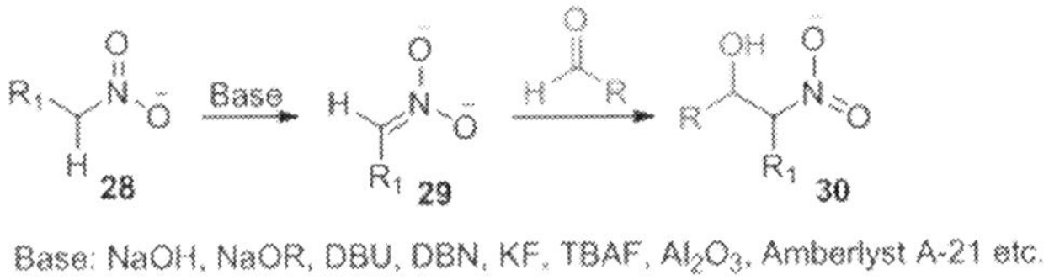
Base
28
29
30
Base: NaOH, NaOR, DBU, DBN, KF, TBAF, Al2O3, Amberlyst A-21 etc.

FIGURE 2.14 The Henry reaction.

using a strong base reacts with simple aldehydes or α,β-unsaturated aldehydes or ketones to form epoxides **35** (Equation 2.3).

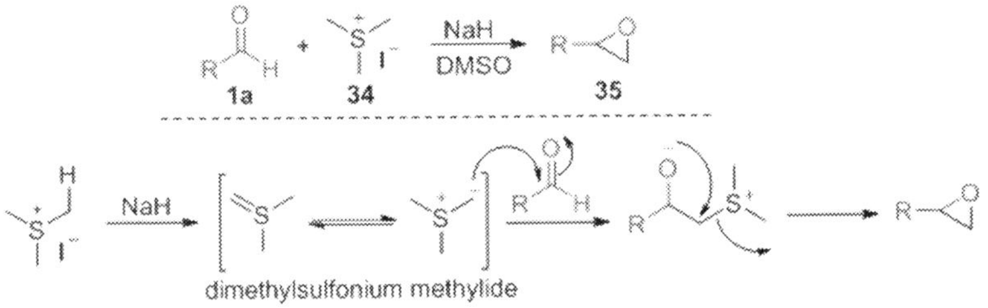
NaH
DMSO
1a
34
35
NaH
dimethylsulfonium methylide

A Corey-Fuchs reaction is carbon homologation of aldehydes to produce terminal alkenes.[13] The reaction involves the use of tetrabromide and triphenylphosphine which converts the given aldehydes into dibromo-olefin (**38a** and **38b**) through a Wittig reaction, which, after treatment with n-BuLi, undergoes lithium-halogen exchange to form the corresponding alkyne **37** after hydrolysis (Fig. 2.15).

Nucleophiles based on organometals posses high degree of nucleophilicity and easily react with aldehydes to form new C—C bonds (Fig. 2.16).[14]

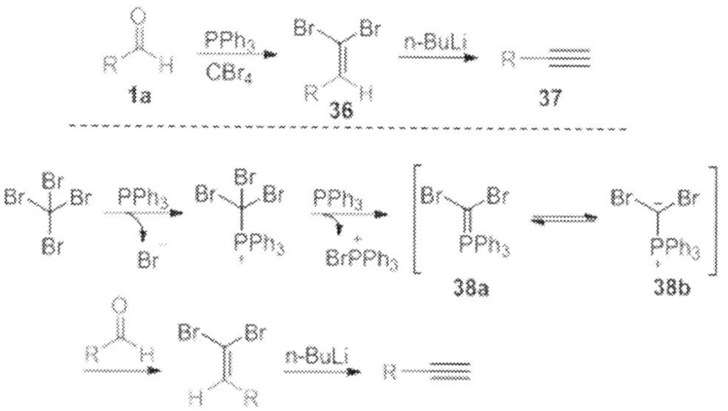

FIGURE 2.15   The Corey-Fuchs homologation of aldehydes.

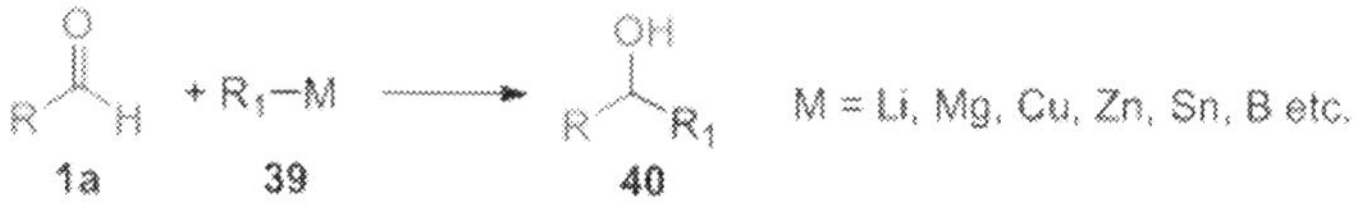

FIGURE 2.16   Nucleophilic addition on aldehydes using organometals.

2.1.1   Imines as intermediates

Imine intermediate is formed during the reaction of aldehydes with primary amines also called Schiff's base. In addition to being important in synthetic chemistry, it is even more important in many biological reactions. Imines are less electrophilic than aldehydes and hence the use of catalysts or lewis acids are necessary to facilitate the reaction.[1]

Some important reactions involving imine as an intermediate are given in Fig. 2.17. An Aza-Henry reaction involves the abstraction of α-hydrogen from nitroalkanes 41 to generate C-nucleophile 42, which can react with masked aldehyde, imine 43 to give α-nitroamines 44 (Fig. 2.18).[15]

An important reaction extending to a wide variety of substitution patterns is a four-component exothermic Ugi reaction between aldehydes 1a, amine 46, carboxylic acid 45, and isocyanide 47 in which α-aminoacyl amide derivative 49 is formed through a stable stable adduct a. The key steps in the mechanism involve the formation of an imine intermediate which then undergoes nucleophilic attack by carboxylic acid to form iminium ion c. The imino carbon of isocyanide ion b and carboxylateion simultaneously react with c to give the final product after rearrangement of acylated isoamide d through acyl-transfer (Fig. 2.19).[16]

Preparation of aziridines 51 through Corey-Chaykovsky's protocol exploits the reaction between less electrophilic imine substrate 43 and more nucleophilic sulfur ylides 50 (Fig. 2.20).[12]

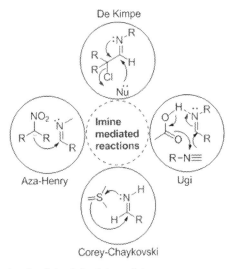

FIGURE 2.17 Reactions involving Imine intermediate.

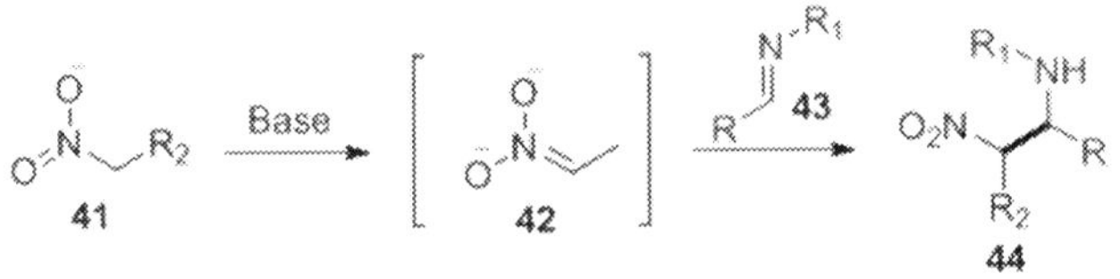

FIGURE 2.18 Aza-Henry reaction of nitro alkanes.

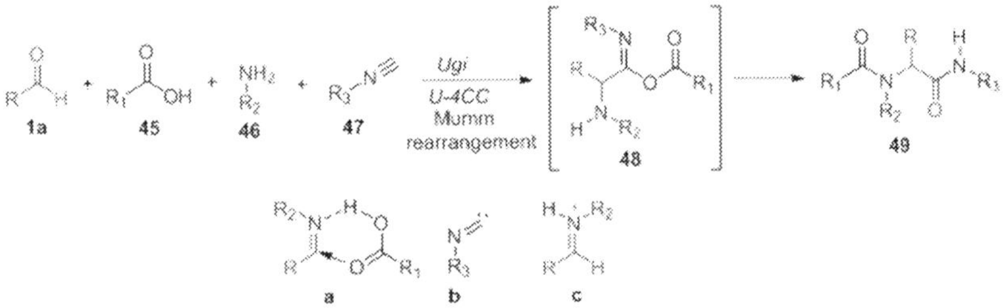

FIGURE 2.19 The Ugi multicomponent reaction.

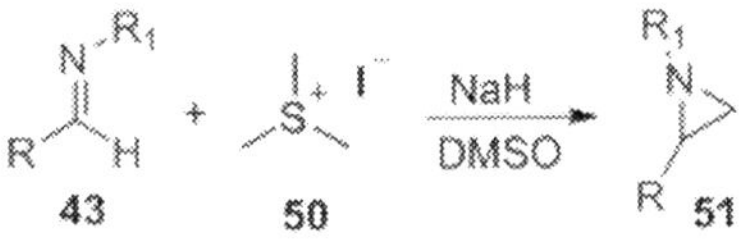

FIGURE 2.20 The Corey-Chaykovsky reaction.

Another method for the preparation of aziridines **53** was introduced by Norbert De Kimpe, which engages imine electrophiles **52** and nucleophilic substrates like α-chloroimines, Grignard reagents, hydrides, and cyanides (Fig. 2.21).[17]

2.1.2 Iminium ion catalyzed reactions

Aldehydes bearing α-hydrogen react with secondary amines to form enamines with the elimination of water molecules. Some of a few important reactions related to iminium ion are Stork enamine synthesis of dicarbonyl compounds (Fig. 2.22). The enamine species **56** is nucleophilic in nature and readily adds to α,β-unsaturated carbonyl compound **54** through conjugate or Michael addition to form 1,5-dicarbonyl compound **57**.[18]

Transamination of the iminium ions is another important reaction in which iminium intermediates like **58** is transformed into imine **59** catalyzed by a secondary amine, usually pyrrolidine (Fig. 2.23).[19]

Like imines α,β-unsaturated iminium ions undergo (2 + 4) cycloaddition with 4π electron systems (e.g., **61**) catalyzed by secondary amine **64** through iminium catalysis (Fig. 2.24).[20]

Mannich reaction (Fig. 2.25) is important in this perspective where iminium ion **56** is generated from the reaction between non-enolizable aldehydes **1a** and secondary amine **55** to form alkylated amine product **66**.

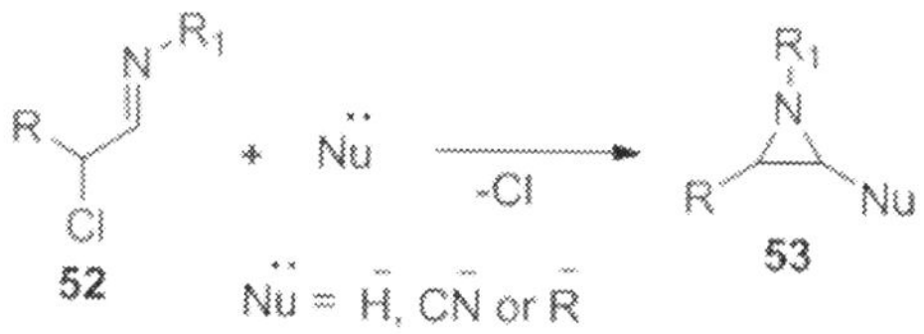

FIGURE 2.21   The Norbert De Kimpe reaction.

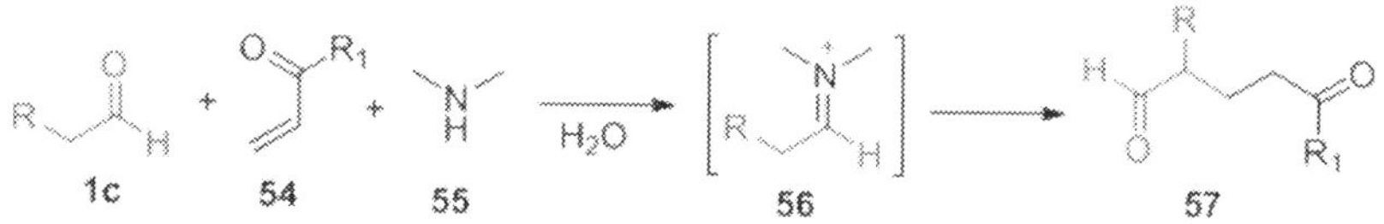

FIGURE 2.22   Stork enamine synthesis.

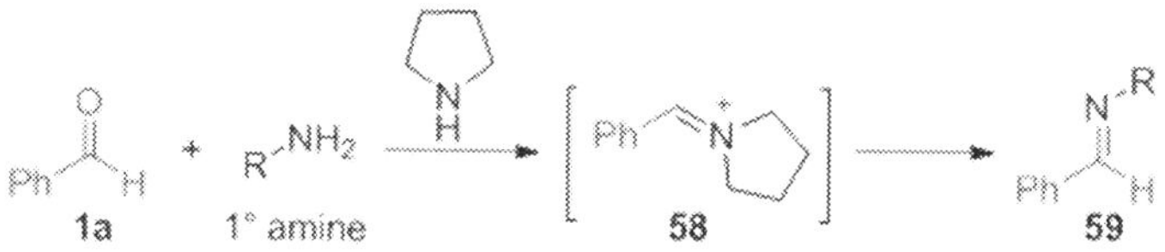

FIGURE 2.23   Formation of enamines through transamination.

Organoboron reagents when used along with Lewis bases generate boron-based nucleophiles which then add to electrophilic carbon with or without any foreign additive. In this context, Petasis reaction or Petasis-Borono Mannich multicomponent reaction proceeds with the formation of an iminium ion **69** formed after the condensation of an aldehyde and primary or secondary amine. The species **69** after the reaction with boronic acid **68** yields secondary or tertiary amine **70** through nucleophilic boronate complex (Fig. 2.26).[21]

2.1.3   Reactivity of 2-oxoaldehydes

As mentioned in Chapter 1, Synthesis and Physical Properties of 2-Oxoaldehydes and 2-Oxoacids, 2-ketoaldehyde constitutes "aldehyde" carbonyl connected to a "2-oxo" group. Therefore, due to the additional electron

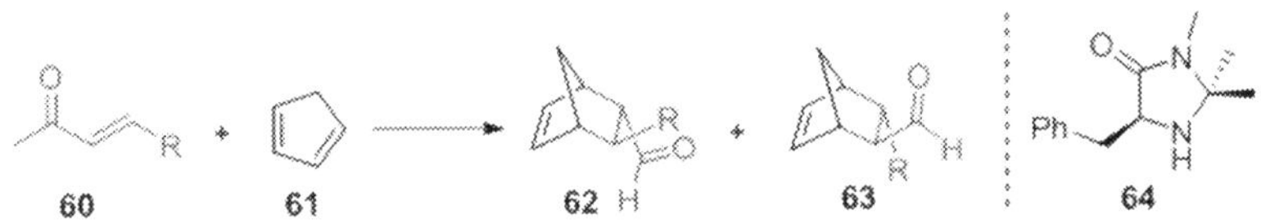

FIGURE 2.24   Cycloaddition of carbonyl compounds through iminium ion.

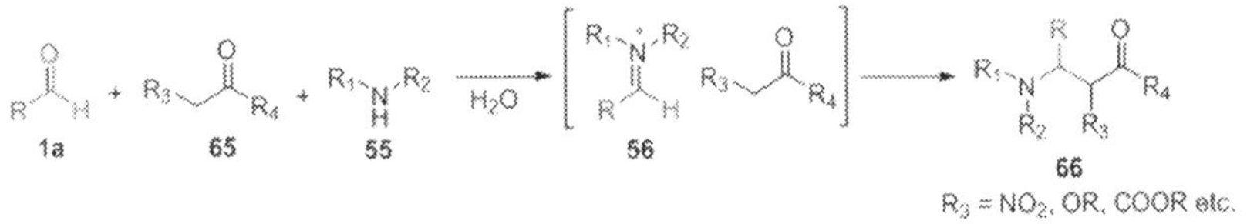

FIGURE 2.25   Formation of alkylated amines via Mannich reaction.

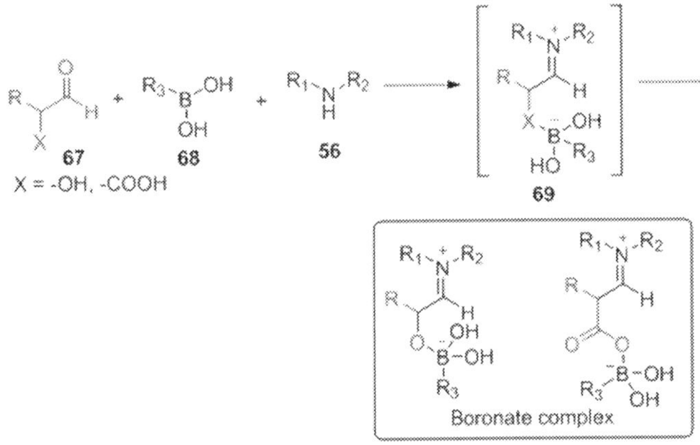

FIGURE 2.26   Formation of tertiary amines through Petasis reaction.

withdrawing 2-oxo carbonyl unit the electrophilicity of the aldehyde unit in 2-oxoaldehyde increases and becomes more labile as compared with the normal aldehyde molecule. Moreover, both of the carbonyl units differ in reactivity and give rise to reactions in three different dimensions in contrast to normal aldehydes and ketones. That means reactions can occur separately at aldehyde and 2-oxo part as well as involving both groups taking part in the reaction simultaneously. This may be fruitful in later but challenging in the former case because many of the reactions that take place at the aldehyde group also take place at a 2-oxo group and vice versa.

Therefore, the limitation of 2-oxoaldehydes as substrates is that these are many times accompanied by side reactions. Hence, mixture of products are formed that may lead to lower yields as for as the expected product is concerned. We have fully donated Chapter 1 for this discussion and concluded that reactions of 2-oxoaldehydes can be broadly classified as[22]:

1. Formation of heterocycles involving the participation of aldehyde group.
2. Formation of heterocycles that contain two hetero atoms involving the participation of both 2-oxo and CHO group.
3. 2Oxoaldehydes and 2-oxoimines act as dienophiles in [4 + 2] and [2 + 2] cycloadditions reactions through C=O and C=N bonds.
4. 2-Oxoaldehyde hydrates as nucleophiles through OH group.

2.1.4   2-Oxoimines as intermediates

Glyoxalimines are generally more electrophilic due to the presence of directly attached carbonyl groups. Therefore, these species react faster and may also act as dienophiles in [4 + 2] and [2 + 2] cycloaddition reactions via C = O and C = N bonds.

The formation of aziridine **72** in a three-component stereoselective aza-Darzens reaction can be directly formed from 2-oxoimines **71** and **74** at lower temperatures by using chiral phosphoric acid **73**. This reaction proceeds without the formation of any transisomer and does not take place when aldimine **43** is used as a substrate (Fig. 2.27).[23]

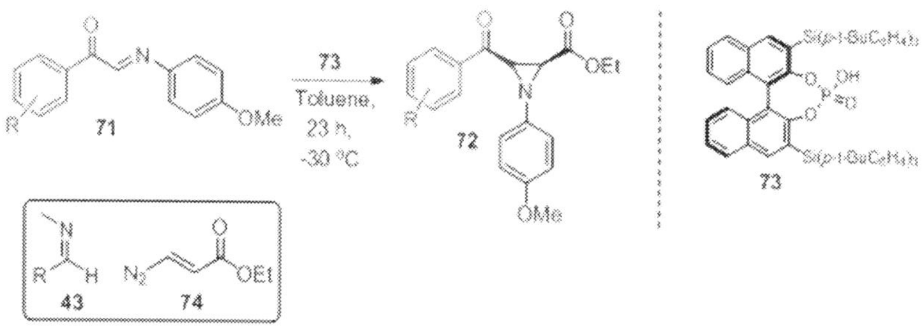

FIGURE 2.27   Synthesis of aziridines through aza-Darzen's reaction.

2-Oxoimines is a reaction center for many transformations. Oxoimine carries the oxidative amidation of weakly nucleophilic substrates like sulfonamides **75**, benzamides **79** and anilines **81** to generate the corresponding keto amides **76**, **80**, and **82**.[24] Similarly, pyrrolidines **85** and **86** are formed during a cycloaddition reaction with azomethine ylide (generated from 2-oxoimine) and N-methylmaleimide **83** or dimethyl fumarate **84**.[25] In addition, a multicomponent Ugi reaction involving insitu 2-oxoimine intermediate produces N-substituted 4-cyano-2,5-dihydro-5-oxopyrrole-2-carboxamides **78** through Ugi adduct adduct **77** formation. (Fig. 2.28).[26]

2.1.5   2-Oxoiminium ion as intermediates

A new and distinct mode of the reactivity of 2-oxoaldehyde and secondary amine was observed with or without the presence of external nucleophiles. The insitu generated 2-oxoiminium ion is highly reactive due to presence of keto group as compared with the iminium ion. 2-oxoaldehydes react with a secondary amine in three different pathways, that is, amino catalytic

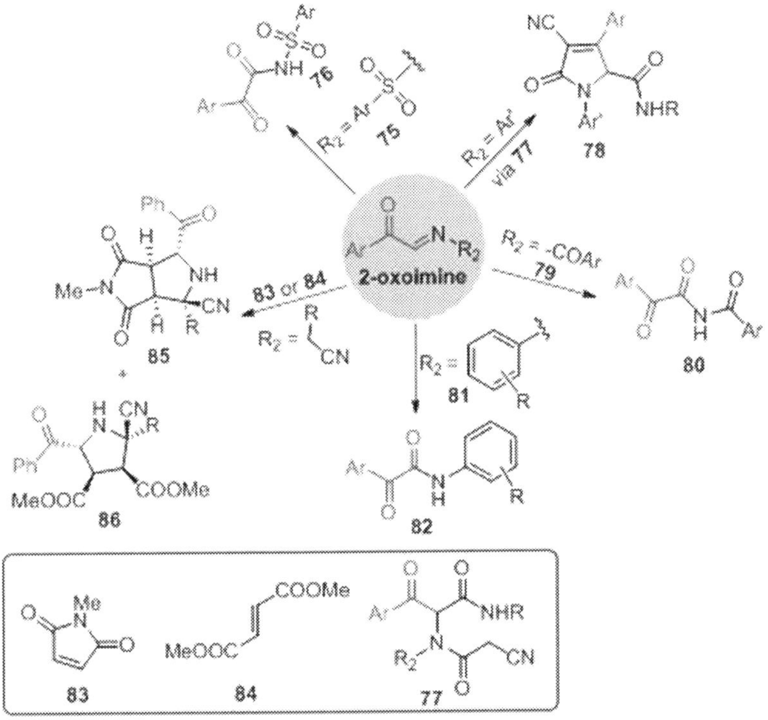

FIGURE 2.28   Different reactions of 2-oxoimine.

pathway, amino-equivalence pathway and amino-equivalence pathway in the presence of external nucleophiles as depicted in Fig. 2.29.

In an amino catalytic pathway, amine acts as a catalyst to initiate the reaction with 2-oxoaldehydes **1** and forms highly reactive 2-oxoiminium ion **II**, followed by the attack of an external nucleophile to generate 1,2-diones **88**. Different nucleophiles were successfully used for the synthesis of different substituted 1,2-diones (Fig. 2.30).

While amino-equivalence pathway works in absence of external nucleophile and produces α-ketoamide **89** through the oxidation of 2-oxoiminium ion **II** in DMSO.[27] Iminium ion **56** generated from normal aldehydes **1a** fails to form any such product under the same conditions. This indicates divergent reactive behavior of 2-oxoiminium ion as compared to the normal iminium ion towards nucleophilic additions (Fig. 2.31).

This distinct mode of reactivity adheres to the 2-oxo group in 2-oxoaldehyde which enhances electrophilicity of aldehyde group such that nucleophilic attack is optimally facilitated that a normal aldehyde cannot achieve.

FIGURE 2.29 Three different pathways of 2-oxoaldehydes and secondary amines.

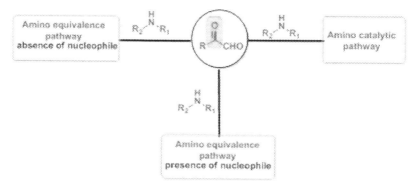

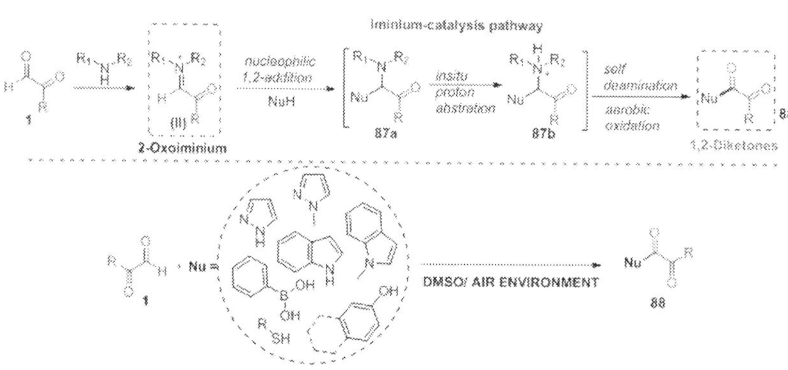

FIGURE 2.30 The amino catalytic pathway of 2-oxoaldehydes and amines to give 1,2-diones.

In the presence of external nucleophiles, 2-oxoiminium ion behaves differently to form different valuable structures.[28] Some examples of amino-equivalence in the presence of external nucleophiles for the synthesis of different heterocyclic compounds are depicted in Fig. 2.32 and are explained in Chapter 3.

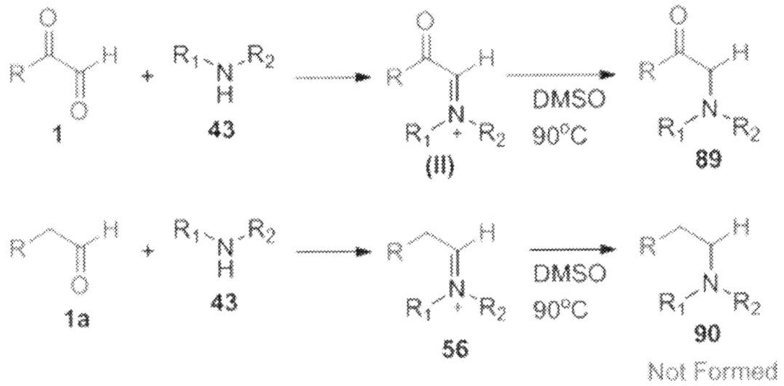

FIGURE 2.31 Synthesis of ketoamides through amino equivalence pathway.

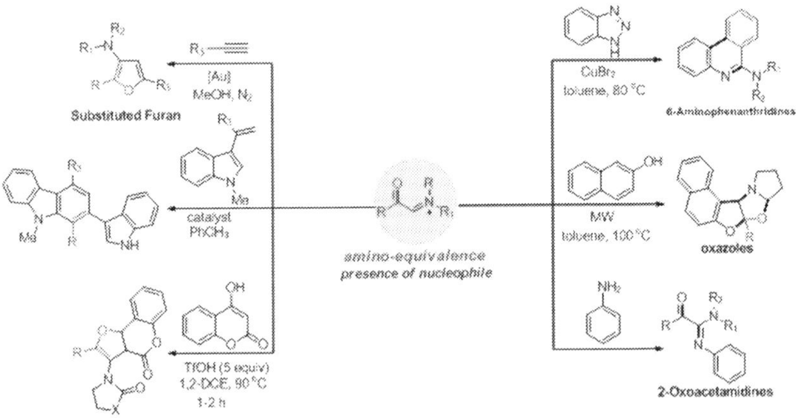

FIGURE 2.32 Different amino-equivalence reactions in the presence of different external nucleophiles.

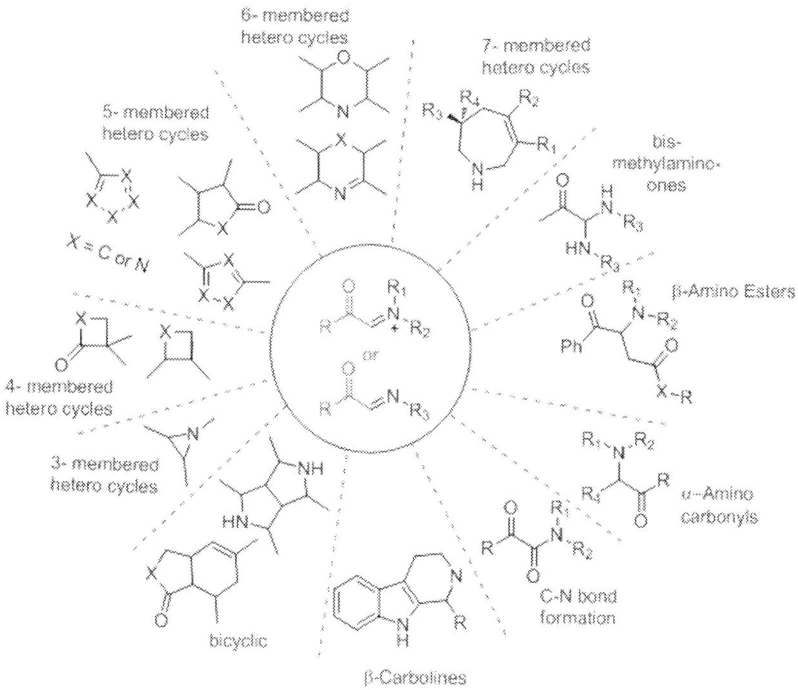

FIGURE 2.33   Applications of 2-oxoiminum and 2-oxoimine.

Furthermore, iminium ion in general and α-carbonyl iminium ion, in particular, are known for their vital role in organo-synthesis and functional group transformations.[22] 2-oxoiminum/2-oxoimine have broad applications for the generation of different heterocyclic compounds of different size as shown in Fig. 2.33.

2.2   Spectral characteristics

2.2.1 ^{1}H and ^{13}C-NMR spectra

Nuclear magnetic resonance (NMR) spectroscopy is an important spectroscopic technique that is commonly used for structure determination in modern organic chemistry. ^{1}H, ^{13}C-NMR, and DEPT spectra of some of the representative examples of 2-oxoaldehydes and different products that were prepared from 2-oxoaldehydes.

2.2.1.1 (E)-1-(2,4-dimethylphenyl)-2-morpholino-2-(phenylimino) ethanone

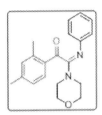

Yellow solid; mp 128−130°C; ^{1}H NMR (400 MHz, CDCl$_3$) δ 7.63 (d, J = 7.9 Hz, 1H), 7.05 (d, J = 7.8 Hz, 1H), 7.00 (t, J = 7.8 Hz, 2H), 6.90 (s, 1H), 6.77 (t, J = 7.4 Hz, 1H), 6.71−6.62 (m, 2H), 3.78 (s, 4H), 3.54 (s, 4H), 2.32 (s, 3H), 2.31 (s, 3H). ^{13}C NMR (125 MHz, CDCl$_3$) δ 195.17, 157.45, 148.33, 144.54, 140.89, 133.16, 132.54, 131.12, 128.25, 126.64, 122.34, 122.31, 66.68, 21.61, 21.42. IR (CHCl$_3$, cm^{-1}) ν; 3385, 2964, 2853, 1667, 1607, 1433, 1212, 981, 769, 544; HRMS (TOF) m/z [M + H]$^+$ Calcdforfor C$_{20}$H$_{22}$N$_2$O$_2$ 323.1760 found 323.1772.

2.2.1.2 ^{1}H-NMR (400 MHz), ^{13}C-NMR, and DEPT (125 MHz) spectra (CDCl$_3$)

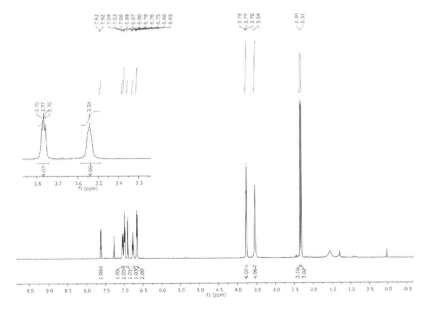

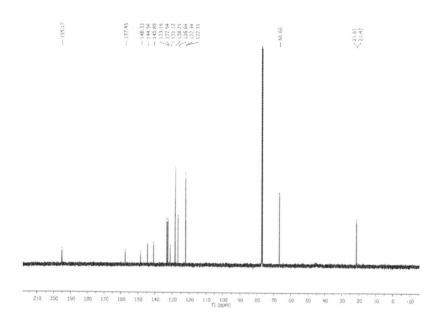

2.2.1.3   (1-(4-hydroxyphenyl)-1H-benzo[d]imidazol-2-yl)(phenyl) methanone

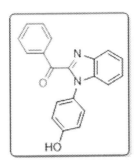

Yellow solid; mp 152–154°C; ¹H NMR (400 MHz, MeOD) δ 8.47 (d, $J = 8.3$ Hz, 1H), 8.03 (d, $J = 8.3$ Hz, 1H), 7.76 (d, $J = 7.4$ Hz, 2H), 7.68 (t, $J = 7.7$ Hz, 1H), 7.53 (t, J = 7.3 Hz, 2H), 7.37 (t, $J = 7.8$ Hz, 2H), 6.86 (d, $J = 8.7$ Hz, 2H), 6.55 (d, $J = 8.7$ Hz, 2H). ¹³C NMR (125 MHz, MeOD) δ 189.5, 155.7, 148.8, 145.6, 136.6, 134.8, 133.9, 130.7, 129.8, 129.10, 129.0, 126.2, 122.8, 119.3, 115.2, 114.6. IR (CHCl₃, cm-1) ν; 3414, 2099, 1649, 1514, 1450, 1216, 1165, 987, 768, 666, 541; HRMS (TOF) m/z [M + H] + Calcd for $C_{20}H_{15}N_2O_2$ 315.1134 found 315.1117.

2.2.1.4   ¹H NMR (400 MHz), ¹³C NMR, and DEPT (125 MHz) spectra (CD₃OD)

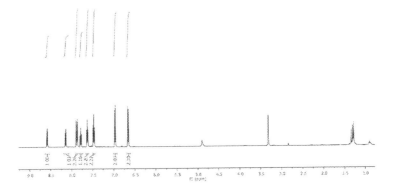

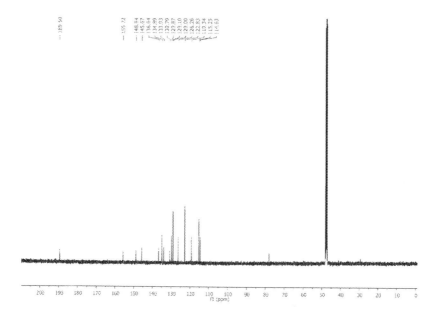

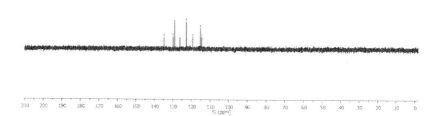

2.2.1.5 2-oxo-N-phenyl-2-(p-tolyl)acetamide

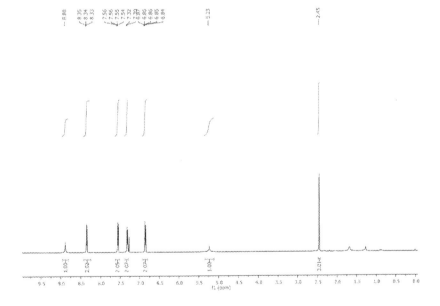

Yellow solid; mp 184−186°C; ^{1}H NMR (400 MHz, CDCl$_3$) δ 8.88 (s, 1H), 8.40−8.25 (m, 2H), 7.62−7.49 (m, 2H), 7.31 (d, $J = 8.5$ Hz, 2H), 6.97−6.76 (m, 2H), 5.23 (s, 1H), 2.45 (s, 3H). ^{13}C NMR (125 MHz, CDCl$_3$) δ 187.0, 159.0, 153.1, 145.9, 131.6, 130.6, 129.8, 129.3, 121.8, 115.9, 21.9. IR (CHCl$_3$, cm^{-1}) ν; 3347, 2960, 1689, 1643, 1340, 982, 750, 583; HRMS (TOF) m/z [M + H]$^{+}$ Calcd for C$_{15}$H$_{14}$NO$_3$ 256.0974 found 256.0992.

2.2.1.6 ^{1}H NMR (400 MHz), ^{13}C NMR, and DEPT (125 MHz) spectra (CDCl$_3$)

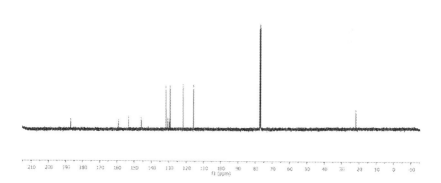

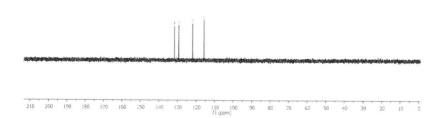

2.2.1.7 Ethyl 3-(1H-benzo[d][1,2,3]triazol-1-yl)-4-oxo-4-(p-tolyl) butanoate

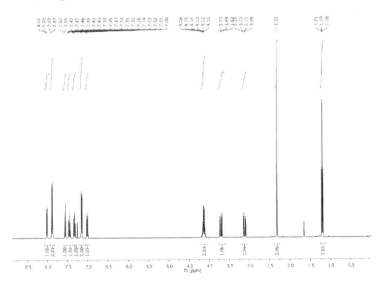

Semi solid; ^{1}H NMR (400 MHz, CDCl$_3$) δ 8.01 (d, J = 8.4 Hz, 1H), 7.88 (d, J = 8.3 Hz, 2H), 7.56 (d, J = 8.4 Hz, 1H), 7.49−7.43 (m, 1H), 7.37−7.30 (m, 1H), 7.15 (d, J = 8.1 Hz, 2H), 7.01 (dd, J = 8.2, 6.3 Hz, 1H), 4.14 (dt, J = 7.1, 3.9 Hz, 2H), 3.70 (dd, J = 16.7, 8.2 Hz, 1H), 3.12 (dd, J = 16.7, 6.3 Hz, 1H), 2.31 (s, 3H), 1.19 (t, J = 7.1 Hz, 3H). ^{13}C NMR (125 MHz, CDCl$_3$) δ 191.40, 169.82, 146.44, 145.40, 132.05, 131.58, 129.63, 128.97, 128.11, 124.26, 120.34, 110.02, 77.21, 61.35, 59.22, 35.16, 21.66, 14.00. IR (CHCl$_3$, cm^{-1}) ν; 3063, 2984, 2927, 2912, 1734, 1693, 1575, 1453, 1371, 1247, 1093, 1024, 778, 743, 505; HRMS (TOF) m/z [M + H] + Calcd for C$_{19}$H$_{19}$N$_3$O$_3$ 338.1505 found 338.1492.

2.2.1.8 ^{1}H NMR (400 MHz), ^{13}C NMR, and DEPT (125 MHz) spectra (CD$_3$OD)

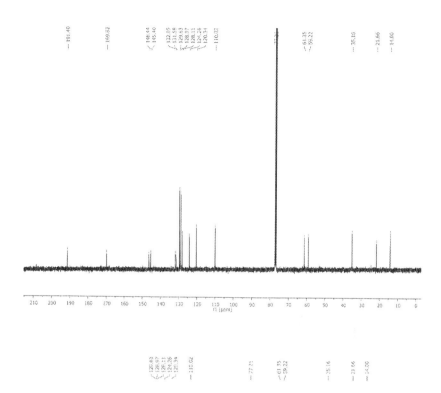

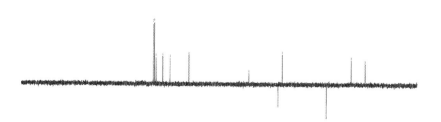

2.3   Reactivity comparison of 2-oxoacids with acids

Carboxylic acids exist in dimeric form through H-bonds and show similarity to alcohols and ketones in certain cases. The carboxyl carbon is sp^2-hybridized as in the case of ketones; therefore, the carboxylic acid group is planar with C−CO and O−CO bond angles of approximately 120 degrees (Fig. 2.34).

All the basic features like conformation and hybridization of 2-oxoacids is same as carboxylic acids however like 2-oxoaldehydes difference arises due to the electronic effect which gives rise to different reactivity in case of 2-oxoacids. Carboxylic acids exist in two conformations. The (E) conformation of a carboxylic acid is less stable by 3−5 kcal mol^{-1} as compared with its (Z) conformer due to electronic effects. 2-oxoacids are governed by the same criteria in which the two adjacent carbonyl groups are maximum apart (θ = 180 degrees), but only the transient conformation is very unfavorable where lone electrons on O atom clash with each other (Fig. 2.35).

In the case of normal carboxylic acids, the lone pair on an oxygen atom in (Z) conformation is aligned to overlap with σ^*C−O of carbonyl oxygen and in its (E) conformation the lone pair of oxygen atom is aligned to overlap with σ^*C−R bond.

Acetic acid dimer

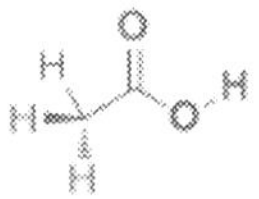

Bond angle	$\theta(°)$	bond length	(pm)
C-C=O	119	C-C	152
C-C-OH	119	C=C	125
O=C-OH	122	C-OH	131

FIGURE 2.34   Bond properties of acetic acid.

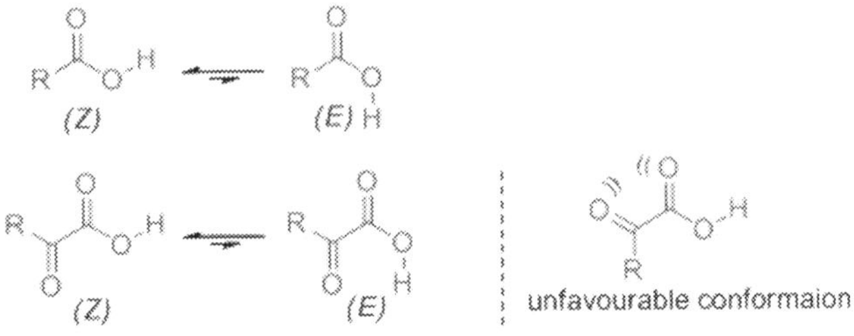

FIGURE 2.35   Favoured conformation in carboxylic acid and 2-oxoacid.

While as the (E) conformation of 2-oxoacid the lone pair of oxygen atoms is aligned to overlap with σ^*C-O bond of a-carbonyl. In addition, the orbital overlap drive in 2-oxoacid is increased due to the electrophilic 2-oxo unit, thereby increasing the magnitude of reactivity toward nucleophiles (Fig. 2.36).

The resonance structure indicates that the rotation about the $C-O$ bond is hindered with rotational barriers $10-12$ kcal mol^{-1} as a measure of the pi bond strength. The filled p-orbital of O atom interacts with π and π^* orbitals in $C=O$ and gives rise to 3-center 4-electron bonding system (Fig. 2.37).

2-oxoacid ion is more stabilized and the enolization step is much faster due to the electron-withdrawing nature of α-carbonyl unit bearing sp-hybridized carbon than that of simple carboxylic acids, which are also reflected in the reactivity of its related species as well.

2.3.1   General reactions of carboxylic acids

Like alcohols, carboxylic acids easily generate anions after deprotonation, which act as nucleophiles in SN^2 reactions and add to the carbonyl group. In

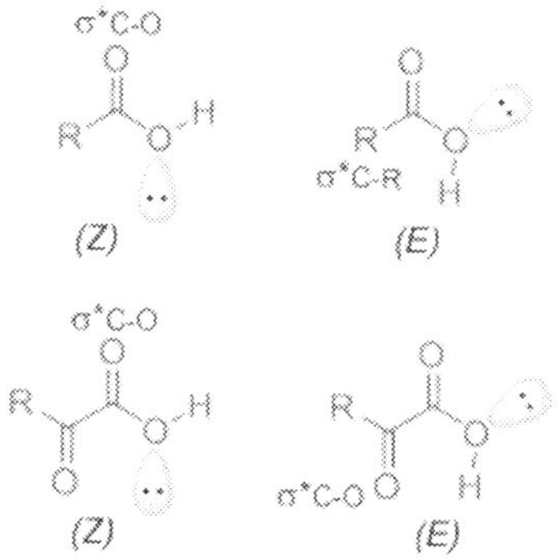

FIGURE 2.36   Lone pair conformations of carboxylic acid and 2-oxoacid.

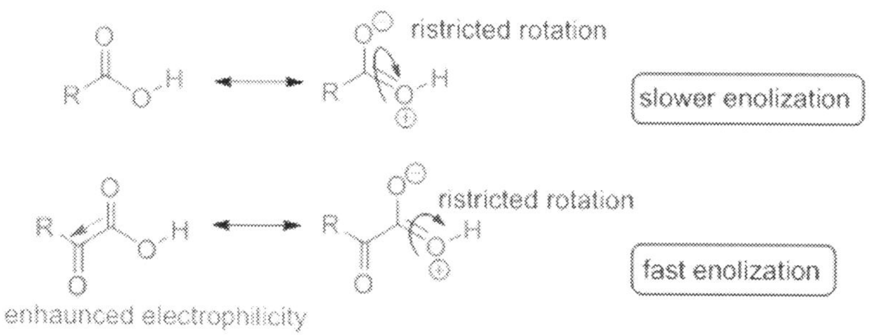

FIGURE 2.37   Rotation about $C-O$ single bond.

addition, carboxylic acids undergo certain reactions neither characteristic of alcohol nor ketones. Carboxylic acids show nucleophilic addition reactions toward carbonyl group similar to ketones. Different types of reactions in carboxylic acids are depicted in Fig. 2.38.

Carboxylic acids are employed as acyl-transfer agents without converting into acid chloride, anhydride, or amide. These reactions are catalyzed by different acids like trifluoroacetic anhydride or trifluoroacetic acid which can be easily recycled. The acylation carried out by carboxylic acids has an advantage over classic Friedel-Crafts acylation using more concentration of environmentally hazardous $AlCl_3$, which needs vigorous conditions giving rise to various side products. In this section, acylation and cyclization, apart from other reactions of carboxylic acids will be briefly discussed.

Acylation by using carboxylic acids **91a** usually follows basic nucleophilic and electrophilic modes and can occur as C-acylation or O-acylation. Direct C-acylation takes place on phenols **92** and naphthols by using carboxylic acids as acylating reagents (Fig. 2.39).[29,30]

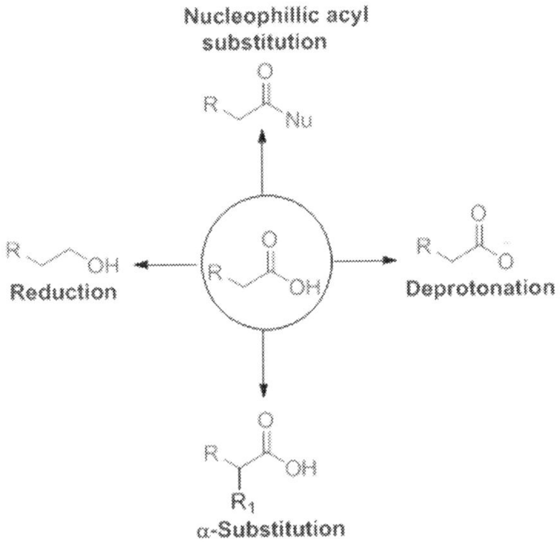

FIGURE 2.38 General reactions of carboxylic acid.

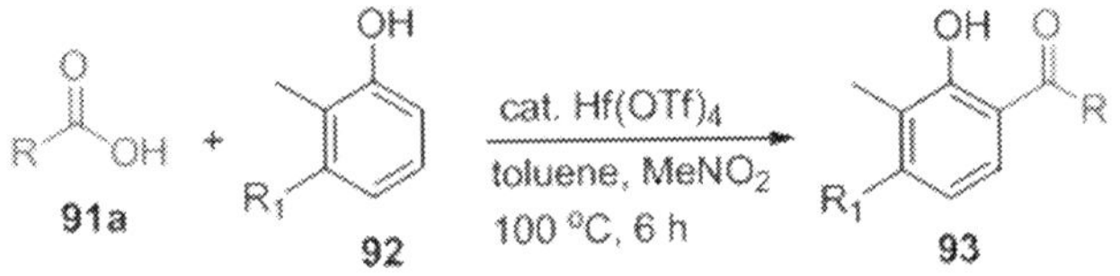

FIGURE 2.39 Hf catalyzed acylation of phenols with acids.

Carboxylic acids **91a** transfer acyl groups to heterocycles like N-substituted pyrrole **94** by simply adding trifluoroacetic acid in order to form C-substituted pyrroles **95** (Fig. 2.40).[31]

Similarly, trifluoroacetic acid catalyzes the acylation of carbazoles **96** in the presence of phosphoric acid. These reactions are simple to occur and do not follow complicated mechanisms (Fig. 2.41).[32]

Carboxylic acid **91a** delivers acyl group to aryl boronic acids **98** through Suzuki cross-coupling in the synthesis of ketones **99**.[33] This process directly forms ketones with 3-substituted benzoic acids which is an advantage over classical Friedel-Crafts acylation. The process is catalyzed by Pd and activators like pivalic acid, dimethyl dicarbonate (DMDC), or 2-chloro-1,3-dimethyl imidazolidinium chloride (DMC), and so on, and follows to complex steps like oxidative addition and reductive elimination during their formation (Fig. 2.42).[34]

Cyclization of carboxylic acid **100** can occur intramolecularly without decarboxylation under simple conditions (Fig. 2.43). Two different intramolecular

FIGURE 2.40 Acid catalyzed acylation of pyrroles.

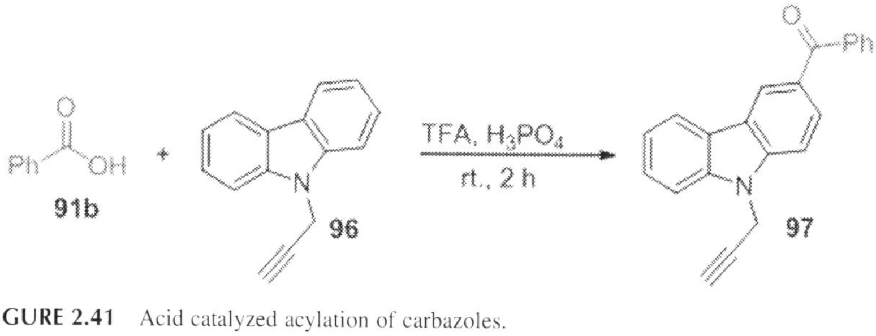

FIGURE 2.41 Acid catalyzed acylation of carbazoles.

R = Ar, vinyl and 1°/ 2°/ 3° alkyl

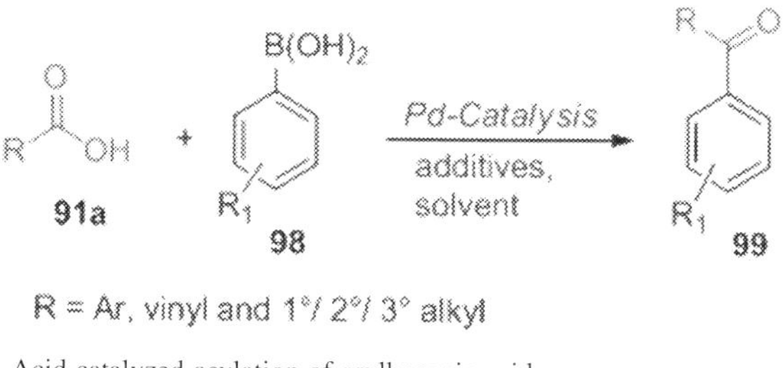

FIGURE 2.42 Acid catalyzed acylation of arylboronic acids.

cyclizations can occur depending upon the chain length by employing trifluoroacetic acid and TfOH to form five and six-membered rings **101** and **102**.[35]

However, when a catalytic amount of weak phosphoric acid is added to thiophene derivative of carboxylic acid **103**, cyclization occurs through nucleophilic attack from C-4 carbon atom of thiophene derivative to yield a bridged compound **104** (Fig. 2.44).[36]

2-phenyl carboxylic acid **105** cyclizes with alkenes **106** under same conditions through nucleophilic addition on TFA in first step followed by the attack over electrophilic carbonyl carbon by alkene to generate a carbocation intermediate **108** which undergoes nucleophilic attack by phenylring to produce 1,2,3,4-tetrahydronaphthalene derivative **107** (Fig. 2.45).[37]

2.3.2 Synthesis of tetrahydronaphthalene from carboxylic acids

2-oxoacid or α-keto acids are biologically significant molecules than normal carboxylic acids, which was discussed in Chapter 1, Synthesis and physical properties of 2-oxoaldehydes and 2-oxoacids. Apart from this, they serve as key starting materials for the formation of C−C, C−N, and C−S bonds. These are employed as a cleaner version of acylating agents instead of hazardous ones like acyl chlorides and other acyl-transfer reagents with carbon dioxide as the only by-product. The reactions of 2-oxoacids can be broadly

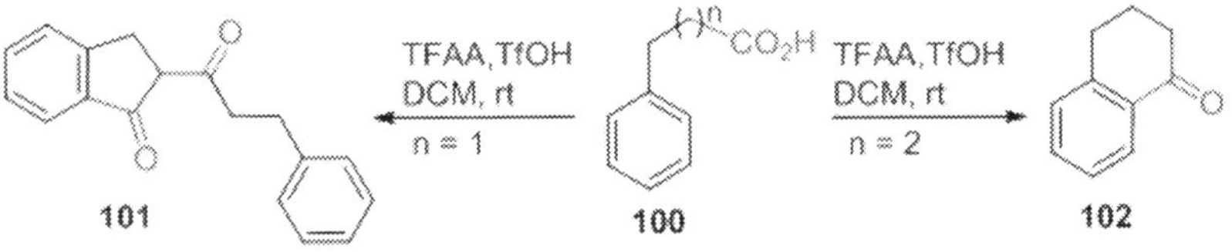

FIGURE 2.43 Acid catalyzed intramolecular cyclization of carboxylic compounds.

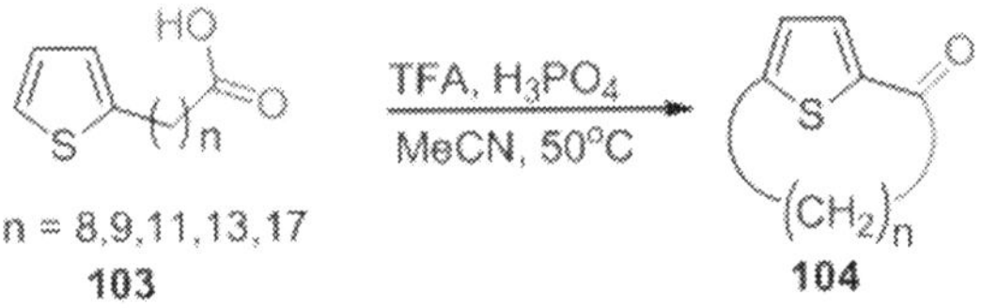

FIGURE 2.44 Acid catalyzed intramolecular cyclization of thiophene.

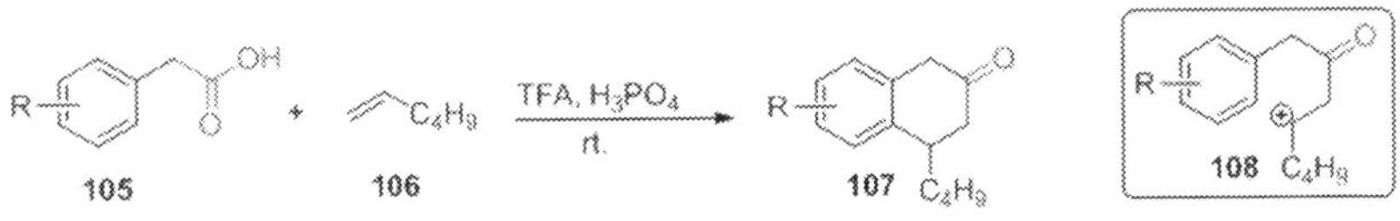

FIGURE 2.45 Acid catalyzed cyclization of.

classified into acyl-transfer (aryl, heteroatom, alkene, or alkyne acyl-transfer) and cyclization reactions (Fig. 2.46).

Although carboxylic acid, acyl chloride, and acetic anhydride are employed as acylating agents along with bases likes DMAP, NEt$_3$, or acids, the problem in using these reagents is their toxicity, sensitivity, storage problems, and difficulty in the recovery of soluble bases or acids. In case of acetic anhydrides, atom economy is the question in which only one acetyl group is utilized.

2-oxoacid conducts the acylation with loss of CO$_2$, called *decarboxylative acylation*, involving *acyl radical* species and is regarded as unique acylating agent for many different kinds of substrates like arenes, heterocyclic alkenes, and alkynes that mostly provide the formation of monoacylated product.[38] For instance, acylation of heteroarenes like pyrazine **109**, pyridine **111**, and quinoline **113** by employing 2-oxoacids along with (NH$_4$)$_2$S$_2$O$_8$ in (H$_2$O/CH$_2$Cl$_2$) produces the corresponding mono- and diacylated derivatives **110, 112, 114** in a regioselective manner (Fig. 2.47).

The acyl group can be directly launched separately through cross-coupling reactions over aryl halides **115** by using 2-oxoacids **91** that leads to the formation of ketones **116**. Different metal catalysts like Cu/Pd, Pd, Ir, and Ni are employed in presence of different chiral ligands like phenanthroline, NMP/quinoline, xantphos, and nixantphos (Fig. 2.48).[39]

Ortho-acylation is the other way of acyl-transfer in aryl substrates. These reactions do not need prefunctionalized coupling partners and costly or sensitive metal catalysts or ligands which may deliver unwanted side products. In addition, 2-oxoacids also promote ortho-acylation after C−H bond activation.

The ortho-acylation of unactivated 2-phenyl pyridine **117** through C-H bond activation catalyzed by Pd(PhCN)$_2$Cl$_2$ is directed by its pyridine ring promoted by Ag$^+$ and S$_2$O$_8^{-2}$ ions (Fig. 2.49).[40]

Indole or its derivatives have a broad biological significance while as some are important pharmaceuticals[41] others like acylated indoles have shown anticancer and anti-HIV potential[42] (Fig. 2.50).

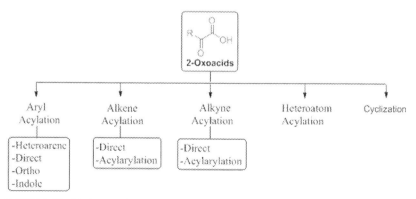

FIGURE 2.46 Various reactions of 2-oxoacids.

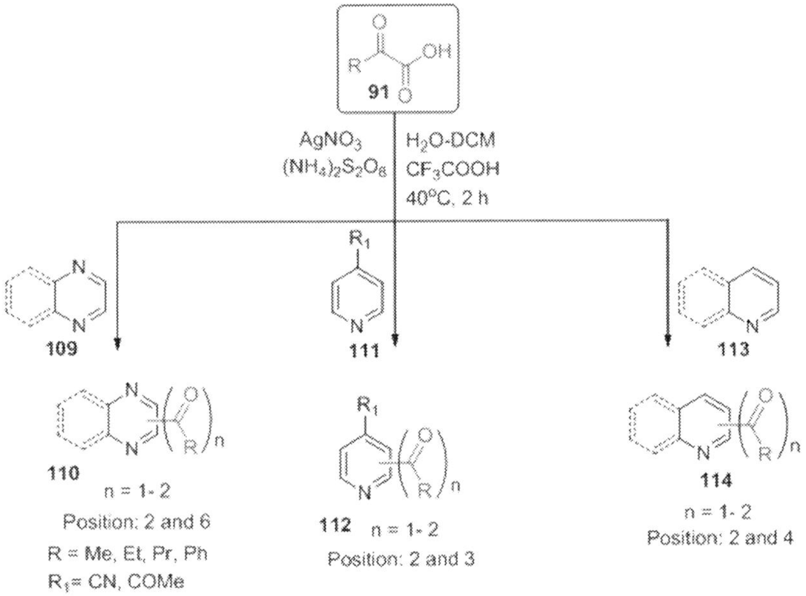

FIGURE 2.47 The decarboxylative acylation of heterocycles by 2-oxoacids.

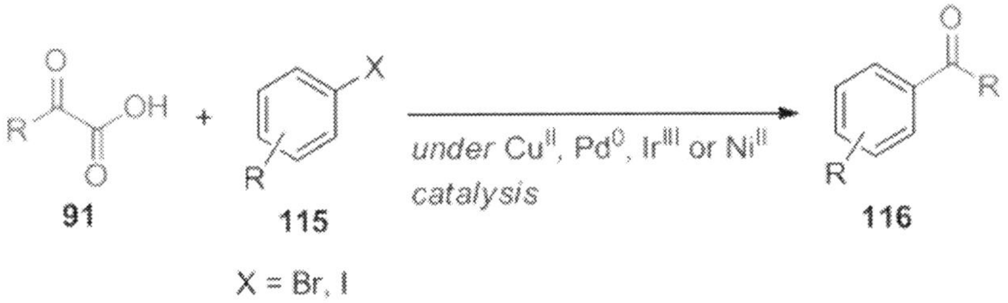

FIGURE 2.48 Metal catalyzed decarboxylative acylation of arylhalides by 2-oxoacids.

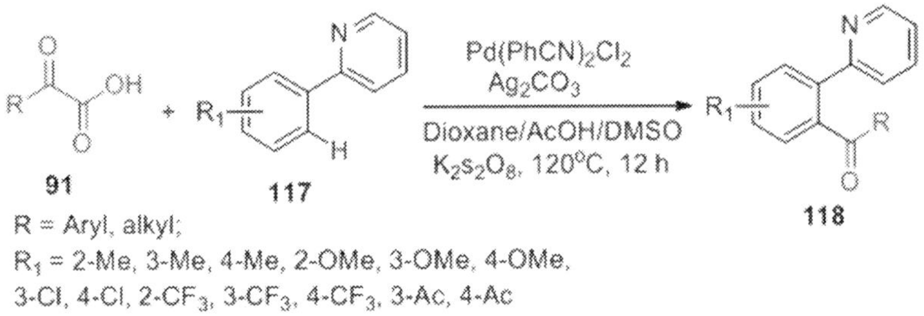

FIGURE 2.49 The decarboxylative ortho-acylation of biaryl compounds.

Although C3 is the most active site for acylation of indole, which is assisted by Cu, Pd, and Ag metal catalysts, 2-oxoacids can be used to transfer the acyl group at C-2 and C-7 in indole. N-substituted indoles **119** easily undergo decarboxylative acylation at C3 position **120** catalyzed by Cu(II) and N-boc protected indoles lose protection to give an acylated product. However, when a free indole or strong electron-withdrawing group is present at N atom, the reaction is not successful (Fig. 2.51).[43]

2-oxoacids are used in the decarboxylative acyl-transfer on alkenes to form α,β-unsaturated ketones following a complex mechanistic pathway. For example, allyl alcohols **123** and allyl carbonates **122** undergo direct acylation

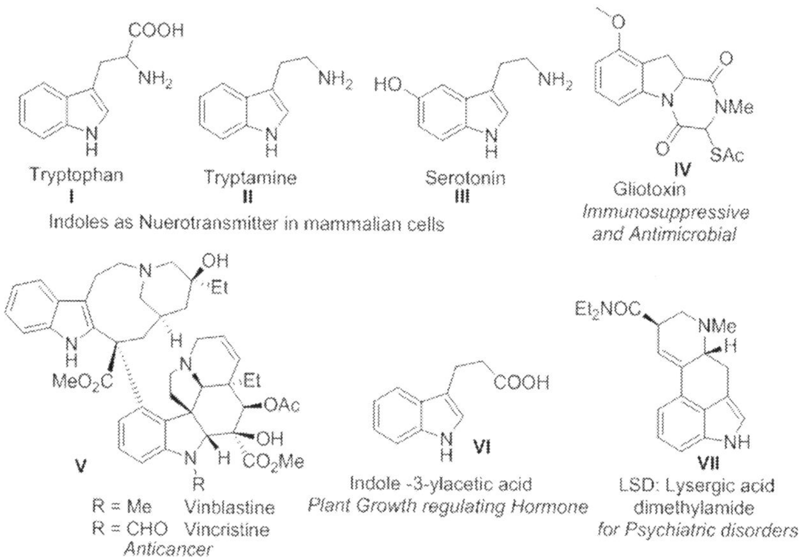

FIGURE 2.50   Some indole based bioactive molecules.

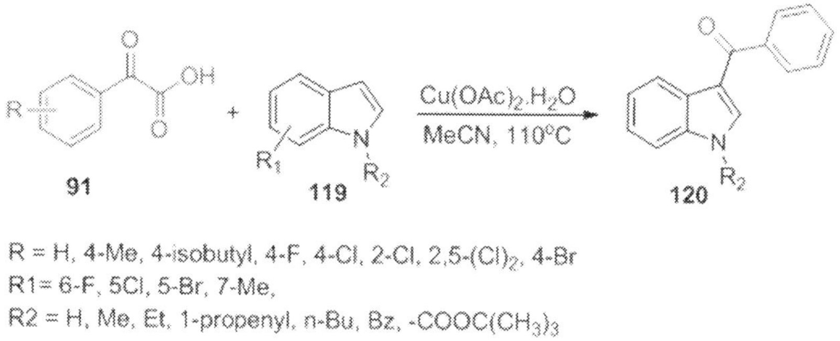

R = H, 4-Me, 4-isobutyl, 4-F, 4-Cl, 2-Cl, 2,5-(Cl)₂, 4-Br
R1= 6-F, 5Cl, 5-Br, 7-Me,
R2 = H, Me, Et, 1-propenyl, n-Bu, Bz, -COOC(CH₃)₃

FIGURE 2.51   Acyl transfer of indoles by 2-oxoacids.

to form aryl enones **121** under similar catalytic systems. Phosphine is necessary for decarboxylation and fetching quantitative yield. The reaction is effected by electron-withdrawing, electron-donating, and sterically hindered groups with no reaction taking place with alkylglyoxylic acids mesitylglyoxylic acid and p-nitrophenylglyoxylic acid (Fig. 2.52).[44]

In a similar way, 2-oxoacid **91** delivers acyl group to alkenes **124** bearing electron-withdrawing groups in a photoredox reaction[45] catalyzed by iridium. The nucleophilic acyl radical adds to the alkene through Micheal addition to form γ-carbonyl derivative. The presence of light and K_2HPO_4 is important for this reaction to occur. The use of cinnamaldehyde and acrylic acid were unsuccessful in product formation (Fig. 2.53).

Although, ynones are obtained from the reaction of toxic carbon monoxide with aryl halides or aryl triflates under Pd catalysis.[46] The better alternative for safer and ecofriendly acylation over acyl chlorides and carbon monoxide is the use of 2-oxoacids (Fig. 2.54). The visible light promotes the decarboxylative acylation of alkynes **126** using hypervalent iodine **128** reagents and Ru(II) catalyst leading to the formation of ynones **127**.

Since amide bonds are the most significant linkages in organic chemistry.[47] Therefore, a very reliable and easy method is the application of 2-oxoacids for delivering acyl group to amino acids, particularly, forming N-acylated amino acids **130** through decarboxylative amidation method by trapping molecular oxygen as oxidant without any metal catalysts (Fig. 2.55).[48]

Apart from acting as an important acyl transferring agent, 2-oxoacids show unique cyclization reactions. These reactions unlike normal carboxylic acids are accompanied by decarboxylation involving acyl radical formation.

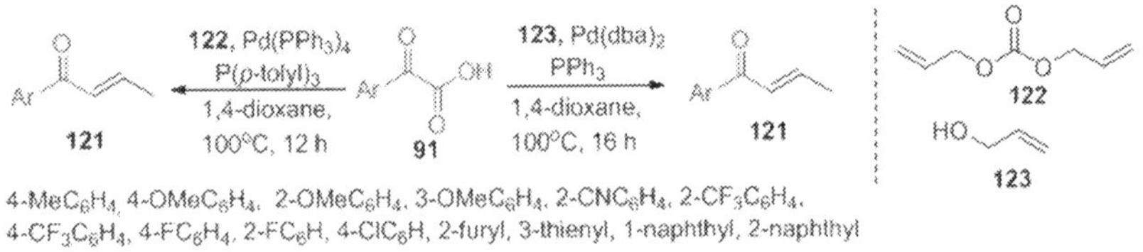
122, Pd(PPh3)4
P(p-tolyl)3
1,4-dioxane,
100°C, 12 h
123, Pd(dba)2
PPh3
1,4-dioxane,
100°C, 16 h
Ar
OH
121
91
121
122
HO
123
4-MeC6H4, 4-OMeC6H4, 2-OMeC6H4, 3-OMeC6H4, 2-CNC6H4, 2-CF3C6H4,
4-CF3C6H4, 4-FC6H4, 2-FC6H, 4-ClC6H, 2-furyl, 3-thienyl, 1-naphthyl, 2-naphthyl

FIGURE 2.52 Acyl transfer of allyl alcohols and allyl carbonates by 2-oxoacids.

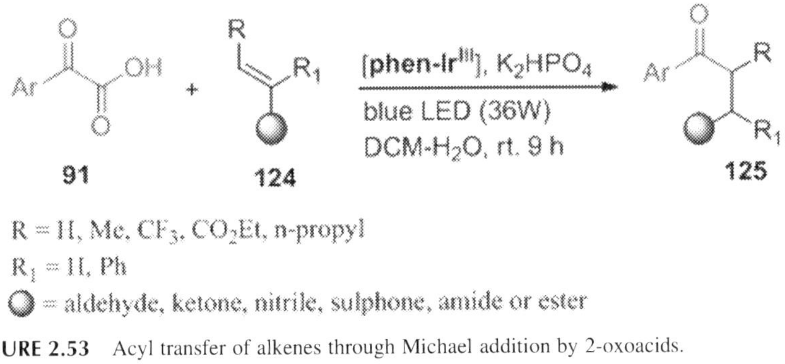
Ar
OH
R
R1
[phen-Ir^III], K2HPO4
blue LED (36W)
DCM-H2O, rt. 9 h
91
124
125
R = H, Me, CF3, CO2Et, n-propyl
R1 = H, Ph
= aldehyde, ketone, nitrile, sulphone, amide or ester

FIGURE 2.53 Acyl transfer of alkenes through Michael addition by 2-oxoacids.

These reactions are catalyzed by metals like Fe(III), Ag(I), and Cu(I) and acids like $BF_3.OEt_2$, TFA, and p-TSA. Many times the reactions take place through dipolar addition but almost all cyclizations are driven by 2-oxo unit present in 2-oxoacids 131–135. Some of the cyclization reactions pertaining to 2-oxoacids are highlighted in Fig. 2.56.

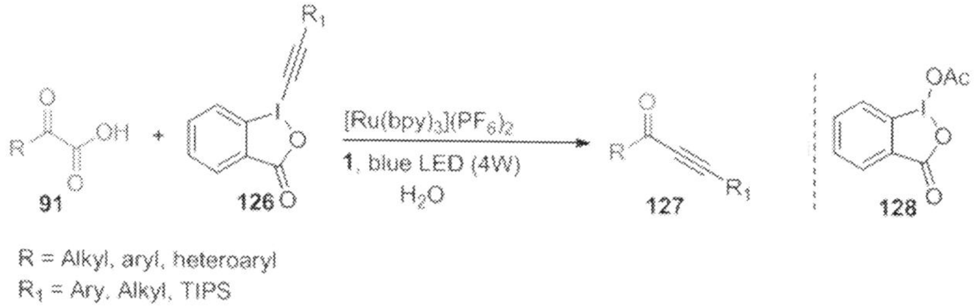

R = Alkyl, aryl, heteroaryl
R_1 = Ary, Alkyl, TIPS

FIGURE 2.54   Light promoted Acyl transfer of ynones by 2-oxoacids.

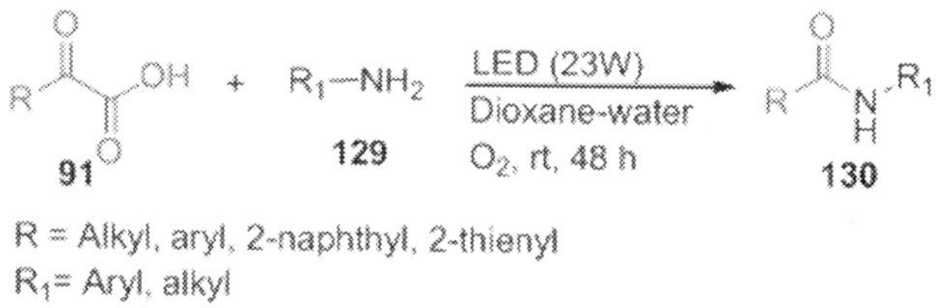

R = Alkyl, aryl, 2-naphthyl, 2-thienyl
R_1= Aryl, alkyl

FIGURE 2.55   Acyl transfer of amines by 2-oxoacids to form amino acids.

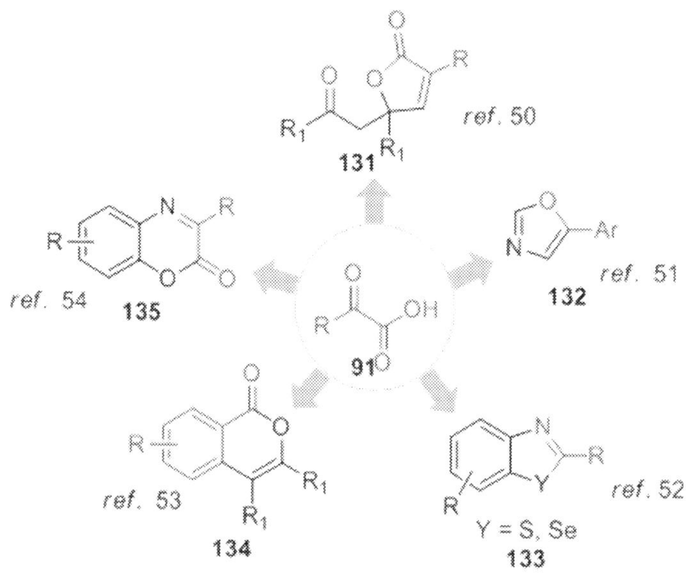

FIGURE 2.56   Cyclization reactions of 2-oxoacids.

These reactions follow complicated mechanisms that have already been discussed in the previous chapter and, therefore, are counted as reactions going in a different way as compared with normal acids where the 2-oxo assisting group is absent.

2.4   Spectral characteristics of 2-oxoaldehydes and 2-oxoacids

2.4.1 ^{1}H and ^{13}C-NMR spectra

2.4.1.1   2-Oxo-2-phenylacetic acid

^{1}H NMR (400 MHz) and ^{13}C NMR (125 MHz) spectra (CDCl$_3$)

^{1}H NMR (400 MHz, CDCl$_3$) δ 9.64 (s, 1H), 8.08 (d, J = 7.8 Hz, 2H), 7.58 (t, J = 7.4 Hz, 1H), 7.41 (t, J = 7.8 Hz, 2H). ^{13}C NMR (101 MHz, CDCl$_3$) δ 185.54, 164.34, 135.53, 131.87, 130.81, 129.12. HRMS (TOF) m/z [M + H] + Calcd for C$_{20}$H$_{25}$O$_8$ 151.0390 found 151.0395.

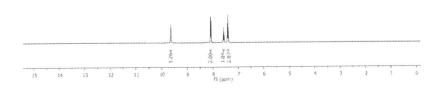

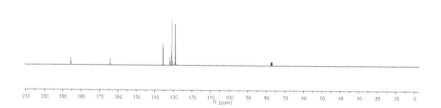

2.4.2 1,2:5,6 di-O-isopropylidene-3-O-oxobenzoyl-α D-glucofuranoside

Isolated as a syrup; ^{1}H NMR (400 MHz, CDCl$_3$) δ 8.06−8.01 (m, 2H), 7.67 (t, J = 7.5 Hz, 1H), 7.50 (t, J = 7.8 Hz, 2H), 5.93 (d, J = 3.7 Hz, 1H), 5.65 (d, J = 3.0 Hz, 1H), 4.64 (d, J = 3.7 Hz, 1H), 4.27 (dd, J = 8.6, 3.0 Hz, 1H), 4.17 (m, J = 8.5, 5.8, 4.6 Hz, 1H), 4.09 (dd, J = 8.7, 6.0 Hz, 1H), 4.01 (dd, J = 8.8, 4.5 Hz, 1H), 1.54 (s, 3H), 1.46 (s, 3H), 1.33 (s, 6H). ^{13}C NMR (101 MHz, CDCl$_3$) δ 185.64, 162.49, 135.16, 132.30, 130.11, 128.93, 112.47, 109.56, 105.36, 83.28, 80.20, 77.27, 72.34, 67.60, 26.92, 26.74, 26.23, 25.23; HRMS (TOF) m/z [M + H]+ Calcd for C$_{20}$H$_{25}$O$_8$ 393.1544 found 393.1547.

2.4.3 ¹H and ¹³C-NMR spectra

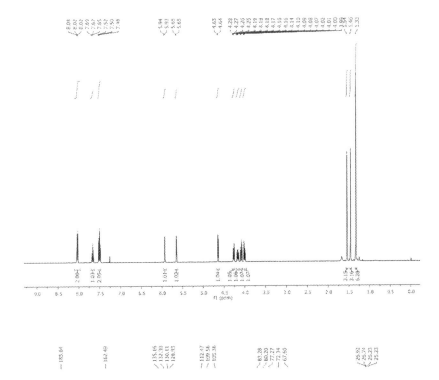

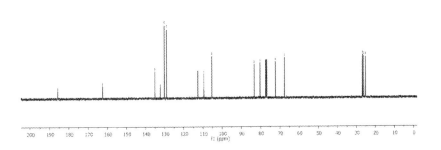

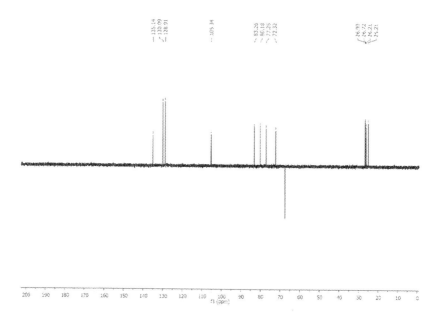

References

1. Erkkila, A.; Majander, I.; Pihko, P. M. *Chem. Rev.* **2007**, *107*, 5416–5470.
2. (a) Appel, R.; Mayr, H. *J. Am. Chem. Soc.* **2011**, *133*, 8240–8251.
   (b) Appel, R.; Chelli, S.; Tokuyasu, T.; Troshin, K.; Mayr, H. *J. Am. Chem. Soc.* **2013**, *135*, 6579–6587.
3. (a) Smith, M. B.; March, J. *Advanced Organic Chemistry*, 5th ed.; Wiley Interscience: New York, 20011218–1223.
   (b) Mahrwald, R. *Modern Aldol Reactions 1,2;* Wiley-VCH: Weinheim, Germany, 20041218–1223.
4. Mukaiyama, T. *Org. React.* **1982**, *28*, 203–331.
5. (a) Okude, Y.; Hirano, S.; Hiyama, T.; Nozaki, H. *J. Am. Chem. Soc.* **1977**, *99*, 3179–3181.
   (b) Takai, K.; Tagashira, M.; Kuroda, T.; Oshima, K.; Utimoto, K.; Nozaki, H. *J. Am. Chem. Soc.* **1986**, *108*, 6048–6050.
   (c) Takai, K.; Kimura, K.; Kuroda, T.; Hiyama, T.; Nozaki, H. *Tetrahedron Lett* **1983**, *24*, 5281–5284.
6. (a) Pudovik, A.; Arbuzov, A. *Dokl. Akad. Nauk. SSSR* **1950**, *73*, 327.
   (b) Rulev, A. Y. *RSC Adv.* **2014**, *4*, 26002–26012.
   (c) Stawinski, J.; Kraszewski, A. *Acc. Chem. Res.* **2002**, *35*, 952–960.
   (d) Ramananarivo, H. R.; Solhy, A.; Sebti, J.; Smahi, A.; Zahouily, M.; Clark, J.; Sebti, S. *ACS Sustain. Chem. Eng.* **2013**, *1*, 403–409.
7. (a) Arundale, E.; Mikeska, L. A. *Chem. Rev.* **1952**, *51*, 505–555.
   (b) Miles, R. B.; Davis, C. E.; Coates, R. M. *J. Org. Chem.* **2006**, *71*, 1493–1501.

8. Belen'Kii, L. I.; Vol'Kenshtein, Yu. B.; Karmanova, I. B. *Russ. Chem. Rev.* **1977**, *46*, 891−903.

9. (a) Tishchenko, V. *J. Russ. Phys. Chem. Soc.* **1908**, *38*, 355.
   (b) Gnanadesikan, V.; Horiuchi, Y.; Ohshima, T.; Shibasaki, M. *J. Am. Chem. Soc.* **2004**, *126*, 7782−7783.

10. (a) Cannizzaro, S. *Justus. Liebigs. Ann. Chem.* **1853**, *88*, 129−130.
    (b) Elderfield, R. C. *J. Chem. Educ.* **1930**, *7*, 594.

11. (a) Wittig, G.; Haag, W. *Chem. Ber.* **1955**, *88*, 1654−1666.
    (b) Maryanoff, B. E.; Reitz, A. B. *Chem. Rev.* **1989**, *89*, 863−927.

12. (a) Corey, E. J.; Chaykovsky, M. *J. Am. Chem. Soc.* **1965**, *87*, 1353−1364.
    (b) Peng, Y.; Yang, J.-H.; Li, W.-D. Z. *Tetrahedron* **2006**, *62*, 1209−1215.

13. (a) Corey, E. J.; Fuchs, P. L. *Tetrahedron Lett* **1972**, *13*, 3769−3772.
    (b) Wang, Z. In Comprehensive Organic Name Reactions and Reagents; *John Wiley & Sons, Inc.*

14. (a) Seyferth, D. *Organometallics* **1984**, *3*, 1775.
    (b) Connelly, N. G.; Geiger, W. E. *Chem. Rev.* **1996**, *96*, 877−910.

15. Noble, A.; Anderson, J. C. *Chem. Rev.* **2013**, *113*, 2887−2939.

16. (a) Ugi, I.; Steinbrückner, C. *Angew. Chem.* **1960**, *72*, 267−268.
    (b) Dömling, A.; Ugi, I. *Angew. Chem. Int. Ed.* **2000**, *39*, 3168−3210.

17. (a) Callebaut, G.; Meiresonne, T.; De Kimpe, N.; Mangelinckx, S. *Chem. Rev.* **2007**, *114*, 7954−8015.
    (b) Denolf, B.; Leemans, E.; De Kimpe, N. *J. Org. Chem.* **2007**, *72*, 3211−3217.

18. (a) Stork, G.; Dowd, S. R. *J. Am. Chem. Soc.* **1963**, *85*, 2178−2180.
    (b) Stork, G.; Brizzolara, A.; Landesman, H.; Szmuszkovicz, J.; Terrell, R. *J. Am. Chem. Soc.* **1963**, *85*, 207−222.

19. (a) Dirksen, A.; Dirksen, S.; Hackeng, T. M.; Dawson, P. E. *J. Am. Chem. Soc.* **2006**, *128*, 15602−15603.
    (b) Morales, S.; Guijarro, F. G.; Ruano, J. L. G.; Cid, M. B. N. *J. Am. Chem. Soc.* **2014**, *136*, 1082−1089.
    (c) Cordes, E. H.; Jencks, W. P. *J. Am. Chem. Soc.* **1962**, *84*, 826−831.

20. Ahrendt, K. A.; Borths, C. J.; MacMillan, D. W. C. *J. Am. Chem. Soc.* **2000**, *122*, 4243.

21. Candeias, N. R.; Montalbano, F.; Cal, P. M. S. D.; Gois, P. M. P. *Chem. Rev.* **2010**, *110*, 6169−6193.

22. Eftekhari-Sis, B.; Zirak, M.; Akbari, A. Arylglyoxals in synthesis of heterocyclic compounds. *Chem. Rev.* **2013**, *113*, 2958−3043.

23. Akiyama, T.; Suzuki, T.; Mori, K. *Org. Lett.* **2009**, *11*, 2445.

24. Anil, K. P.; Nagaraju, M.; Deepika, S.; Ram, A. V.; Qazi, N. A. *Eur. J. Org. Chem.* **2015**, 3577−3586.

25. Tsuge, O.; Uneo, K.; Kanemasa, S.; Yorozu, K. *B. Chem. Soc. Jpn.* **1986**, *59*, 1809.

26. Bossio, R.; Marcaccini, S.; Pepino, R. *Synthesis* **1994**, 765.

27. Mupparapu, N.; Khan, S.; Battula, S.; Kushwaha, M.; Gupta, A. P.; Ahmed, Q. N.; Vishwakarma, R. A. *Org. Lett.* **2014**, *16*, 1152−1155.

28. (a) Battula, S.; Kumar, A.; Gupta, A. P.; Ahmed, Q. N. *Org. Lett.* **2015**, *17*, 5562−5565.
    (b) Battini, N.; Battula, S.; Kumar, R. R.; Ahmed, Q. N. *Org. Lett.* **2015**, *17*, 2992−2995.
    (c) Kumar, A.; Battini, N.; Kumar, R. R.; Athimoolam, S.; Ahmed, Q. N. Air-assisted 2-oxo-driven dehydrogenative α, α-diamination of 2-oxo aldehydes to 2-oxo acetamidines. *Eur. J. Org. Chem.* **2016**, 3344−3348.
    (d) Li, J.; Liu, L.; Ding, D.; Sun, J.; Ji, Y.; Dong, J. *Org. Lett.* **2013**, *15*, 2884−2887.

    (e) Huang, Y.-W.; Li, X.-Y.; Fu, L.-N.; Guo, Q.-X. *Org. Lett.* **2016**, *18*, 6200–6203.

    (f) Babu, V. N.; Murugan, A.; Katta, N.; Devatha, S.; Sharada, D. S. *J. Org. Chem.* **2019**, *84*, 6631–6641.

29. Huang, H.; Song, C.; Chang, J. *Prog. Chem.* **2019**, *31* (1), 1–9.

30. Kobayashi, S.; Moriwaki, M.; Hachiya, I. *Tetrahedron Lett.* **1996**, *37* (24), 4183–4186.

31. Kakushima, M.; Hamel, P.; Frenette, R.; Rokach, J. *J. Org. Chem.* **1983**, *48*, 3214.

32. Kolli, S. K.; Prasad, B.; Babu, P. V.; Ashfaq, M. A.; Ehtesham, N. Z.; Raju, R. R.; Pal, M. *Org. Biomol. Chem.* **2014**, *12*, 6080.

33. (a) Gooßen, L.; Ghosh, K. *Angew. Chem. Int. Ed.* **2001**, *40*, 3458–3460.

    (b) Gooßen, L.; Ghosh, K. *Eur. J. Org. Chem.* **2002**, 3254–3267.

    (c) Kakino, R.; Narahashi, H.; Shimizu, I.; Yamamoto, A. *Bull. Chem. Soc. Jpn.* **2001**, *30*, 1242–1243.

34. (a) Kwon, Y. B.; Choi, B. R.; Lee, S. H.; Seo, J. S.; Yoon, C. M. *Bull. Korean Chem. Soc.* **2010**, *31*, 2672–2674.

    (b) Pathak, A.; Rajput, C.; Bora, P.; Sharma, S. *Tetrahedron Lett* **2013**, *54*, 2149–2150.

    (c) Garcia-Barrantes, P.; McGowan, K.; Ingram, S.; Lindsley, C. *Tetrahedron Lett* **2017**, *58*, 898–901.

    (d) Wu, H.; Xu, B.; Li, Y.; Hong, F.; Zhu, D.; Jian, J.; Pu, X.; Zeng, Z. *J. Org. Chem.* **2016**, *81*, 2987–2992.

    (e) Si, S.; Wang, C.; Zhang, N.; Zou, G. *J. Org. Chem.* **2016**, *81*, 4364–4370.

35. Kim, J.; Shokova, E.; Tafeenko, V.; Kovalev, V. *Beilstein J. Org. Chem.* **2014**, *10*, 2270.

36. Galli, C.; Illuminati, G.; Mandolini, L. *J. Org. Chem.* **1980**, *45*, 311.

37. Gray, A. D. :; Smyth, T. P. *J. Org. Chem.* **2001**, *66*, 7113.

38. Fontana, F.; Minisci, F.; Barbosa, M. C. N.; Vismara, E. *J. Org. Chem.* **1991**, *56*, 2866–2869.

39. (a) Gooßen, L. J.; Rudolphi, F.; Oppel, C.; Rodríguez, N. *Angew. Chem. Int. Ed.* **2008**, *47*, 3043–3045.

    (b) Ji, Y.; Yang, X.; Mao, W. *Appl. Organomet Chem.* **2014**, *28*, 678–680.

    (c) Zuo, Z.; Ahneman, D.; Chu, L.; Terrett, J.; Doyle, A. G.; MacMillan, D. W. C. *Science* **2014**, *345*, 437–440.

    (d) Noble, A.; MacMillan, D. W. C. *J. Am. Chem. Soc.* **2014**, *136*, 11602–11605.

    (e) Cheng, W.-M.; Shang, R.; Yu, H.-Z.; Fu, Y. *Chem. Eur. J* **2015**, *21*, 13191–13195.

40. Li, M.; Ge, H. *Org. Lett.* **2010**, *12*, 3464–3467.

41. (a) Sundberg, R. J. *The Chemistry of Indoles;* Academic Press: New York, 1996.

    (b) Casapullo, A.; Bifulco, G.; Bruno, I.; Riccio, R. *J. Nat. Prod.* **2000**, *63*, 447–451.

    (c) Garbe, T. R.; Kobayashi, M.; Shimizu, N.; Takesue, N.; Ozawa, M.; Yukawa, H. *J. Nat. Prod.* **2000**, *63*, 596–598.

    (d) Bao, B.; Sun, Q.; Yao, X.; Hong, X.; Lee, C.; Sim, C. J.; Im, K. S.; Jung, J. H. *J. Nat. Prod.* **2005**, *68*, 711–715.

    (e) Ertl, P.; Jelfs, S.; Mühlbacher, J.; Schuffenhauer, A.; Selzer, P. *J. Med. Chem.* **2006**, *49*, 4568–4573.

42. (a) Nicolaou, I.; Demopoulos, V. J. *J. Med. Chem.* **2003**, *46*, 417–426.

    (b) Barreca, M. L.; Ferro, S.; Rao, A.; De Luca, L.; Zappala, M.; Monforte, A.-M.; Debyser, Z.; Witvrouw, M.; Chimirri, A. *J. Med. Chem.* **2005**, *48*, 7084–7088.

    (c) Wu, Y.-S.; Coumar, M. S.; Chang, J.-Y.; Sun, H.-Y.; Kuo, F.-M.; Kuo, C. C.; Chen, Y.-J.; Chang, C.-Y.; Hsiao, C.-L.; Liou, J.-P., et al. *J. Med. Chem.* **2009**, *52*, 4941–4945.

43. Yu, L.; Li, P.; Wang, L. *Chem. Commun.* **2013**, *49*, 2368–2370.

44. Grünberg, M. F.; Gooßen, L. J. *J. Organomet Chem.* **2013,** *744,* 140−143.

45. Wang, G.-Z.; Shang, R.; Cheng, W.-M.; Fu, Y. *Org. Lett.* **2015,** *17,* 4830−4833.

46. Wu, X.-F.; Neumann, H.; Beller, M. *Chem. Soc. Rev.* **2011,** *40,* 4986−5009.

47. (a) Arthur, G. The Amide Linkage: Selected Structural Aspects in Chemistry, Biochemistry, and Materials Science; Wiley: Interscience, 2000.

     (b) Wieland, T.; Bodanszky, M. *The World of Peptides: A Brief History of Peptide Chemistry;* Springer, 1991.

     (c) Pattabiraman, V. R.; Bode, J. W. *Nature* **2011,** *480,* 471−479.

48. Wang, Q.; Zhang, X.; Fan, X. *Org. Biomol. Chem* **2018,** *16,* 7737−7747.

Chapter 3

Applications of 2-Oxoaldehydes

3.1 Introduction

2-Oxoaldehydes (AGs) possess both ketone and aldehyde, functional groups
with different reactivity, that play an important role in the synthesis of various important synthons and biologically important compounds. AGs (α-keto
aldehydes) in particular are among the most attractive precursors that are
used for the synthesis of different heterocyclic compounds and also to
develop carbon-heteroatom bond formation methods.[1] Reactions of AGs
were performed in three different ways: (1) reactions take place from the
more reactive aldehyde group and AGs provided one atom in the ring of the
heterocycles to produce a benzoyl substituent, (2) reactions take place from
the keto group of the AG without affecting the aldehydic group, (3) reaction
take place from the participation of both aldehyde and ketone groups of AG
to provide two atoms of the heterocyclic rings. Frequently, the aldehyde
group of AG reacts rapidly with different nucleophiles, which then undergo
cyclization either by the aldehyde group residue, which provides one atom of
heterocycle along with producing an aroyl substituent on the obtained rings
or by ketone of AG to provide two atoms of the heterocyclic rings. In addition, AGs and AG-imines can act as dienophiles in [4 + 2] and [2 + 2] cycloaddition reactions via $C=O$ and $C=N$ bonds respectively. In some cases,
AG-hydrates act as nucleophiles via the oxygen atom of the OH group. In
this chapter, we are particularly focusing on the application of AGs for the
construction of different valuable structures or medicinally important scaffolds. There have been many published articles on different reactions of
AGs and their derivatives, such as allylation, arylation, Cannizzaro, Henry,
Mannich, reductive amination, reductive coupling with dienes, and Wittig
reactions.[2]

In addition to some common heterocyclic compounds, other uncommon
heterocycles such as pyrrolizidine, indolizidine, furofuran, dioxophospholane,
β-carboline, benzoxathiin, fused heterocycles, and some seven-membered
heterocycles such as azepine, diazepine, and oxazepane are also reported
starting with AGs and their derivatives. Furthermore, AGs have been
explored against several reactions such as cyclo condensation, cycloaddition,
Pictet − Spengler, Ugi − Wittig, Ugi cyclo condensation, Aldol-
Paal − Knorr, and Wittig dehydrative cyclization Diels−Alder cycloaddition

Chemistry of 2-Oxoaldehydes and 2-Oxoacids. DOI: https://doi.org/10.1016/B978-0-12-824285-8.00004-2

reactions (Fig. 3.1). The unique reactivity of AGs is valuable for the formation of new molecules which otherwise will be either difficult or not possible to synthesize. This may include simple to complicated ring systems, aliphatic systems, carbocycles, and O, N, and S containing heterocycles and so on. In most of the case, a new bond formation like C−C, C−N, C−O, C−S, and C−P will take place and some of the representative structures are shown in (Fig. 3.2).[3]

In this chapter, we will discussing three different aspects of the reactions of AGs separately, that is, reactions from the aldehydic group, reactions from the keto group, and reactions from both aldehyde and keto group. In addition, the divergent reactivity of AGs is more firmly validated by various types of reactions involving addition, substitution, multicomponent, click, and pericyclic reactions, which will form the core subject of this chapter. This chapter will highlight the reactivity of arylglyoxal or its hydrate form in the synthesis of various molecules or substrates directly or through an intermediate including their different aspects of chemical modes (Fig. 3.3).

3.2 Reactions with aldehyde group

3.2.1 Synthesis of oxiranes

Oxirane is the three-membered cyclic ether and simplest oxygen-containing saturated heterocycle commonly called epoxide. Oxirane is present in natural products like cryptophycin A and B, which exhibits anticancer properties,[4] azinomycins A and B,[5] triptonid,[6] epoxomicin,[7] and psorospermin (from *Psorospermum febrifugum*), which shows activity against drug-resistant leukemia and AIDS-related lymphoma.[8,9] Other oxirane containing biologically

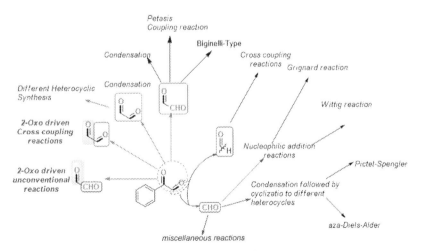

FIGURE 3.1 Different types of reactions in 2-oxoaldehydes.

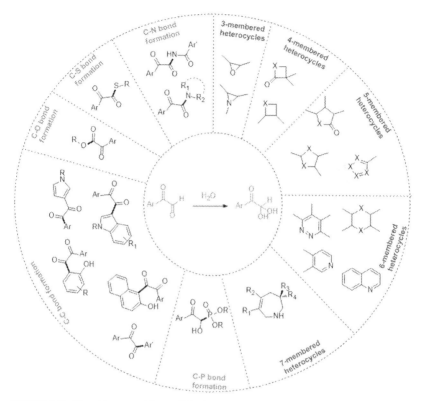

FIGURE 3.2   Chemistry around 2-oxoaldehyde.

active molecules show antiinflammatory,[10] immunosuppressive,[11] and antitumor[12] activities. Oxirane is a strained ring susceptible to various nucleophiles, ring-opening, or rearrangement reactions[13] and hence they are counted among the most important intermediates in organic synthesis. While as **2** is the starting material for the synthesis of protease inhibitor atazanavir and HIV protease inhibitor amprenavir[14] **3** is the necessary intermediate in the synthesis of Food and Drug Administration (FDA)-approved antibiotic levofloxacin (Fig. 3.4).

Another common example is the epoxide hydrolysis necessary in the detoxification of toxic substances in animals. The benzo[a]pyrene is a carcinogen, found in cigarettes, soot, and barbecued meat is detoxified in the liver by its conversion to diol epoxide, which is then hydrolyzed by epoxide hydrolase enzyme to give a soluble tetrol, which is easily soluble in water and is excreted out from the body (Fig. 3.5).

In this context, some of the important reactions are the synthesis of epoxides via epoxidation of alkenes,[15] dehydrochlorinations of chlorohydrins,[16] and preparation from carbonyl compounds using sulfur ylides.[17]

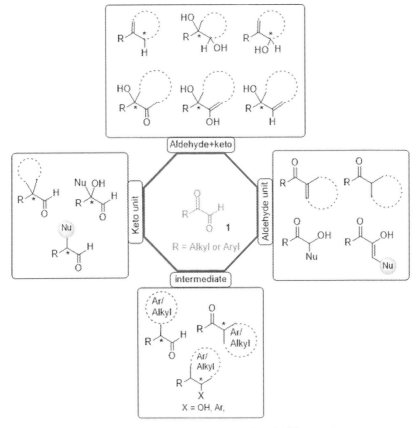

FIGURE 3.3   Various modes of reactions in 2-oxoaldehydes via different units.

R₁ = Me, R₂ = Cl Cryptophycin A
R₁ = Me, R₂ = H Cryptophycin B
R₁= R₂ = H   Arenastatin A

Psorospermin
Anti-leukemic

FIGURE 3.4   Biologically active molecules containing oxirane moiety.

An example of the reactivity of AG is explained by the synthesis of oxiranes. Oxiranes are prepared by reacting AGs **1a** with phenacyl bromide **4** in the presence of 10% NaOH solution in water: dioxane solvent mixture for 15 min at room temperature that leads to the formation of *syn*-dibenzoyloxirane **5** in low yields[18] (Fig. 3.6).

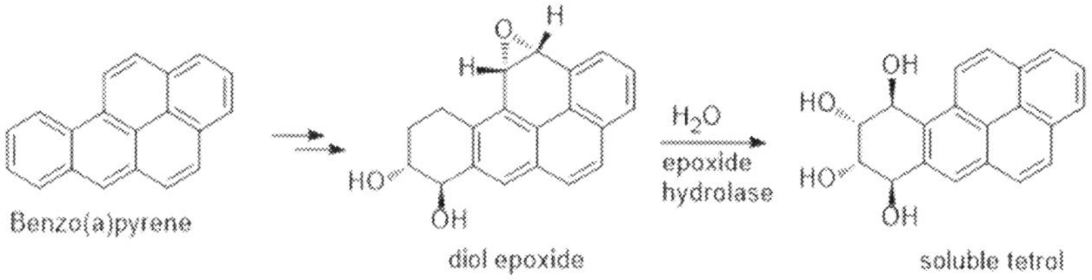

FIGURE 3.5 Example for detoxification of toxic substances in animals.

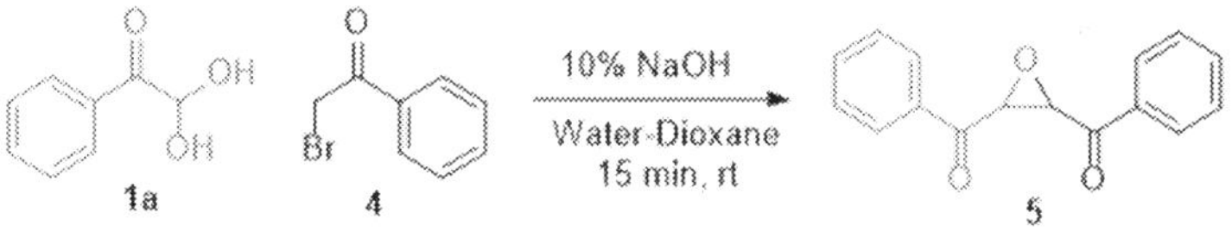

FIGURE 3.6 NaOH-catalyzed synthesis of oxiranes.

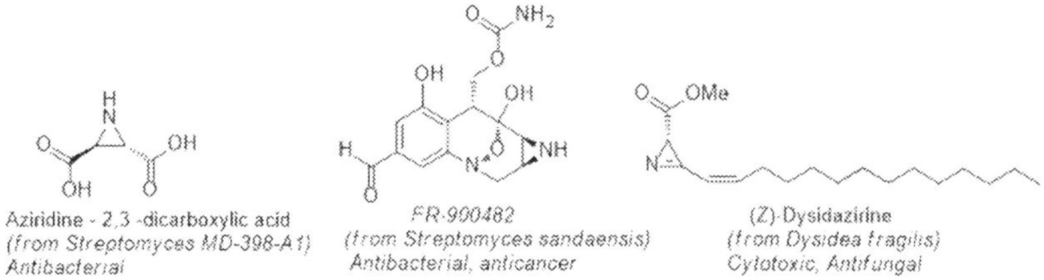

Aziridine - 2,3 -dicarboxylic acid
(from *Streptomyces MD-398-A1*)
Antibacterial

FR-900482
(from *Streptomyces sandaensis*)
Antibacterial, anticancer

(Z)-Dysidazirine
(from *Dysidea fragilis*)
Cytotoxic, Antifungal

FIGURE 3.7 Biologically active molecules containing aziridines scaffold.

3.2.2 Synthesis of aziridines

Aziridines, on the other hand, are three-membered saturated heterocycles that contain nitrogen atoms in the ring. Aziridines are utilized as substrates for the synthesis of various scaffolds.[19] Beside their presence in many natural products[20,21] they are also used in the synthesis of diverse nitrogen-containing building blocks[22] such as chiral amino acids,[23] indolizidines, alkaloids,[24] tetrahydropyridines that involve ring-expansion and ring-opening reactions.[25] They are also present as a key structural unit in some synthetic pharmaceuticals and are found in biologically active compounds[26] like antitumorials,[27,28] and antibiotics like aziridine-2,3-dicarboxylic acid,[29] FR-900482[30] and (Z)-dysidazirine[31] which are microbial metabolites (Fig. 3.7). Therefore the synthesis of aziridine, oxirane, and other hetero- and carbocyclic molecules by using AG as a substrate reflects the scale of its divergent reactivity.

Arylglyoxal **1** and *p*-anisidine **6** reacts in the presence of a catalytic amount of **7** and $MgSO_4$ in toluene at room temperature for 1 h to form in situ arylglyoxal-imine **8**, which undergoes aziridination by the addition of α-diazoacetate **9** at $-30°C$ to furnish the substituted *cis*-aroylaziridine **10** in

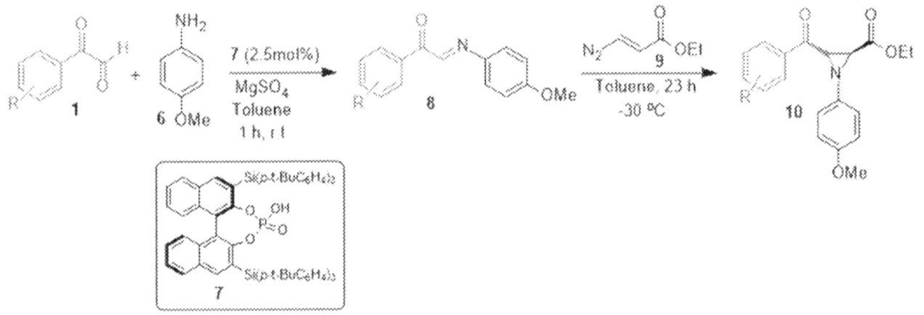

FIGURE 3.8 Chiral phosphoric acid-catalyzed synthesis of aziridines.

$92\%-97\%$ ee.[32] The three-component reaction is catalyzed by a chiral phosphoric acid **7** and proceeds through enantioselective and stereoselective aza-Darzens reaction (Fig. 3.8) without the formation of any *trans*-isomer.

Aldimine derived from benzaldehyde and *p*-anisidine did not lead to any product formation. Aziridine is formed in a similar three-component reaction with AGs after they are treated with diphenylmethylamine and α-diazoacetate in the presence of 4 Å Ms in hexane at room temperature. The reaction is catalyzed by Yb(OTf)$_3$ and proceeds stereoselectively to yield *syn* isomer as a major product.[33]

3.2.3 Synthesis of ketoamides

Ketoamides are 2-oxoamides that are formed when an additional carbonyl group is bonded adjacent to the carbonyl group of amide and are found in various natural products and biologically active molecules.

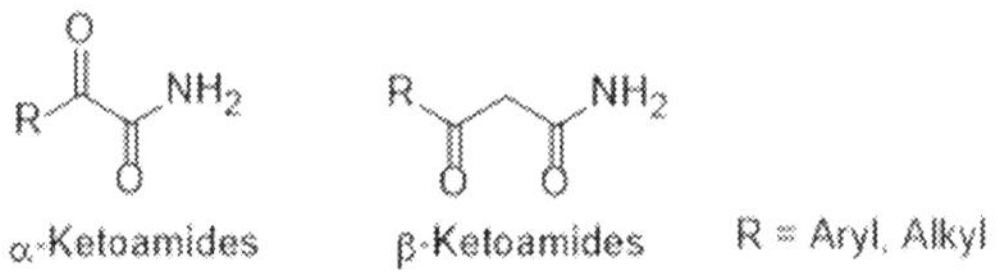

AGs **1** have also been applied for the synthesis of different substituted α-keto amides **12** in different conditions. When 2-oxoaldehydes **1** reacts with secondary amines **11** like morpholine, pyrrolidine, and N, N-dimethylamine in the presence of DMSO, which acts both as a solvent and delivers O atom to the product. 2-Oxoaldehydes acts like a ligand which directs and eases the formation of iminium ion. Moreover, this metal-free oxidative amidation reaction works well over both aryl- as well as aliphatic 2-oxaldehydes (Fig. 3.9).[34]

The reactivity of AGs toward oxidative amidation has been reported in which three different reagent systems have been successfully employed. AGs react with primary amines **15** in three different optimized conditions to

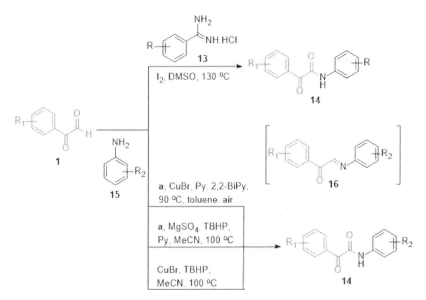

FIGURE 3.9 Metal-free oxidative amidation reaction.

FIGURE 3.10 Different examples of oxidative amidation.

produce the desired product α-carbonylamides **14** in good yields. The three
optimized conditions are (1) CuBr, pyridine, 2,2-biPy in toluene at 90°C. (2)
TBHP, MgSO₄, Py in acetonitrile at 100°C, and (3) CuBr, TBHP, in acetonitrile at 100°C.[35]

In contrary to this, α-ketoamides **14** are formed as a result of the reaction
between AGs **1** and benzamidine hydrochlorides **13** in the presence of iodine
and DMSO (Fig. 3.10). The reaction works well with acetophenone; but fails
with benzamides.[36]

As evident from earlier examples, the in situ generated imines **16** from
AG **1** and primary amine **15** is a reaction center wherefrom many molecules
can be synthesized. One such example is the reactivity of 2-oxoimine **16**
generated from 2-oxoaldehydes and primary amine with SeO₂ and pyridine.
The intermediate **16** reacts with different weakly nucleophilic amines like
anilines **15**, sulfonamides **17**, and benzamides **19** that leads to the formation
of α-carbonylamides **14**, α-carbonylsulfonamides **19**, and α-carbonylimides
20, respectively (Fig. 3.11). DMSO plays a dual role of solvent as well as an
oxidant. These reactions suggest primary reports on oxidative amidation

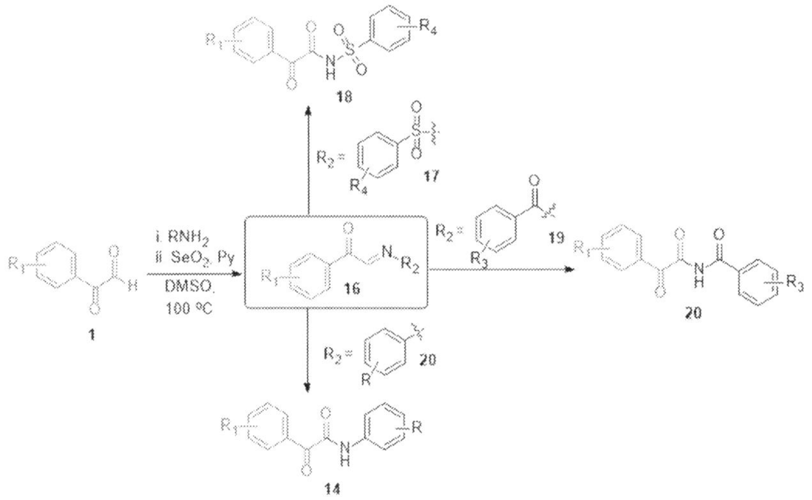

FIGURE 3.11   Reactivity of 2-oxoimine for the synthesis of ketoamides.

sulfonamides **17** and benzamides **19** while exploiting AG to form the corresponding keto amides.[35]

An important reaction pertaining to AGs is the formation of C-C bonds by cross-coupling reaction through iminium ion **21**. Iminium ion, generated from AGs **1** with pyrrolidine, may lead to two different products through different activation pathways. In an iminium equivalence pathway, aerobic oxidation of iminium ion occurs simply by heating AG **1** and cyclic amines **11** such as pyrrolidine, piperidine, morpholine, and diethylamine in DMSO to yield corresponding α-keto amides **12** at 80°C.

In other iminium catalysis activation mode, the direct addition of nucleophiles like indole, pyrrole, phenol, and 2-naphthol takes place on iminium ion **21**, which undergoes deamination followed by aerobic oxidation to form 1, 2-diketones **22** (A−D) in good to moderate yields (Fig. 3.12). Such a difference in the reaction arises due to the presence of adjacent C = O unit in the iminium ion.[37]

Arylglyoxals **1** reacts with primary amines, isocyanides **23**, and carboxylic acids in Ugi fashion to form β-keto amides **24** on Wang resin as solid support. Compound **24** can further be cyclized into imidazoles **25** in the presence of ammonium acetate in AcOH at 100°C followed by the reaction with 10% TFA in DCM at 23°C (Fig. 3.13).[38]

Further, AGs **1** reacts with benzyl isocyanide **26** under Passerini reaction to form β-ketoamide **27** followed by the Beirut reaction with benzofuroxanes **28** and CaCl₂ with a catalytic amount of ethanolamine in MeOH gave 1,4-di-N-oxide-quinoxaline-2-carboxamide **29**, which is potent against *Mycobacterium tuberculosis* strain H37Rv (Fig. 3.14).[39]

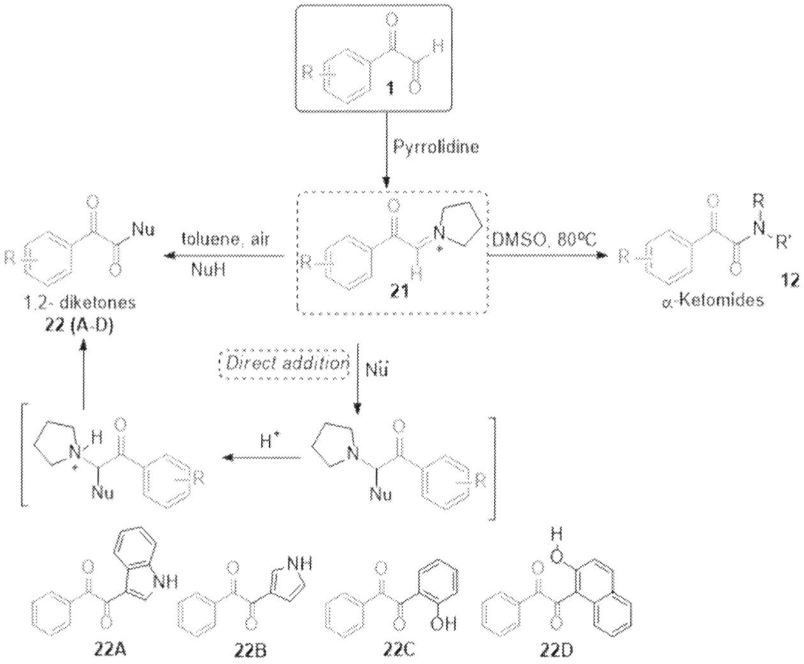

FIGURE 3.12 Cross-coupling reaction through iminium ion.

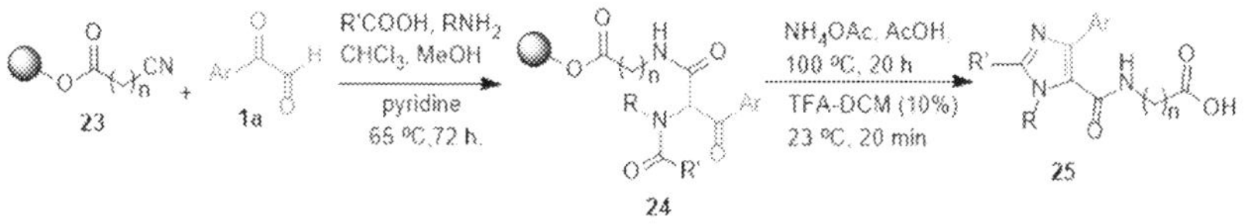

FIGURE 3.13 Synthesis of 3-ketoamides.

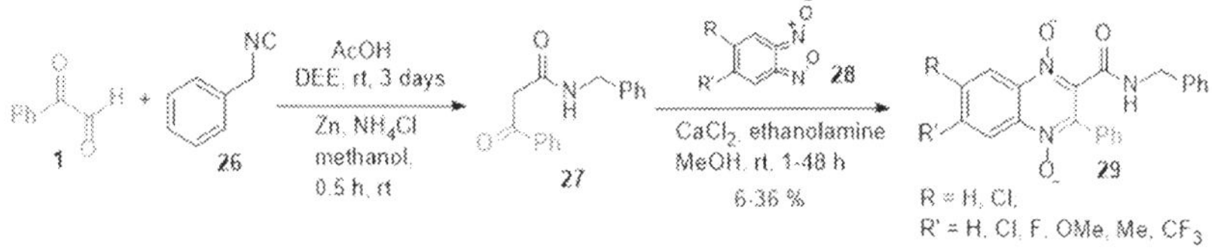

FIGURE 3.14 Synthesis of 1,4-di-N-oxide-quinoxaline-2-carboxamide.

AG-hydrate **1a** reacts in a multicomponent reaction (MCR) with benzoic acids (isophthalic or terephthalic acid) **30**, cyclohexylisocyanide **31** and *n*-,butylamine **32** in MeOH at room temperature to form α-amido-β-ketoamide **33** via an Ugi product intermediate, which further undergoes cyclization with in situ generated ammonia delivered from $(NH_4)_2CO_3$ to yield *bis*-imidazole **34** in good to moderate yields (Fig. 3.15).[40]

3.2.4 Synthesis of imidazoles

Imidazoles are important scaffolds and are present in various natural products and biologically active compounds. A green and efficient method was reported by Khalili and coworkers[41] for the synthesis of aroylimidazoles **35** in two isomeric forms (**35a** and **35b**) by conducting a reaction in water between AG-hydrate **1** and an excess amount of NH_4OAc at room temperature for 30 − 45 min in 48% − 86% yields (Fig. 3.16). The ratio of two isomers of aroylimidazoles **35** was determined by measuring NH signal intensities in ^{1}H-NMR.

Arylglyoxal-hydrate can easily undergo a condensation reaction with 3-hydroxyamino-2-butanone oxime **36** in MeOH to form α-aroylnitrone **37** which consequently cyclizes to lose water molecule upon heating in acetic acid and produces 1-hydroxyimidazoles **39** in 46%−84% yields via **38** (Fig. 3.17).[42]

In addition, a regioselective selenium dioxide promoted reaction between AGs **1** and amino acid alkyl ester hydrochlorides **40** for the synthesis of imidazoles **41** was described in the presence of a base (Fig. 3.18). The in situ

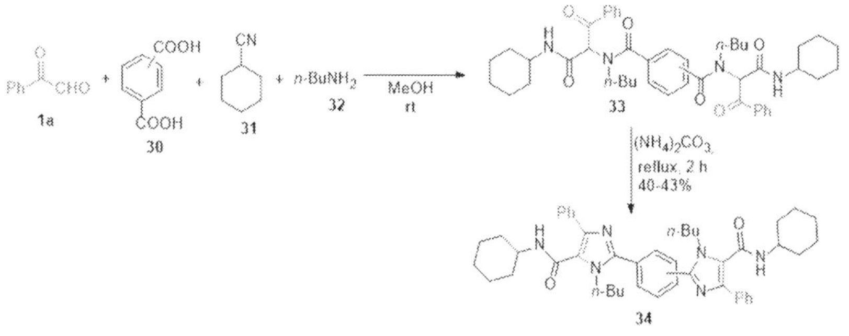
1a
30
31
+ n-BuNH2
32
MeOH
rt
33
(NH4)2CO3
reflux, 2 h
40-43%
34

FIGURE 3.15 Synthesis of α-amido-β-ketoamide.

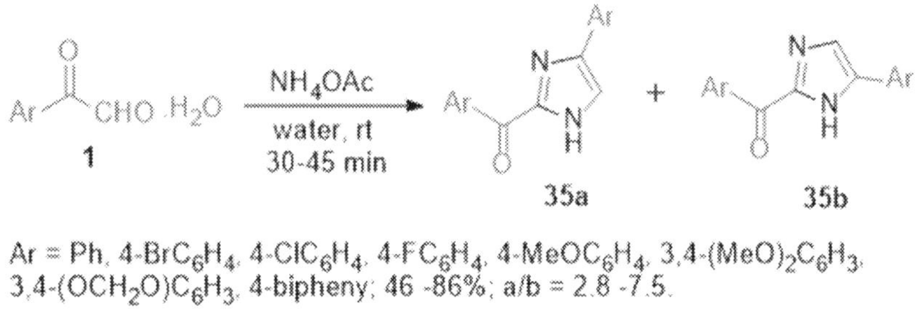
NH4OAc
water, rt
30-45 min
1
35a
35b
Ar = Ph, 4-BrC6H4, 4-ClC6H4, 4-FC6H4, 4-MeOC6H4, 3,4-(MeO)2C6H3,
3,4-(OCH2O)C6H3, 4-bipheny; 46 -86%; a/b = 2.8 -7.5.

FIGURE 3.16 Synthesis of aroylimidazoles.

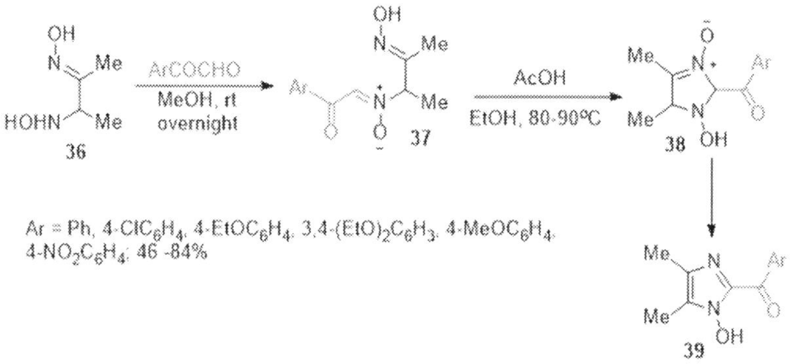

FIGURE 3.17   Synthesis of 1-hydroxyimidazoles.

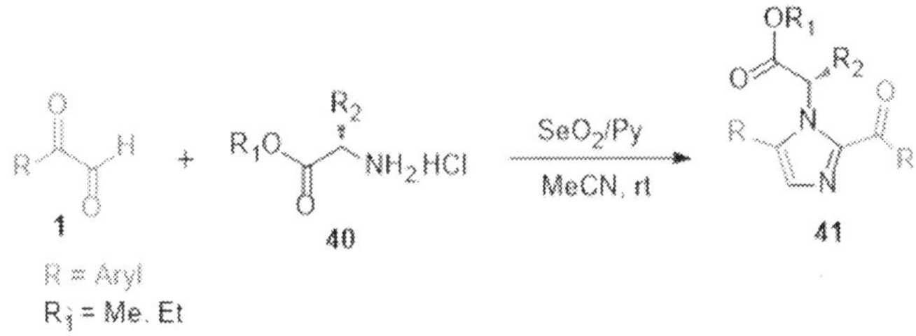

FIGURE 3.18   Selenium dioxide-promoted synthesis of imidazoles.

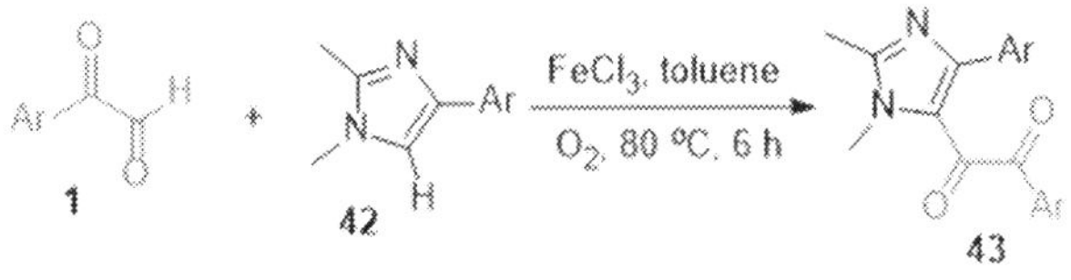

FIGURE 3.19   FeCl$_3$-catalyzed cross-dehydrogenative coupling reaction.

formed intermediate ArCOCHN$_1$N$_2$ installs overall two N atoms from amino acids into the amino acid substituted imidazoles. The reaction has a wide scope over various natural/nonprotein amino acids and peptide substrates.[43]

Furthermore, a cross-dehydrogenative coupling of AGs **1** with imidazole derivative **42** leads to the dicarbonyl functionalization of imidazole hetero-cycles in the presence of FeCl$_3$ to form substituted imidazoles **43** (Fig. 3.19). The reaction is feasible under aerial oxidation in toluene at 80°C and obtains the desired product in 87% yield.[44]

3.2.5   Synthesis of azetidines and β-lactams

Azetidine is the N-containing four-membered heterocycles with two confor-mational forms **44** and **45**. The equilibrium between the two conformers lies toward the right while as the lone pair and H atom on N prefers axial and equatorial position, respectively.

The azetidine unit is predominant in various naturally occurring molecules like nicotianamine,[45] medicanine,[46] and forms an essential part of many active nucleosides, which bear antifungal antibiotic properties.[47] Moreover, it is present in melagatran, a thrombin inhibitor.[48]

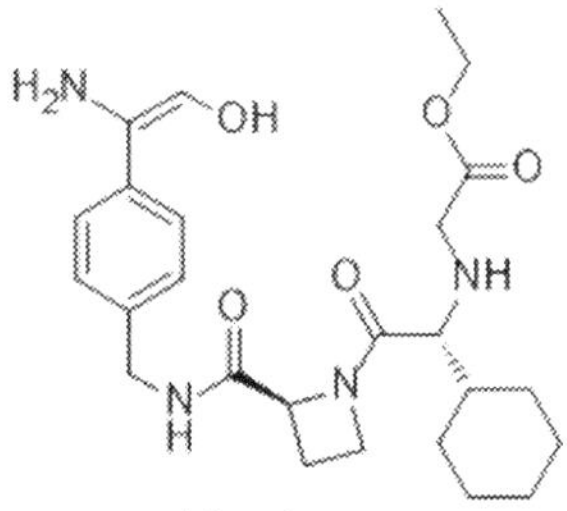

Ximelagatran

On the other hand, β-lactams are 2-azetidinones and form a basic structural core in antibiotics. Apart from antibiotic activity, they also show different medicinal properties like antiinflammatory arylglyoxalents,[49] antidepressants,[50] anticancer,[51] antimicrobials,[52,53] cholesterol absorption inhibitors, and antitubercular.[54] β-lactams are used as synthons for the synthesis of different naturally occurring molecules of biological significance (Fig. 3.20).[55]

The azetidine derivatives **48** were made by heating amide derivative of menthol **46** with AGs **1** in toluene to give perhydrobenzoxazines **47**. Perhydrobenzoxazines **47** was transformed finally into the azetidine derivatives **48** in a five-step sequence that involves O-benzylation and reductive cleavage of the O−C bond followed by oxidation and subsequent tosylation (Fig. 3.21).[56]

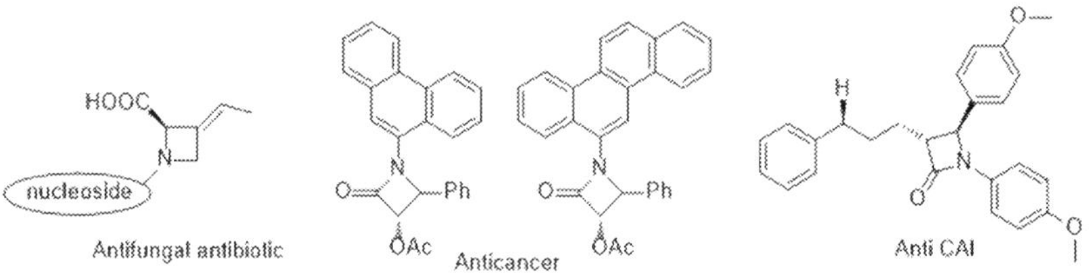

FIGURE 3.20 Biologically active molecules containing azetidine scaffold.

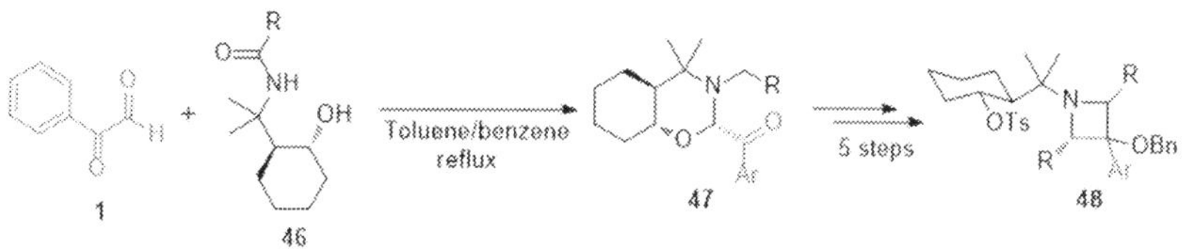

FIGURE 3.21 Synthesis of azetidine derivatives.

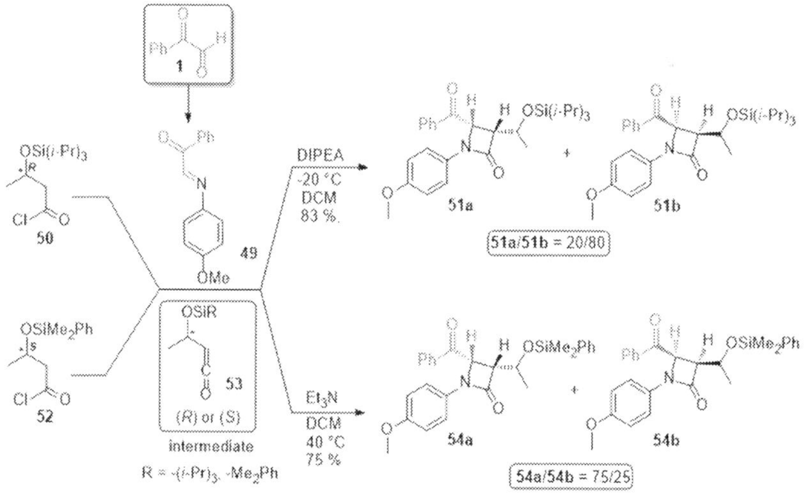

FIGURE 3.22 Temperature-dependent reactions for the synthesis of azetidine derivatives.

Further, a different case of a temperature-dependent reaction involving 2-oxaldehyde **1** with acid chloride **50** and **52** through 2-oxaldehyde imine **49** is the formation of antibiotic intermediate carbapenem.[57] A reaction was performed with isopropylsilane substituted acid chloride **50** with imine **49** in the presence of DIPEA in DCM at −20°C to obtain two isomeric products of 4-benzoylazetidin-2-ones **51a** in major and **51b** in minor quantity through ketene intermediate **53**. Interestingly, [2 + 2] cycloaddition reaction takes place with acid chloride **52** when treated with **49** in the presence of TEA at 40°C after 24 h through **53** to produce carbapenem **54a** in major and **54b** in minor (Fig. 3.22).[58]

AG imine **55** reacts with allyl phosphate **56** stereoselectively in a carbonylation process catalyzed by a Pd-catalyst, tertiary amine (c-Hex$_2$NMe), and CO to produce vinyl azetidin-2-ones **57**. The product is formed in a *syn*-selective manner and requires high pressure of 30 kg cm^{-2} of CO for 0.5 mmol of AGs-imine in THF for 5 h.[59] The vinyl azetidin-2-ones **57** can also be transformed into another derivative of azetidin-2-ones **58** in the presence of a catalytic amount of Co(acac)$_2$ and PhSiH$_3$ under the oxygen atmosphere (Fig. 3.23).

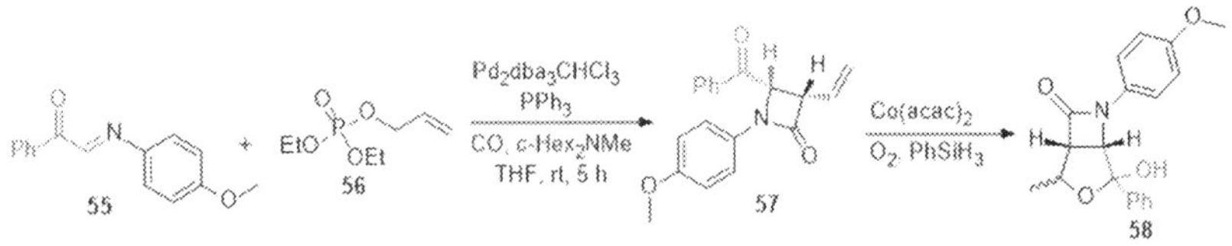

FIGURE 3.23   Synthesis of azetidin-2-ones.

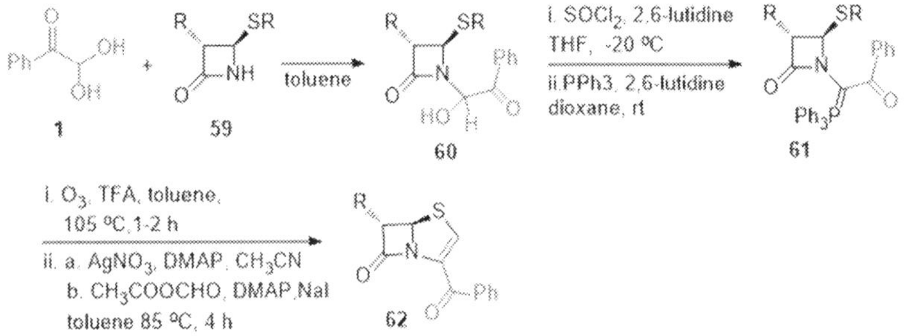

FIGURE 3.24   Synthesis of benzoylpenems in a multistep process.

3.2.6   Synthesis of azetidinones

Pearson and coworkers reported the reaction of AGs **1** with azetidinones **59** for the synthesis of benzoylpenems **62** in a multistep process. In the first step, AGs **1** reacts with azetidinones **59** to produce compound **60** after that the azeotropic mixture is removed by reaction with $SOCl_2$ and 2,6-lutidine followed by the addition of triphenylphosphine and 2,6-lutidine in dioxane furnishes phosphorane derivative **61** which was finally converted into the corresponding penem ring **62** after ozonolysis and then the addition of silver nitrate and DMAP followed by formylation by acetic formic anhydride (Fig. 3.24).[60]

3.2.7   Synthesis of β-lactones

β-Lactones are attractive intermediates in natural product and polymer synthesis like malyngolide and grandinolide.[61]

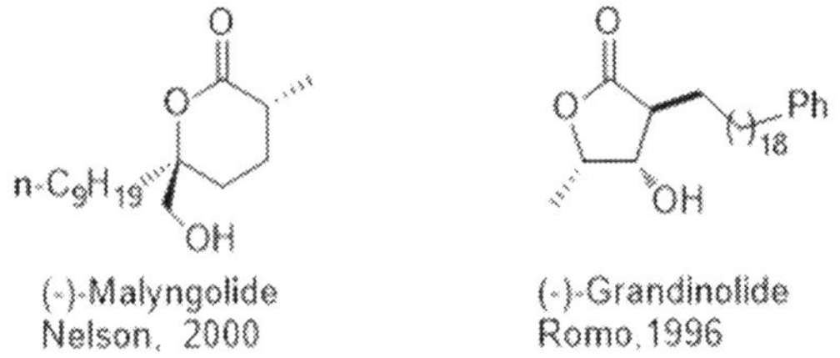

(-)-Malyngolide
Nelson, 2000

(-)-Grandinolide
Romo,1996

AG **1** also forms β-lactones **64a** and **64b** when reacted with ketenes **63** and in situ generated carbene source from a triazolium salt **66** in [2 + 2] enantioselective cycloaddition reaction in THF at room temperature. AGs bearing both electron-donating and electron-withdrawing groups undergo [2 + 2] cycloaddition to form β-lactones **64** in excellent yields with high enantio- and diastereoselectivity (Fig. 3.25). Meanwhile, N-heterocyclic carbene **67** was generated in situ from the action of Cs_2CO_3 on triazolium salt **66** in THF at room temperature in 1 h.[62]

3.2.8   Synthesis of pyrrolidines

While proline is a well-known proteogenic amino acid, pyrrolidines are unique building blocks in organic synthesis[63] and are found in various naturally occurring compounds,[64] bioactive molecules,[65] and organocatalysts.[66] More than 60 FDA-approved drugs are based on the pyrrolidine core structure.[67] Pyrrolidine is present in advanced pharmaceutical clinical candidates as a DPP-IV inhibitor (e.g., sitagliptin),[68] NK3 receptor antagonists,[69] and anticoagulants[70] (Fig. 3.26).

A variety of pyrrolidines are formed when 2-ketoaldehydes **1** reacts with amine **68** in chloroform to give N-(a-cyanoalkyl)- and N-(α-cyanoaryl) imines **69**. These imines are exploited for the synthesis of pyrrolidines **71** and **72** when azomethine ylide **70** reacts with dimethyl fumarate **74** or N-methyl-maleimide **73** (Fig. 3.27).[71]

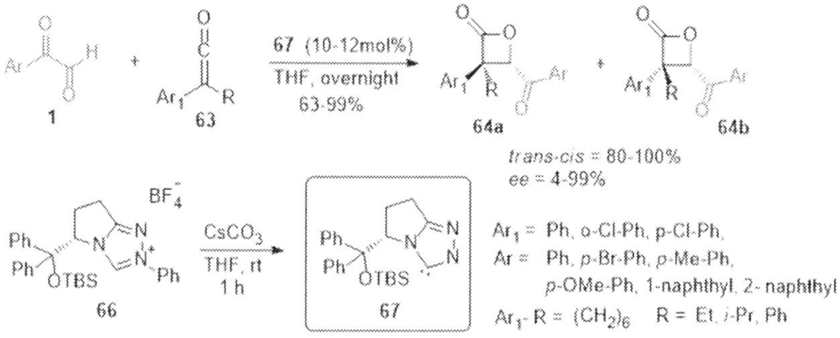

FIGURE 3.25   [2 + 2] Enantioselective cycloaddition reaction.

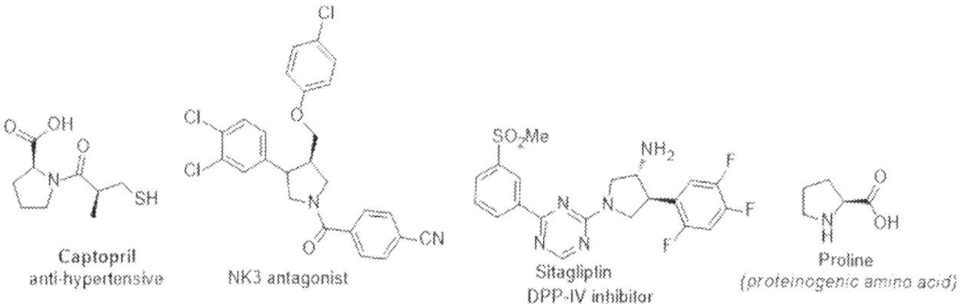

FIGURE 3.26   Biologically active molecules containing pyrrolidines scaffold.

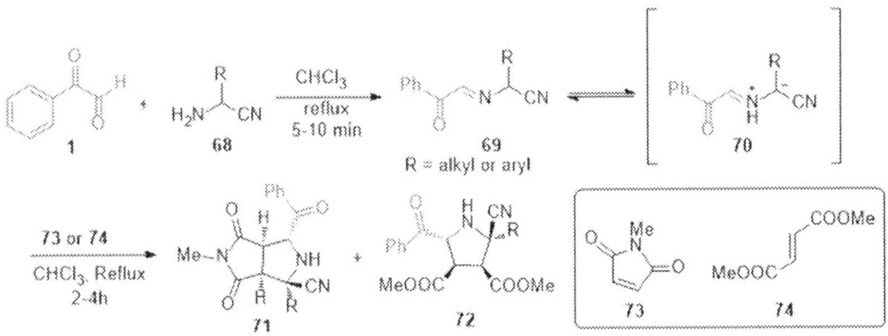

FIGURE 3.27   Synthesis of pyrrolidine derivatives.

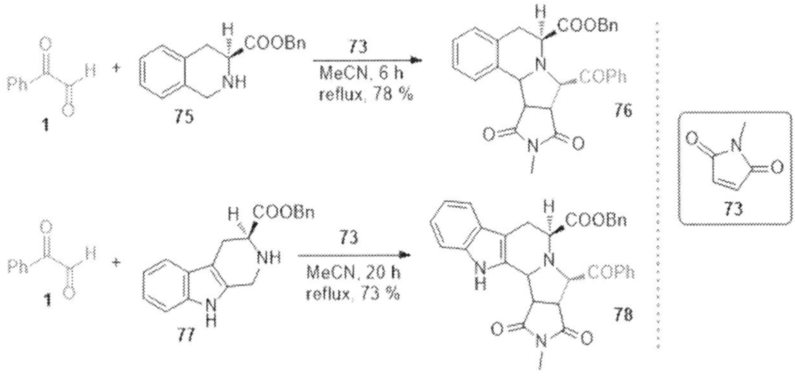

FIGURE 3.28   Synthesis of pyrroloisoquinoline and indolzinoindole.

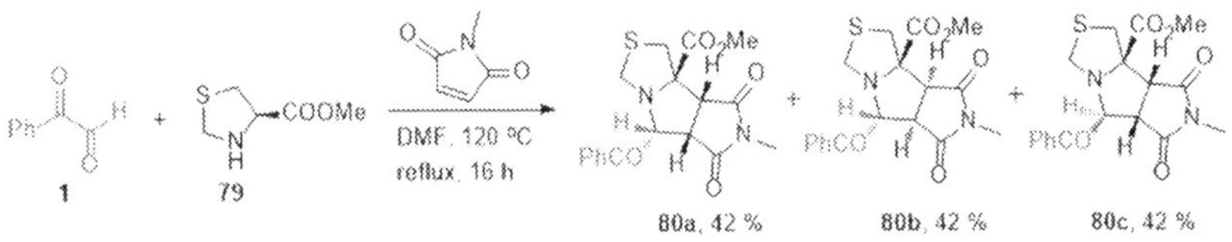

FIGURE 3.29   Formation of different isomers.

This reaction has been applied for the synthesis of pyrroloisoquinoline **76** and indolzinoindole **78** by the reaction of AGs with *N*-methylmaleimide **73** and its subsequent reaction with isoquinoline **75** and **77** when refluxed in MeCN for 6 h (Fig. 3.28).[72]

However, when AG **1** is refluxed with **79** at 120°C for 16 h in DMF, **80a** was formed in major quantity and other two isomers **80b** and **80c** in minor concentration (Fig. 3.29).

Yong Yuan and co-workers demonstrated access to functionalized pyrrolidones from AGs, α-angelica lactone, and anilines through a multicomponent process in a simple acidic environment. In addition to the greener and scalable approach, the reaction proceeds without any solvent in H_2SO_4 (30 mol %) in 1 h up to 75% yields.[73]

3.2.9   Synthesis of pyrrolizidine

Phenylglyoxals also show the pericyclic mode of reactions in which AG reacts with proline **81** and β-nitrostyrene **82** to yield pyrrolizidine **83** in a 1,3-dipolar addition via an intermediate **84b** formed in situ from the decarboxylation of **84a** (Fig. 3.30). The reaction proceeds smoothly to form a single regioisomer with an overall 80% yield.[74]

3.2.10   Synthesis of tetrahydrofuran

Tetrahydrofuran and dihydrofuran form the basic structural unit of many naturally occurring scaffolds like gambieric acid A and ciguatoxin,[75] goniocin,[76] and some biologically active molecules, for example, lasalocid A (X537A)[77] (Fig. 3.31). Owing to their importance in synthetic chemistry[78] various strategies have been launched for their synthesis[79] based on cycloaddition and cyclization reactions.

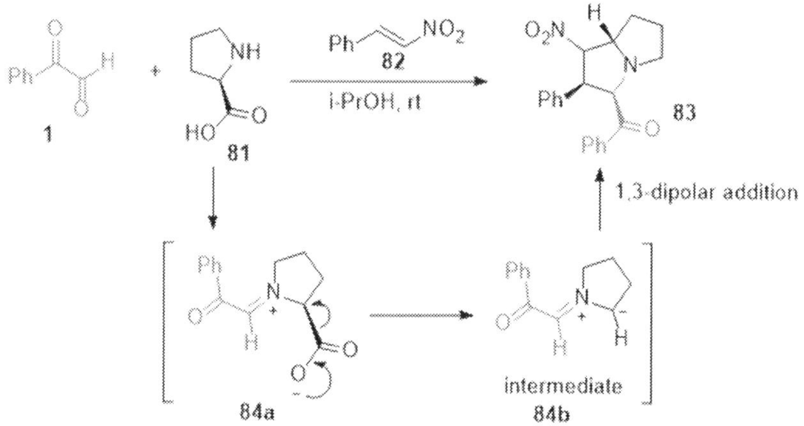

FIGURE 3.30   Synthesis of pyrolizidine derivatives.

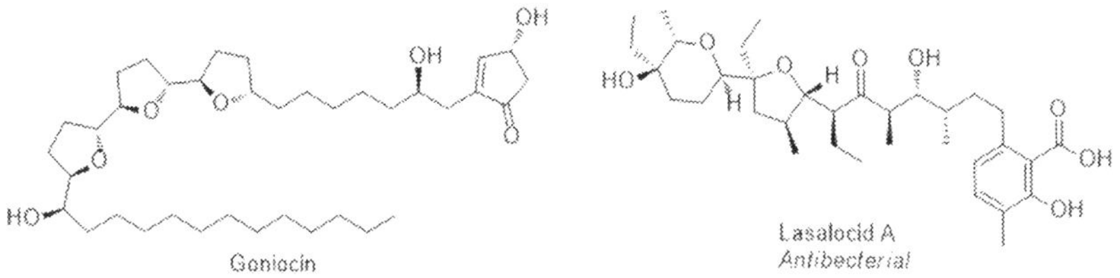

Goniocin

Lasalocid A
Antibecterial

FIGURE 3.31   Biologically active molecules containing tetrahydrofuran moiety.

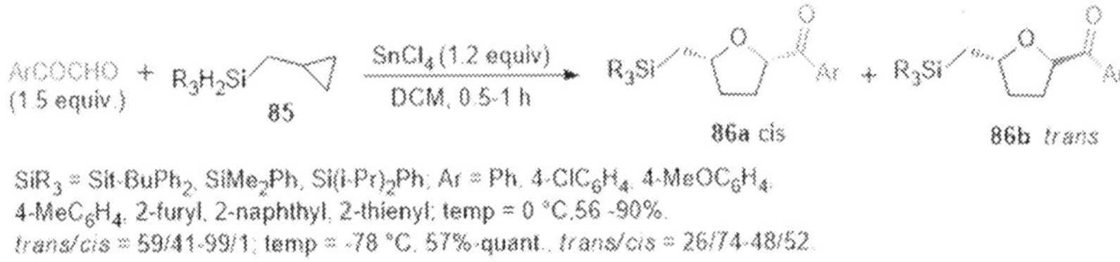
ArCOCHO
(1.5 equiv.)
R3H2Si
85
SnCl4 (1.2 equiv)
DCM, 0.5-1 h
R3Si
O
Ar
86a cis
86b trans
SiR3 = Sit-BuPh2, SiMe2Ph, Si(i-Pr)2Ph; Ar = Ph, 4-ClC6H4, 4-MeOC6H4,
4-MeC6H4, 2-furyl, 2-naphthyl, 2-thienyl; temp = 0 °C,56 -90%,
trans/cis = 59/41-99/1; temp = -78 °C, 57%-quant., trans/cis = 26/74-48/52.

FIGURE 3.32 Synthesis of 2-silylmethyl substituted tetrahydrofurans.

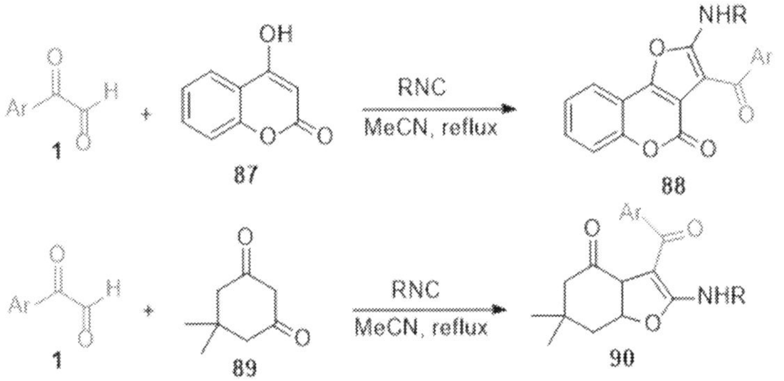
OH
Ar
H
O
1
87
RNC
MeCN, reflux
NHR
88
89
90

FIGURE 3.33 Synthesis of furocoumarins and benzofurans.

2-Silyl methyl-substituted tetrahydrofurans (**86a** and **86b**) were synthesized by Fuchibe et al.[80] via [3 + 2] cycloaddition reactions of cyclopropyl methyl silanes **85a** and AGs. Different reaction conditions were optimized by using different Lewis acids in different solvents and $SnCl_4$ in DCM was found as best as far as the yields of the desired products are considered. *Cis* **86a** and *trans* **86b** isomers of the products are formed and the stereoselectivity of the products is dependent on the reaction temperature. *Trans*-isomer **86b** was formed in major quantity when the reaction was carried out at 0°C and *cis* isomer **86a** was formed at −78°C in major quantity (Fig. 3.32).

3.2.11 Synthesis of furocoumarins and benzofurans

Furocoumarins **90** and benzofurans **88** are prepared from a reaction between arylglyoxals **1** with 4-hydroxycoumarin **87** and dimedone **89** in the presence of isocyanides in MeCN in very efficient yields (Fig. 3.33).[81] Enolate formation is the main step in the formation of both heterocycles.

3.2.12 Synthesis of dioxolanes

1,3-Dioxolanes are important scaffolds as far as their biological activities are concerned and are present in various biologically active compounds.[82] The

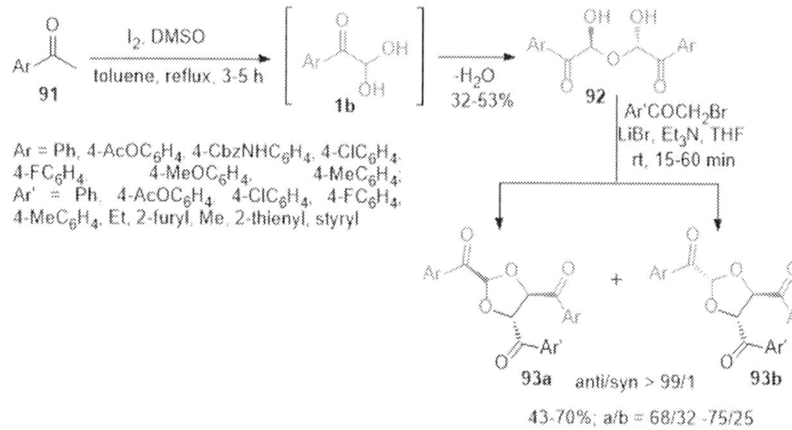

FIGURE 3.34   Synthesis of 2,4,5-triaroyl-1,3-dioxolanes.

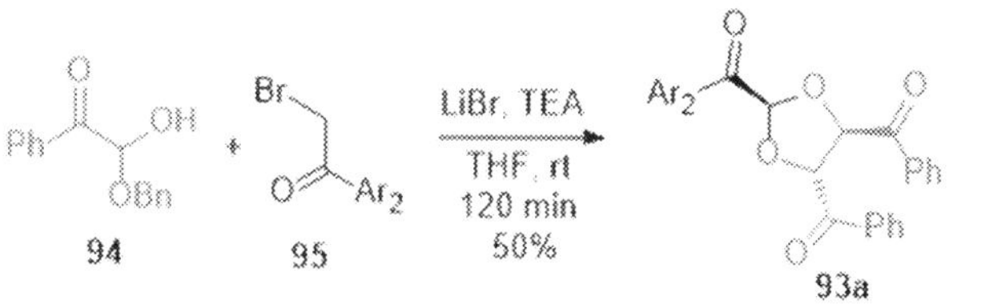

FIGURE 3.35   Synthesis of anti dioxolanes.

dioxolane nucleosides show anticancer activity and also inhibit the herpes simplex viruses.[83] There are various synthetic procedures available for the synthesis of dioxazoles depending on the starting material, catalyst and conditions such as starting from tartaric acid, carbohydrate derivatives,[84] dicarbonyl compounds,[85] carbonyl ylides, and 2-hydroxyethyl vinyl ethers are commonly used.

The case where arylglyoxal-hydrate is generated as intermediate is discribed[86] in which the arylglyoxal-hydrate was reported by Shao and Li for the synthesis of 2,4,5-triaroyl-1,3-dioxolanes **93a** and **93b** from hemialdals **92**, which was prepared from the direct oxidation of acetophenones **91** in presence of the catalytic amount of I_2 in DMSO.[87] In this reaction dehydration of 2-oxoaldehyde hydrate occurs to form hemialdal intermediate **92** which subsequently reacts with α-bromoacetophenone and LiBr in basic conditions to produce trisubstituted 1,3-dioxolanes **93a** stereoselectively in 43%−70% yield (Fig. 3.34).

This reaction was carried out in Dean−Stark assembly to eliminate any fraction of water produced which otherwise may hamper the reaction. On the other hand, when only one hydroxyl group is present in hemialdal **94**, the reaction selectively proceeds to form anti dioxolanes **93a** in 35%−61% yields under similar reaction conditions (Fig. 3.35).[88]

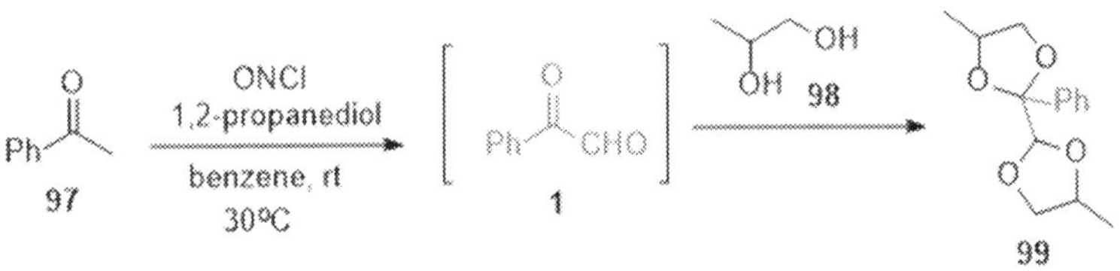

FIGURE 3.36   Synthesis of bisdioxolane.

Perilloxin (I)   Tenual (II)   Tenucarb (III)
(Perilla frutescens)   (Asphodeline tenuior)

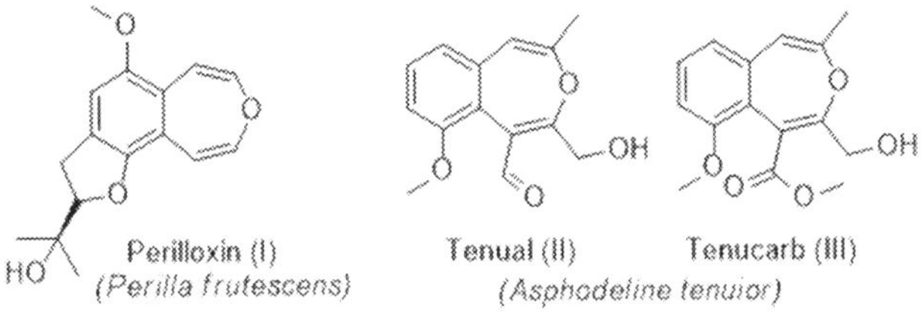

FIGURE 3.37   Biologically active molecules containing benzoxepine scaffolds.

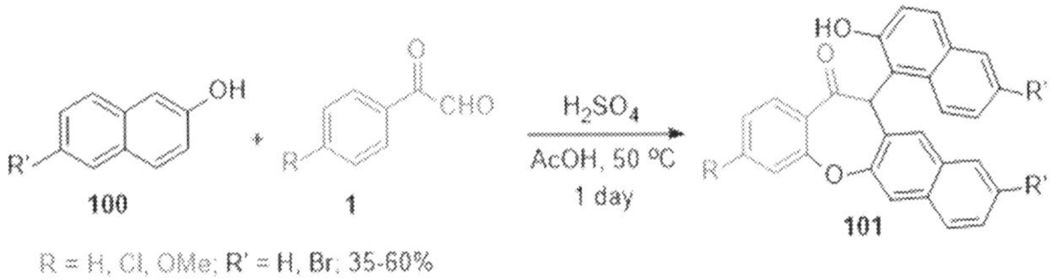

R = H, Cl, OMe; R' = H, Br; 35-60%

FIGURE 3.38   Synthesis of benzoxepine derivatives.

In addition, Manning et al. reported[89] the synthesis of bisdioxolane **99** from acetophenone **97** and nitrosyl chloride in the presence of 1,2-propanediol in benzene at room temperature. Nitrosyl chloride is slowly added to the mixture of acetophenones in benzene and 8 equivalent of 1,2-propanediol is added to generate PG in in situ, which undergoes reaction with 1,2-propanediol **98** to afforded bisdioxolane **99** in 36% yield (Fig. 3.36).

3.2.13   Synthesis of benzoxepine

Benzazepines are heterocyclic chemical compounds consisting of a benzene ring fused to an oxepine ring. The bicyclic ring system of the benzoxepines is a part of certain naturally occurring compounds like peilloxin (I) and tenual (II) and tenucarb (III) (Fig. 3.37).[90]

In contrast, when AGs are heated with 2-naphthol **100** in acetic acid and H[+] source seven-membered ring containing product **101** was formed

(Fig. 3.38). The product formation can be explained on the basis of a bulky naphthol ring that prevents the Friedel−Craft semiacetalization reaction.[91]

3.2.14   Synthesis of oxazole

Arylglyoxal-hydrate also reacts with S_4N_4 to yield **102** in 10%−33% along with oxazole **103** via aryglyoxime imine as an intermediate. The substitution pattern of arylglyoxals is supposed to determine the isomer of the product that is formed during the reaction.[92] When part of the aryl oxaldehyde are phenyl, 4-MeOC$_6$H$_4$, 4-MeC$_6$H$_4$ or 4-CNC$_6$H$_4$, 2-aroyl-5-aryloxazoles are formed and when arylglyoxal substrates contain phenyl, 2-thienyl, 4-MeOC$_6$H$_4$, 4-BrC$_6$H$_4$, 4-NCC$_6$H$_4$, 4-MeC$_6$H$_4$ or 4-ClC$_6$H$_4$, **102** is formed (Fig. 3.39).

3.2.15   Synthesis of oxazolidine

The dihydro- and tetrahydro-1,3-azoles are called oxazoline and oxazolidine. In addition to their presence in certain natural compounds like quinocarcin[93] they are constituted in medicinally active scaffolds, for instance, L-159463 and L-159692 are antibacterial molecules.[94]

In asymmetric catalysis, they are useful as ligands (**a−d**) and can coordinate with metal through either O or N atom. Copper(II) complexes of oxazolidine

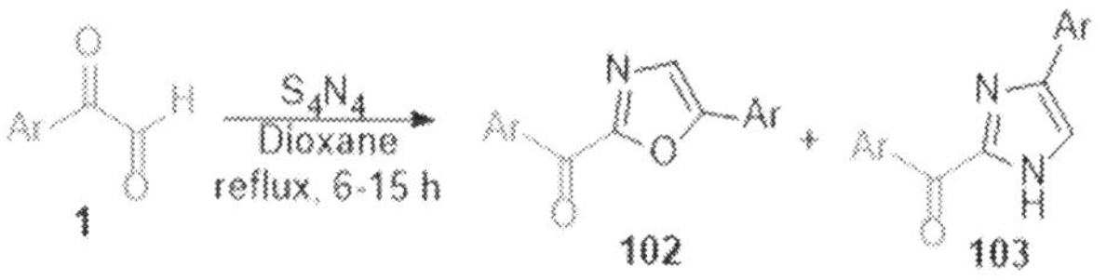

FIGURE 3.39   Synthesis of oxazole derivatives.

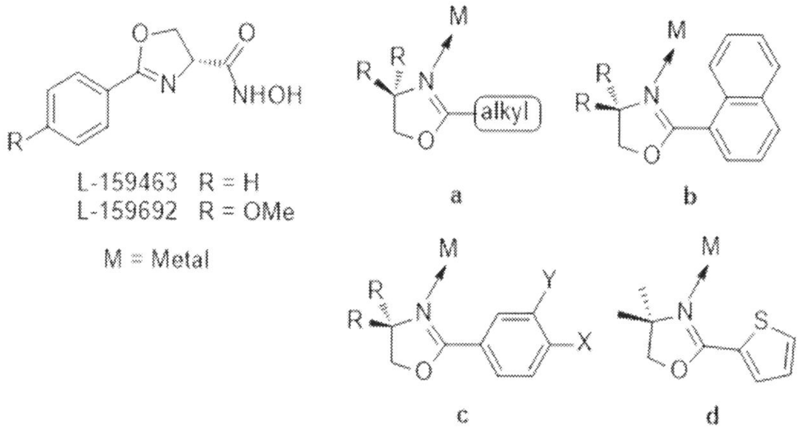

FIGURE 3.40   Biologically active molecules and useful ligands.

ligands derived from hydroxyquinoline carboxaldehyde show catalytic activity that mimics superoxide dismutase (SOD). These molecules were also tested for calf thymus DNA interactive experiments in the presence of the copper(II) complexes using UV–visible spectroscopy.[95] 2-Aryl aldehydes therefore can play a beneficial role through divergent reactivity in addition to other developments in oxazolidine and oxazoline chemistry that exist (Fig. 3.40).[96]

Chiral auxiliaries are important substrates that can be exploited for the stereoselective synthesis of complex molecules. In this perspective, AG-hydrate reacts with (R)-piperidin-3-ol **104** and **106** (S)-prolinol to form a bicyclic compound **105** and a fused ring system **107** respectively (Fig. 3.41). Both reactions occur under mild and dry conditions with excellent yields.

3.2.16 Synthesis of thiazolidines

The synthesis of substituted thiazolidine **109** involves the condensation reaction of aminothiol (X = SH) **108** with PG in refluxing CH_3CN. The product 2-benzoylthiazolidine **109** was obtained as a mixture of two diastereoisomers. The same concept was used for N-methyl phenyl glycinol (X = OH) to give morpholinone under the same conditions (Fig. 3.42).[97]

Thiazolidines **111** is formed in the diastereomeric ratio when arylglyoxal is treated with an ester of L-cysteine **110** in protic solvents in mild basic conditions with R/S in 74%/68% yields (Fig. 3.43).[89] With phenyl glyoxal, (S)-isomer was formed in major quantities; however, with more electron-donating groups like 4-methoxyphenylglyoxal, (R)-isomer is predominantly formed.

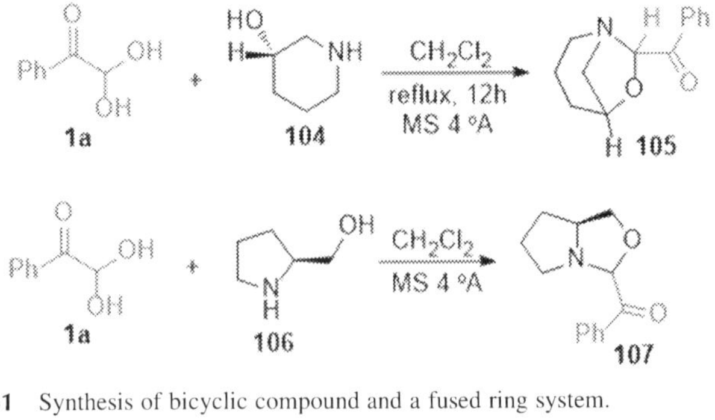

FIGURE 3.41 Synthesis of bicyclic compound and a fused ring system.

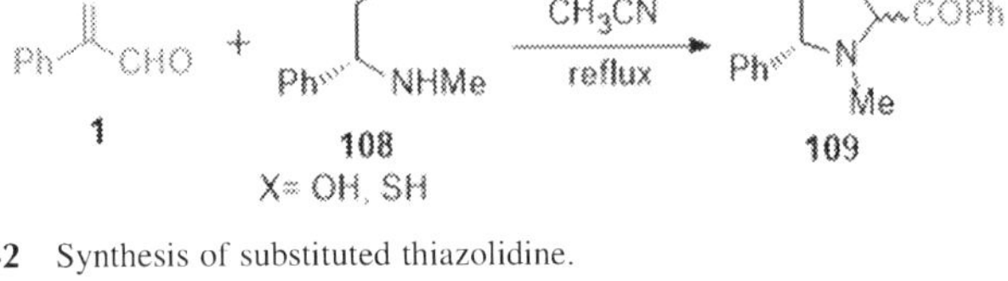

FIGURE 3.42 Synthesis of substituted thiazolidine.

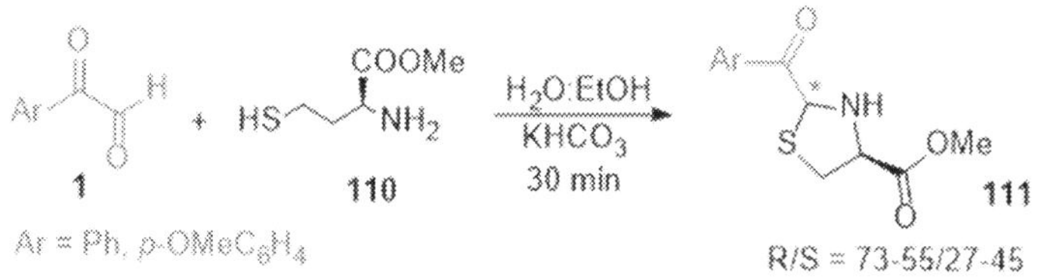

FIGURE 3.43 Synthesis of diastereomeric forms of thiazolidines.

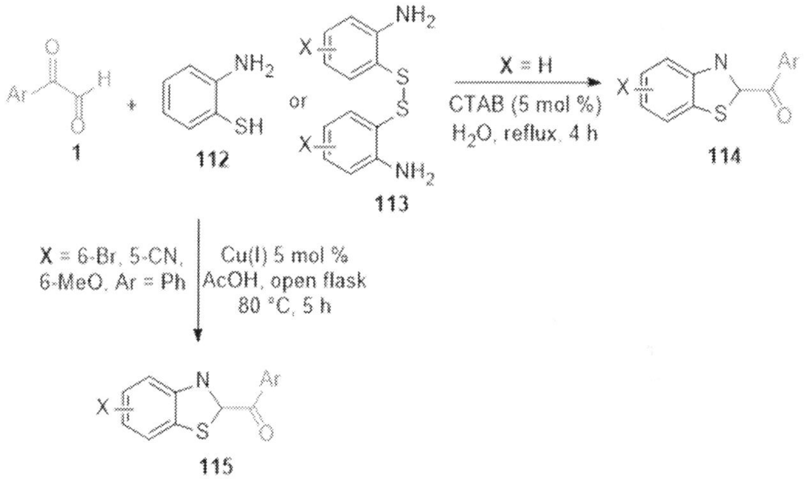

FIGURE 3.44 Synthesis of substituted benzothiazole.

3.2.17 Synthesis of benzothiazole

Benzothiazole **114** is formed when arylglyoxal condenses with 2-amino thiophenol **112** and its dimer disulfide arylamines **113** by using catalytic quantities of cetyltrimethylammonium bromide in water. The reaction occurs in an open flask and uses air from the atmosphere as an oxidant for the oxidation of thiazoline to obtain thiazole **114** in 80%−85% yields.[98] However, when 6-Br, 5-CN, 6-OMe, and Ph are part of the S-S dimer **113** then under Cu (I) catalysis in AcOH, the reaction proceeds to form **115** in 66%−85% yields (Fig. 3.44).[99]

A diluted and cold solution of AG-hydrate and thiosemicarbazide **116** react to form thiosemicarbazones **117**, which on heating with $FeCl_3$ through nucleophilic addition of S over α-carbon of hydrazide moiety gives two products depending upon the substitutions.[100] It is noteworthy that when $R_3 = H$, thiadiazole **118** was obtained, and when $R_3 = Me$ and $R_2 = H$ thiadiazoline **119** was formed, which can be easily freed from benzoyl group through base hydrolysis (Fig. 3.45).[101]

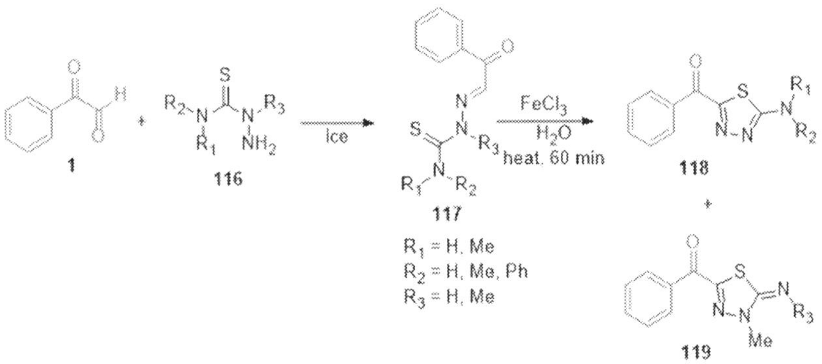

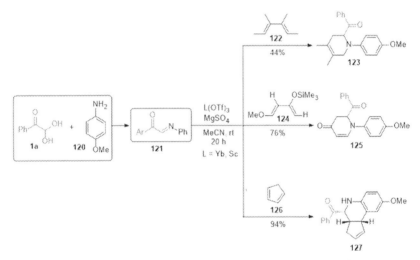

FIGURE 3.45 Synthesis of thiadiazole and thiadiazoline derivatives.

FIGURE 3.46 Lanthanide metal triflates-catalyzed synthesis of tetrahydropyridine.

3.2.18 Synthesis of tetrahydropyridine

Tetrahydropyridine heterocycles bear a wide range of applications in pharmaceuticals and synthetic intermediates.[102,103] Among the important reactions, the aza-Diels−Alder reaction is widely used to obtain substituted piperidines.[104]

AGs **1a** and para methoxy toluene **120** form AGs-imine **121** which subsequently reacts with 3 different dienes **122, 124, 126** through aza-Diels-Alder reaction to form three types of products **123, 125** and **127**. The reaction is catalyzed by lanthanide metal triflates like Yb(OTf)$_3$ and Sc(OTf)$_3$ along with MgSO$_4$ in MeCN. AGs-imine **121** acts as a diene source in case of cyclopentadiene (Fig. 3.46).[105]

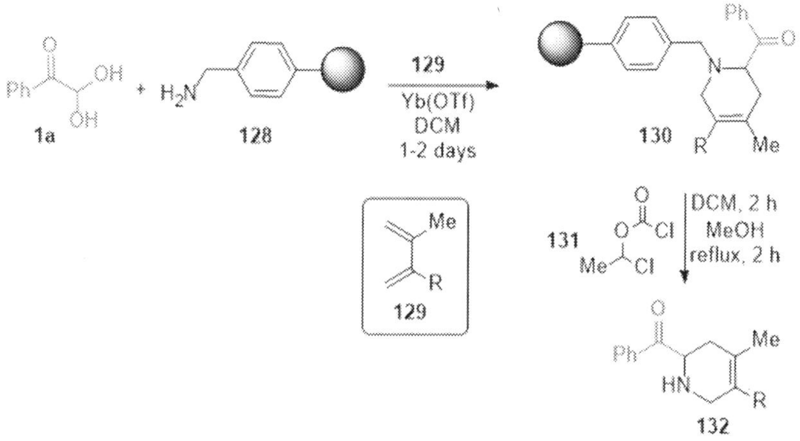

FIGURE 3.47 Solid-phase synthesis of tetrahydropyridine derivatives.

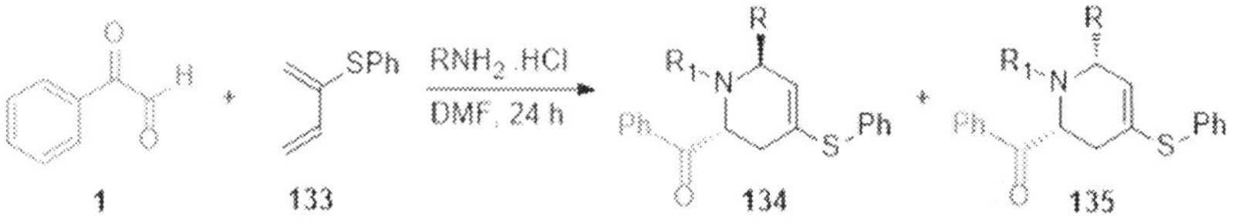

FIGURE 3.48 Aza-Diels−Alder reaction for the synthesis of tetrahydropyridine analogs.

Solid-phase synthesis of tetrahydropyridine derivatives **130** was reported in a similar type of aza-Diels−Alder reaction between AG-hydrates, diene **129**, and benzylamine **128** anchored to a solid support by using Yb(OTf)$_3$ in CH$_2$Cl$_2$. The solid resin support was cleaved by methyl chloroformate **131** and, as an advantage, very pure forms of the product **132** are obtained (Fig. 3.47).

AGs also reacts with the diene **133** through aza-Diels−Alder reaction to form tetrahydropyridine analogs **134** and **135** when stirred in DMF for 24 h. The reaction is dependent on temperature and type of the substituent R attached to the diene. It has been investigated that when R = H and R$_1$ = methyl, benzyl, phenyl, 63%−95% of the product **134** is formed at room temperature. However, when R = Me and R$_1$ = benzyl, and the temperature is reduced to 0°C the overall yield of **135** is 60% with *trans* as a major product (Fig. 3.48). Further reduction in temperature yields **135** as *trans* isomer with increased yields.[106]

3.2.19 Synthesis of tetrahydroquinolines and quinolines

Quinolines and tetrahydroquinolines are prominent in naturally occurring compounds such as helquinoline, L-689560, aspernigrin B[107] and show broad-spectrum activities such as antibiotics, for example, Ofloxacin and levofloxacin (Fig. 3.49).[108]

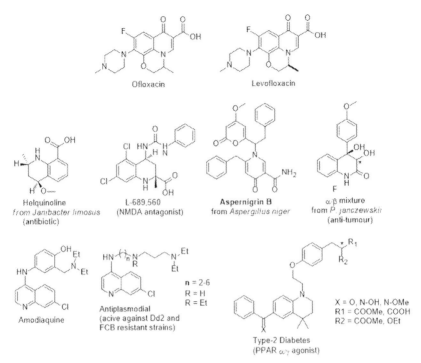

FIGURE 3.49 Biologically active molecules containing tetrahydroquinolines and quinolines moiety.

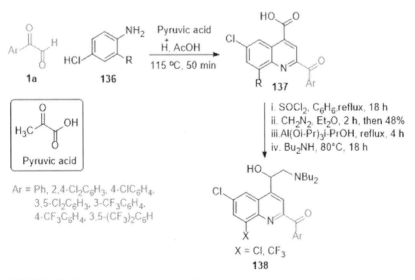

FIGURE 3.50 Synthesis of quinoline derivatives.

Their derivatives show a range of other biological activities including antimalarial, antioxidant, antibacterial, fungicidal, pesticidal, antidepressive, antiinflammatory, and antidiabetic activities.[109] Compound **F** exists in two diastereomeric forms and shows activity against human tumor cell lines, β-isomer is more potent against human ovarian carcinoma (SKOV-3).[110] Arylglyoxals have also been employed for the synthesis of quinolines and tetrahydroquinolines.

Arylglyoxal results in the formation of quinoline derivative **137** when a solution of aniline **136**, pyruvic acid, and arylglyoxal-hydrate was heated to 115°C for 50 min in AcOH/H$_2$SO$_4$. The reaction proceeds via arylglyoxal-imine intermediate. The product was further transformed into **138** in four steps followed by auto-oxidation in 59% yield. Some of the derivatives have shown antimalarial potential (Fig. 3.50).[111]

The hydrate form of arylglyoxal is also a key reactant for the Pictet–Spengler synthesis of tetrahydroisoquinoline derivatives **140** when reacted with aryl ethylamine **139** (Fig. 3.51). The reaction works well when aqueous HCl (3N) is used and the mixture is refluxed for 5 h to yield 25%−75% of the desired product.[112] Acids like acetic acid, trifluoroacetic acid, and p-TsOH were not beneficial for the synthesis of desired product.

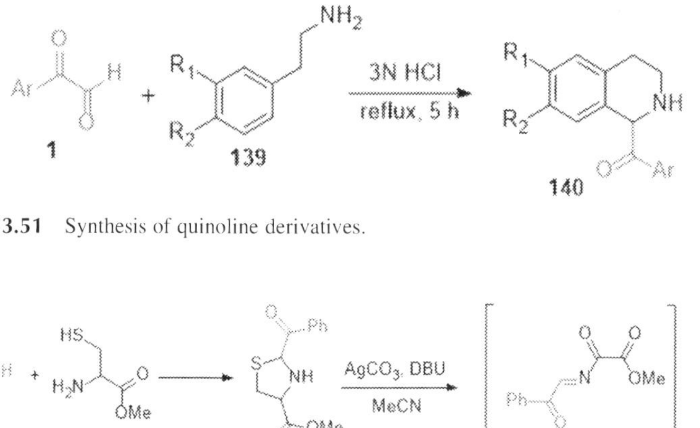

FIGURE 3.51 Synthesis of quinoline derivatives.

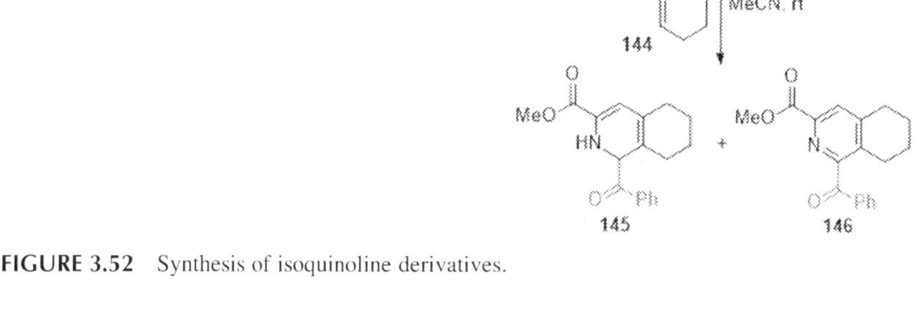

FIGURE 3.52 Synthesis of isoquinoline derivatives.

The same concept for the reactivity of AGs can be described for the synthesis of isoquinoline **145** and **146** with methyl ester of cysteine **141**. The thiazolidine **142** formed in the first step undergoes oxidation with silver carbonate and base to form the aza intermediate **143**. Subsequent addition of pyrrolidinocyclohexene **144** to **143** gave the corresponding products **145** and **146** in 35% and 45% yields, respectively (Fig. 3.52).[113]

3.2.20   Synthesis of furoquinolines

3, 4-Dihydrofuran **148** is another very important system that forms furoquinolines **149** and **150** by using the salen-AlCl complex catalytic system **151**. AGs-imine is formed during the reaction, which adds to 3,4-dihydrofuran **148** in an aza-Diels−Alder fashion. The reaction is completed in 7 h in MeCN and two diastereomeric products **149** and **150** are obtained during the reaction (Fig. 3.53).[114]

3.2.21   Synthesis of pyrimidine

The reactivity of AGs is sometimes sensitive to solvent systems; in this context, AG reacts with propane diamine **152** to form pyrimidine derivative **153** in 89% yields when the reaction was carried in the ether as a solvent. However, when the solvent system was switched to ethanol and heated to

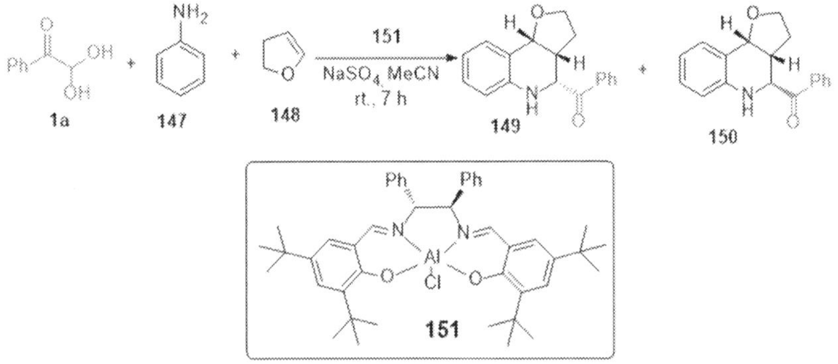

FIGURE 3.53   Synthesis of furoquinolines.

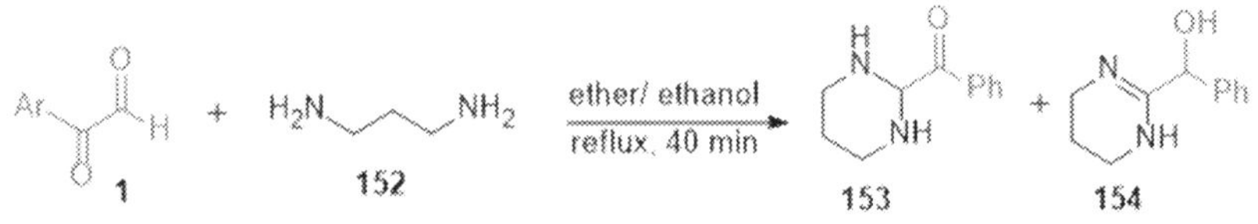

FIGURE 3.54   Synthesis of pyrimidine derivatives.

65°C, along with **153**, the additional product **154** was also produced in 17% overall yield (Fig. 3.54).[115]

In contrast, tetrahydropyrimidine analog **157** are obtained when arylglyoxal-hydrates react with urea **155** and 1,3-dicarbonyl compound **156** under Lewis acid catalysis through Biginelli reaction (Fig. 3.55). Appreciable yields (59%–73%) were obtained when ZnCl$_2$ was employed in refluxing ethanol. The reaction can also be promoted by microwave (MW) irradiation with ZnCl$_2$ and AlCl$_3$ on silica in a short time, but the product is formed in lesser quantity with 26%–42% yields.[116]

3.2.22 Synthesis of pyran

A hetero Diels–Alder reaction of AGs was also described with various other dienes like **158** to form a pyrene ring **160** in presence of Pd or Pt catalyst.[117] Out of other chiral ligands, (S)-BINAP **161** is proved to be a very effective and selective ligand, which not only furnished high yields but generated the desired product in 33%–99% enantiomeric excess (Fig. 3.56). Various other diene systems like cyclohexadiene **162** and diethyl 3,4-dimethylenecyclopentane-1,1-dicarboxylate **163** were also used[118] in hetero Diels–Alder reaction with AGs under a different catalyst [(S) MeObiphepPd(NCAr)$_2$(SbF$_6$)$_2$] CH$_2$Cl$_2$ at 0°C with 55%–80% product formation with 98%–99% enantiomeric excess and 98:2 diastereomeric ratio.[119]

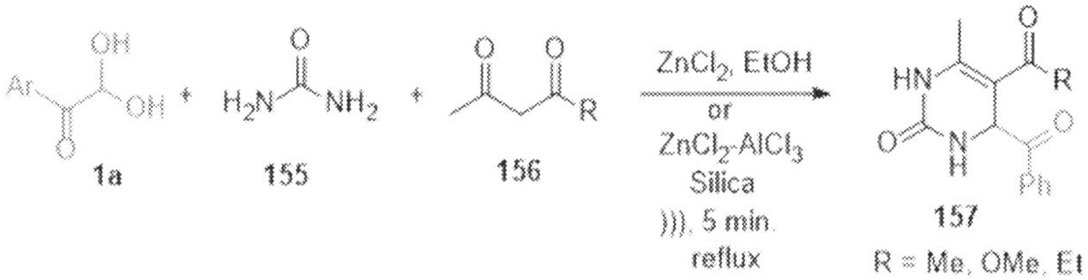

FIGURE 3.55 Synthesis of tetrahydropyrimidine analogs.

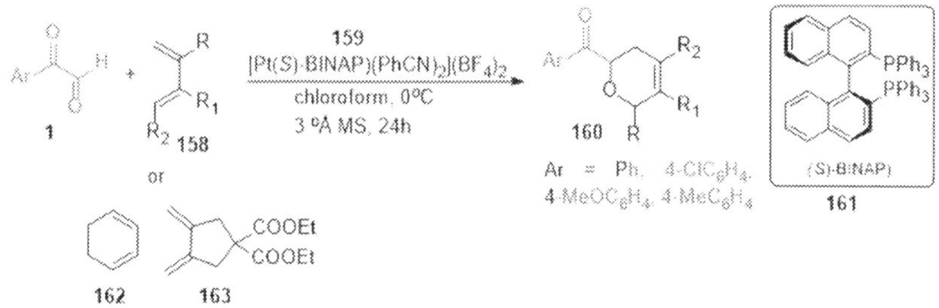

FIGURE 3.56 Synthesis of pyrane derivatives.

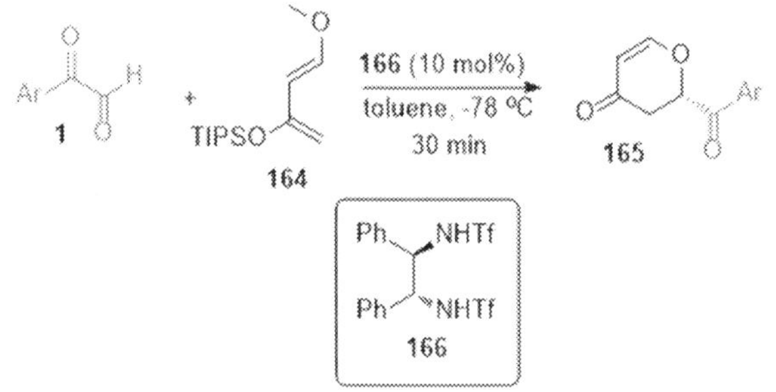

FIGURE 3.57 Synthesis of pyrone derivatives.

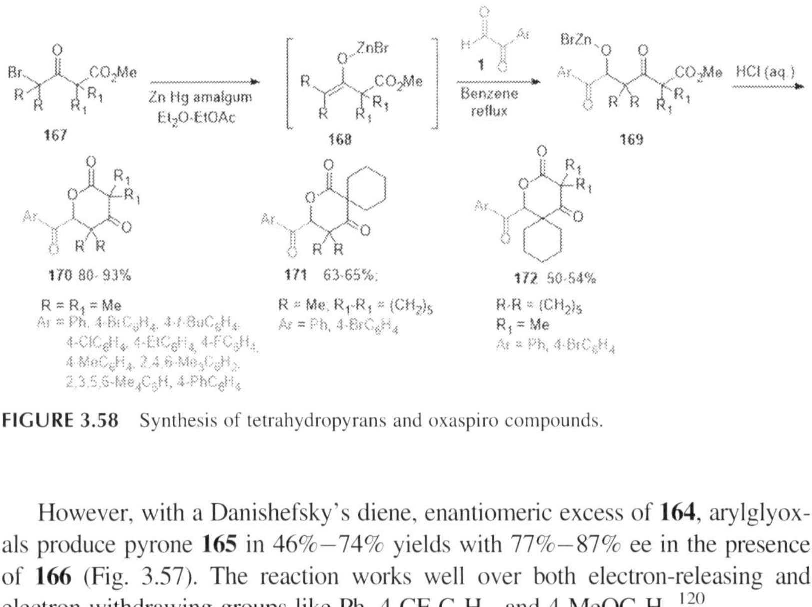

FIGURE 3.58 Synthesis of tetrahydropyrans and oxaspiro compounds.

However, with a Danishefsky's diene, enantiomeric excess of **164**, arylglyoxals produce pyrone **165** in 46%−74% yields with 77%−87% ee in the presence of **166** (Fig. 3.57). The reaction works well over both electron-releasing and electron-withdrawing groups like Ph, $4\text{-}CF_3C_6H_4$, and $4\text{-}MeOC_6H_4$.[120]

Arylglyoxal also reacts through enolates; a very important reaction in this context is the Reformatsky reaction [121] between methyl pentanoate derivative ($R = R_1 = Me$) **167** and arylglyoxal **1**. Zn amalgam generates the required Zn-Br enolate ion **168** that reacts with arylglyoxal in refluxing benzene to form **169**, which cyclizes to tetrahydropyrans **170** and oxaspiro compounds **171** and **172** under the acid hydrolysis by HCl (5%) at 0°C (Fig. 3.58).

Experimentally, methyl pentanoate derivative **167** was added slowly to the activated Zn fillings followed by the addition of $HgCl_2$ in catalytic quantity in diethyl ether and ethyl acetate (1:3 ratio). Furthermore, the addition of arylglyoxal in the reaction mixture and refluxing for 30 min, followed by acid hydrolysis formed the desired products **170, 171,** and **172**.[122]

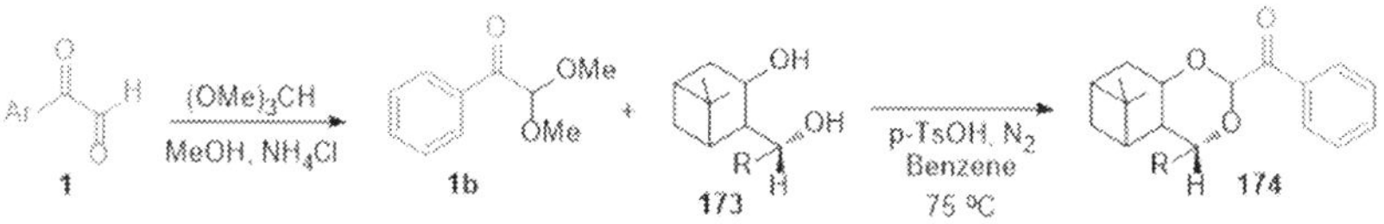

FIGURE 3.59   Synthesis of dioxane derivatives.

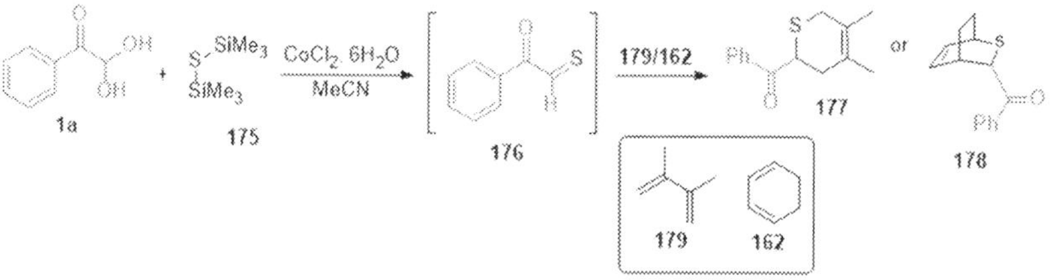

FIGURE 3.60   Synthesis of thiopyran systems.

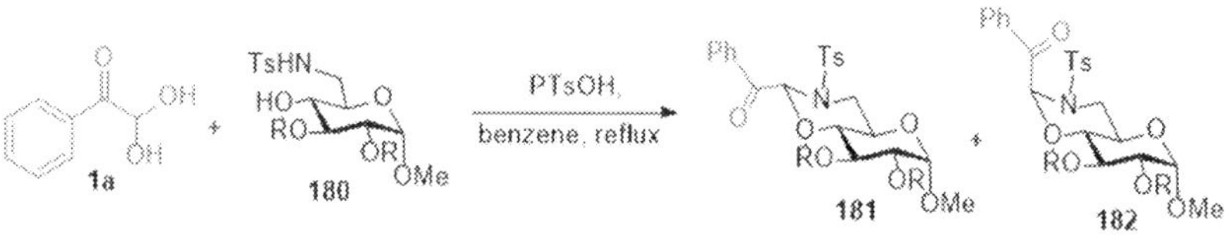

FIGURE 3.61   Synthesis of oxazine derivatives.

3.2.23   Synthesis of dioxane

The reactivity of arylglyoxal acetals **1b** with a pinane derivative **173** for the synthesis of dioxane derivative **174** has been reported by using p-toluene sulfonic acid in benzene (Fig. 3.59). The product so formed can be transformed into carbinols using a Grignard reagent.[123]

A very important reaction of arylglyoxal-hydrates is the formation of thioaldehyde **176** when they react with bis(trimethylsilyl) sulfide **175**, which can undergo [2 + 4] Diels—Alder reaction with either **179** or **162** to give thiopyran systems **177** and **178**. The reaction occurs in very mild conditions in MeCN and cobalt chloride hydrate to obtain the product with *endo* selectivity in 85% or 89% yields.[124] Thiopyran systems occur widely in biological compounds[125] and are used to synthesize many heterocycles.[126] They also find applications in material science for electronic devices (Fig. 3.60).[127]

3.2.24   Synthesis of 1,3-oxazines

1,3-Oxazines are present in pharmacologically active scaffolds that show numerous biological activities like analgesic, antitubercular, anticancer, anti-HIV, antihypertensive, antibiotic, antithrombotic, and anticonvulsant. Some oxazines are useful in the development of photochromic compounds.[128–137] Owing to their importance, arylglyoxlals have been exploited for the synthesis of various oxazines.

The hydrate form of AG also shows reactivity toward D-glucose derivative **180** at the tosyl protection of the primary alcohol and free secondary hydroxyl group at C4 to produce oxazines in α/β mixture **181** and **182**, which can undergo nucleophilic attack or even reduction at carbonyl unit to form 2-benzoyl-1,3-oxazines in a stereoselective manner (Fig. 3.61).[138]

Under the same conditions (1R)-(+)-camphor **183** reacts with AGs to form oxazines **184** and **185** which can be diastereoselectively reduced to form chiral alcohols (Fig. 3.62).[139]

AGs also reacts with naturally occurring amino alcohol, L-ephedrine **186** to form chiral oxazolidine **187** in toluene by treatment of AG-hydrate with L-ephedrine **186**. Other solvents like DEE and EtOH were used in the presence of molecular sieves at 20°C. The reaction occurs regiospecifically to give diastereomeric forms of 2-benzoyloxazolidine **187**, which can produce morpholine **188** through rearrangement when kept for many days under ambient conditions (Fig. 3.63).[140]

The reactivity of AGs is also explored by its condensation with (-)-8-aminomenthol **189** to form oxazine **190** in efficient yields simply by dissolving the reagents in CH_2Cl_2.[141] Oxazine is also used as a precursor for the synthesis of some morpholine derivatives like **191** (Fig. 3.64).[142]

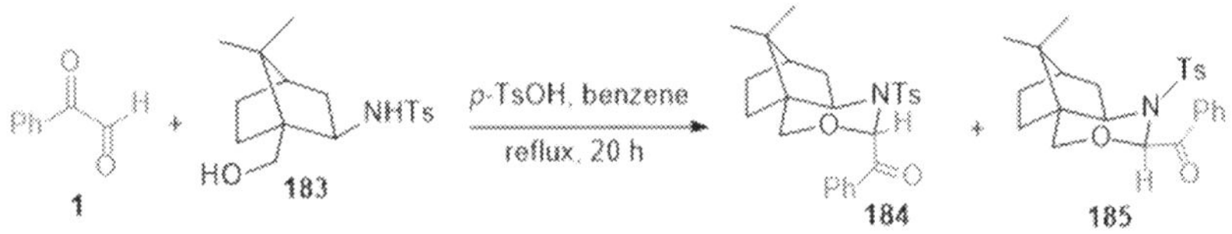

FIGURE 3.62 Synthesis of substituted oxazines.

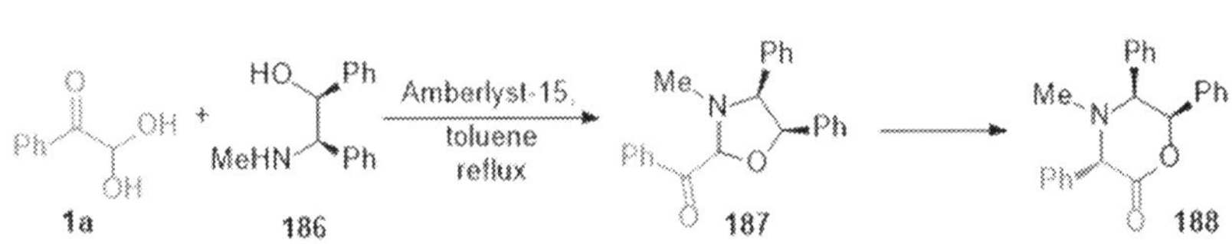

FIGURE 3.63 Synthesis of chiral oxazolidine derivatives.

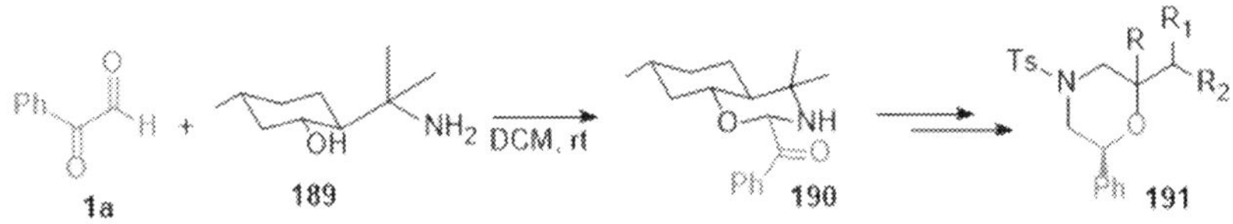

FIGURE 3.64 The condensation reaction of 2-oxoaldehydes with (-)-8-aminomenthol.

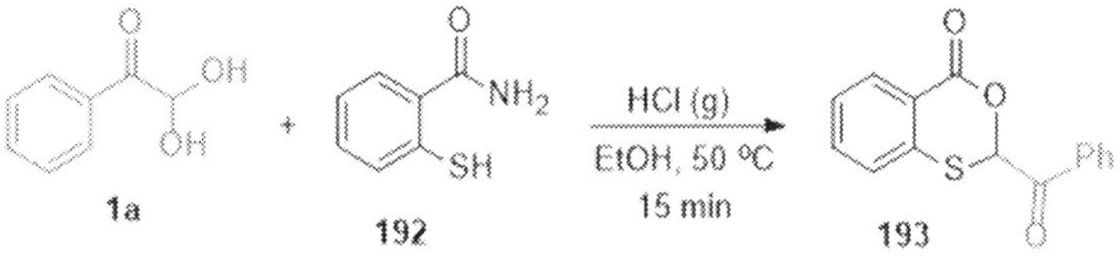

FIGURE 3.65 Synthesis of benzoxathiin derivatives.

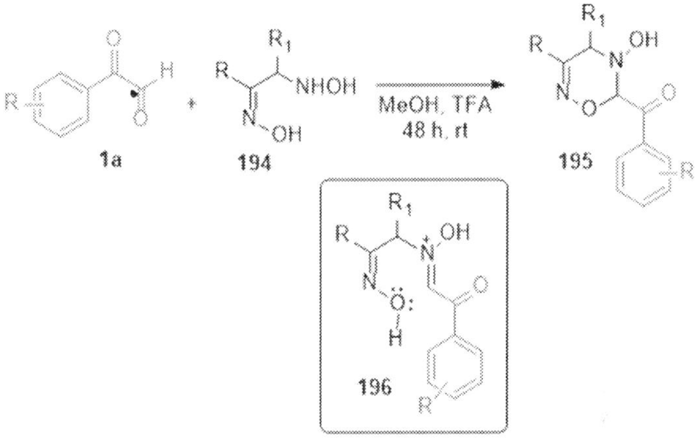

FIGURE 3.66 Synthesis of oxadiazine analogs.

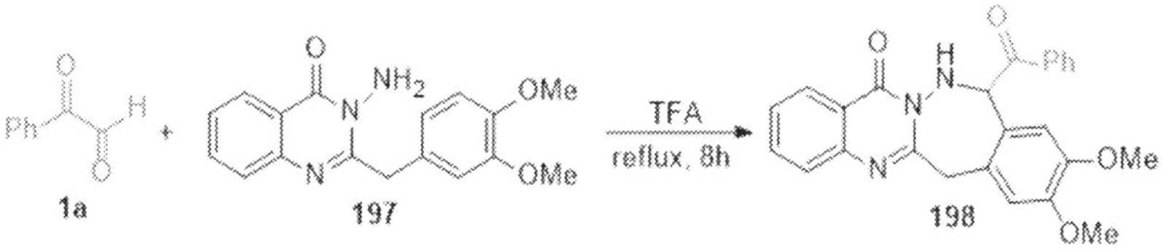

FIGURE 3.67 Synthesis of oxadiazine analogs.

AGs hydrates react with 2-amidothiols **192** in the presence of pure gaseous HCl (g) in ethanol to form benzoxathiin derivative **193** in 28% yield (Fig. 3.65).[143]

Arylglyoxal react with oximes **194** (where R = Ph, 4-ClC$_6$H$_4$; R$_1$ = H, Me; Ar = Ph, 4-BrC$_6$H$_4$, 4-ClC$_6$H$_4$, 4-MeOC$_6$H$_4$) in MeOH and trifluoroacetic acid to produce oxadiazines **195**. The reaction proceeds slowly within 48 h and passes through intermediate **196** (Fig. 3.66). The addition of oxygen to the iminium ion furnishes the required product in 54%−97% yields.[144]

Another instance for the formation of the seven-membered ring by AGs is through Pictet−Spengler condensation with 3-aminoquinazolinone **197** to form benzodiazepine derivative **198** (Fig. 3.67). The reaction can also be carried out by using HCl-H$_2$O in dioxane at 80°C for 8 h in 65% yields.[145]

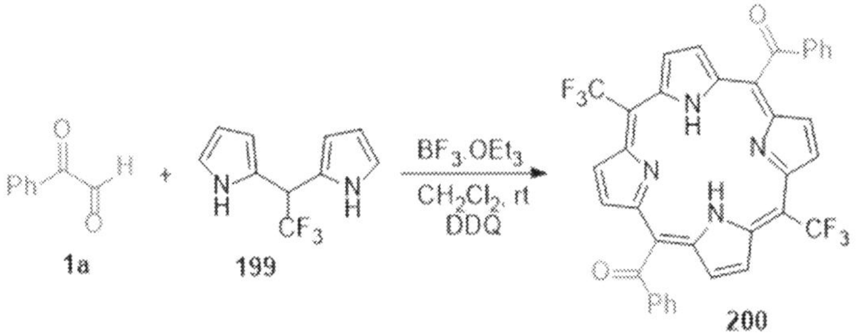

FIGURE 3.68   Synthesis of trans-substituted porphyrin.

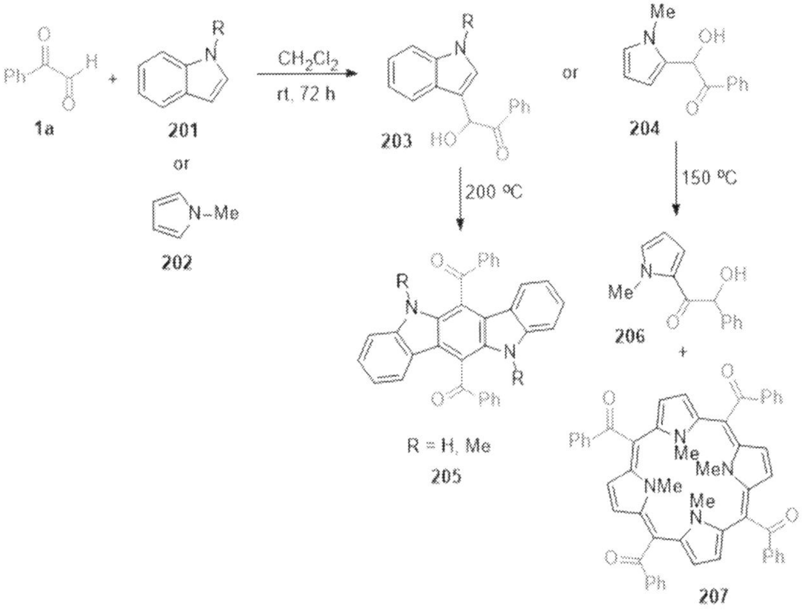

FIGURE 3.69   Synthesis of porphyrin stems.

3.2.25   Synthesis of porphyrins

Porphyrins have a vital role in photosynthesis,[146] electronics,[147–152] and DNA binding arylglyoxalents.[153] Apart from other known strategies for the preparation of porphyrins,[154] 2-oxoaldehyde **1a** hydrates have been employed for the synthesis of *trans*-substituted porphyrin **200** by reaction with dipyrromethane **199** in 14% yield (Fig. 3.68).[155]

Another example for the synthesis of porphyrins stems from 2-oxoaldehyde hydrate, indole **201**, and *N*-methyl pyrrole **202** to form benzoins **203** and **204**.[156] After heating pyrrole **204** at 150°C, in addition to the formation of **206**, polymerization takes place to form porphyrin **207**. However, when indole **203** derivative (R = H) is heated at 200°C, the dimer

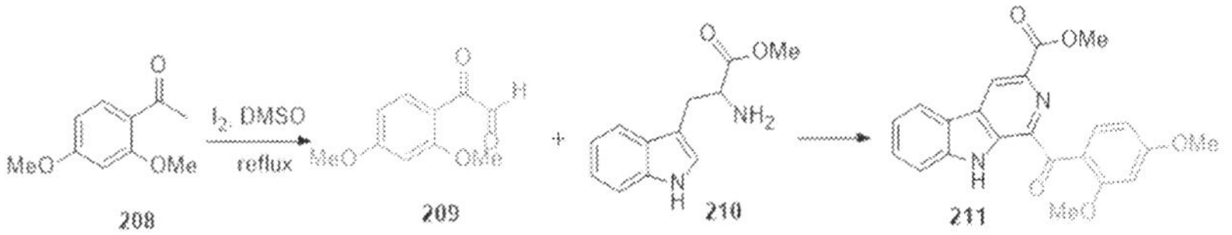

FIGURE 3.70   Synthesis of β-carboline derivatives.

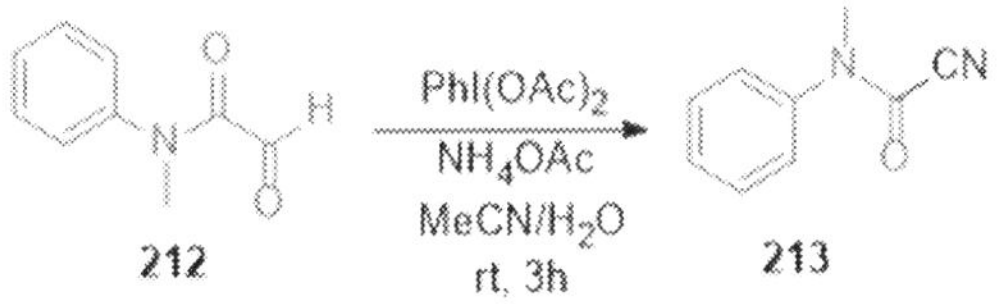

FIGURE 3.71   Synthesis of cyanoformamides.

205 was obtained in lesser yields as compared to when the reaction was performed with *N*-methylindole **203** (R = Me) in which **205** were formed in higher yields (Fig. 3.69).[157]

3.2.26   Synthesis of β-carboline

Another example of the reactivity of arylglyoxal through Pictet–Spengler fashion is explored through the synthesis of a marine natural compound, β-carboline in a multicouple domino reaction between arylglyoxal, tryptophan methyl ester, or tryptamine. The reaction has been eventually applied for the synthesis of merinacarboline (A and B), pityriacitrin, eudistomin Y1, pityriacitrin B, and fascaplysin analogs.

The arylglyoxal **209** was generated when 2, 4-dimethoxy acetophenone **208** was treated with iodine in DMSO. The resulting compound reacts with tryptophan methyl ester **210** and later cyclizes into **211** at 90°C for 3 h in 85% overall yield (Fig. 3.70).[158]

In another way, the derivative of AG **212** reacts under oxidative conditions with (diacetoxyiodo) benzene PhI(OAc)$_2$ as an oxidant and NH$_4$OAc as a nitrogen source to produce cyano formamides **213**. The reaction occurs in very simple conditions in MeCN/H$_2$O (v/v = 4/1) at room temperature for 3 h (Fig. 3.71). The reaction can tolerate electronic effects of AGs and groups like biphenyl, thienyl, benzyl, and vinyl in ecofriendly reaction conditions.[159]

3.2.27   Synthesis of carboxylic acid

Aryl, heteroaryl, or aliphatic α-oxoaldehydes **1** undergoes cleavage at C–C bond to lose the carbonyl group at C2 and oxidation of carbonyl group at C1 to form carboxylic acids **214**. The reaction is performed in DMSO solvent, which acts as an oxygen source and iodine. The reaction is supposed to pass via a solvated complex[160] **215** and works very well in both MW (in 5 min)

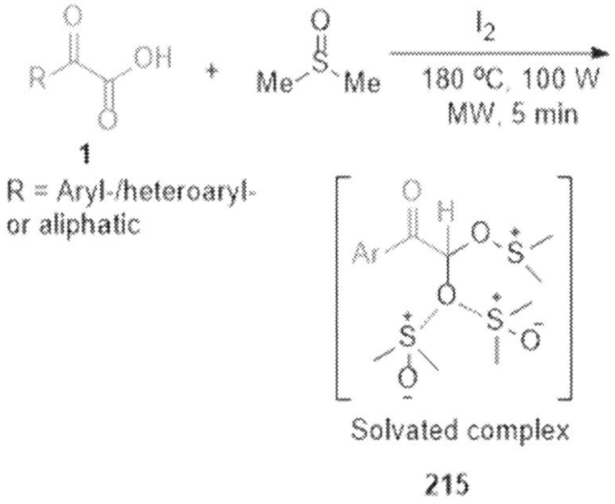

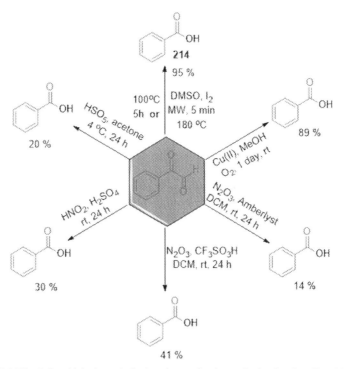

FIGURE 3.72   Synthesis of carboxylic acids.

FIGURE 3.73   2-Oxoaldehyde as the basic substrate for the synthesis of carboxylic acids under variable conditions.

and heating conditions at 180°C and 100°C, respectively (Fig. 3.72). The synthesis of aryl carboxylic acids **214** from arylglyoxal as a substrate or an intermediate are summarized in Fig. 3.73.[161]

3.2.28 Synthesis of hydroxyphenylbutenone

Direct alkenylation of arylglyoxal takes place with 3-vinylindoles **215** to yield α-hydroxyl alcohols **216** catalyzed by chiral phosphoric acid catalyst H8-BINOL **217** with an E/Z ratio of 87:13 and up to 90% ee (2.62). The product formed serves as the basic precursor for the synthesis of chiral tetrahydrocarbozol-2-ones and indole molecules. ¹H-NMR study revealed that the function of the catalyst is the activation of phenylglyoxal hydrate and to maintain the stereoselectivities in a dissymmetric coordination manner (Fig. 3.74).[162]

Arylglyoxal reacts with alkenes **228** catalyzed by Pd(II)-BINAP **220** and produces **229** with ee = 93.8%. Both *para-* and *ortho*-methyl-substituted substrates increase the selectivity of the products owing to the unique chiral environment created by Pd(II)-BINAP catalyst (Fig. 3.75). In addition, the Pd(II)-BINAP is stable and does not go through any conformational changes and can be recycled more than 20 times with enantioselective efficacy still intact.[162]

Yong Yuan and coworkers demonstrated an approach for the synthesis of functionalized pyrrolidones **223** from AGs, α-angelica lactone **222**, and anilines **221** through a multicomponent process in simple acidic conditions. In addition to being a greener and scalable approach, the reaction proceeds without any solvent in H_2SO_4 (30 mol%) in 1 h and gives up to 75% yields (Fig. 3.76).[163]

The reactivity of arylglyoxal is also demonstrated by its reaction with aromatic fused ring azulene **224** and cyclohexane-1,3-dione **225** in a MCR to

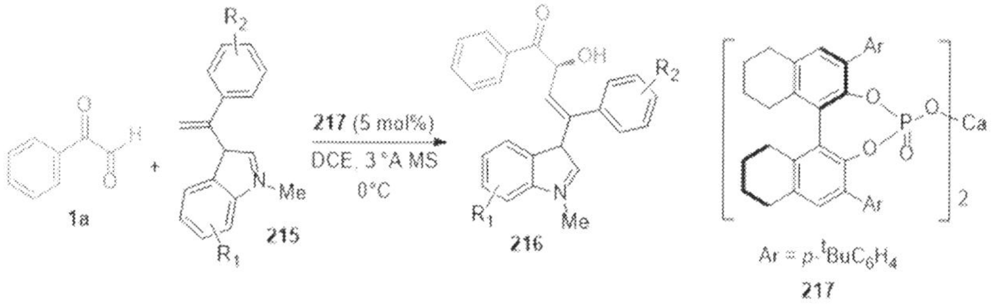

FIGURE 3.74 Synthesis of hydroxyphenylbutenone derivatives.

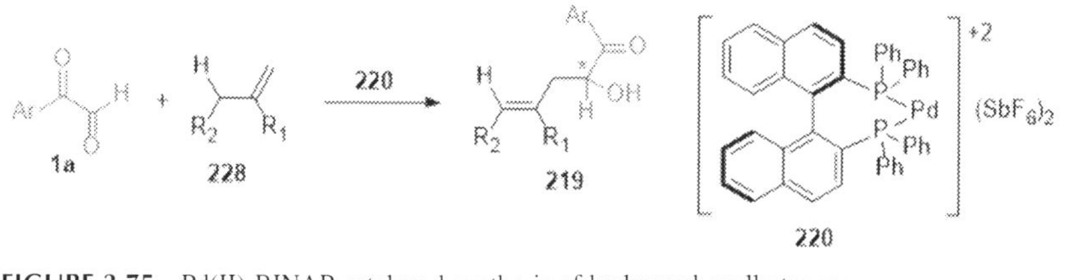

FIGURE 3.75 Pd(II)-BINAP-catalyzed synthesis of hydroxyphenylbutenone.

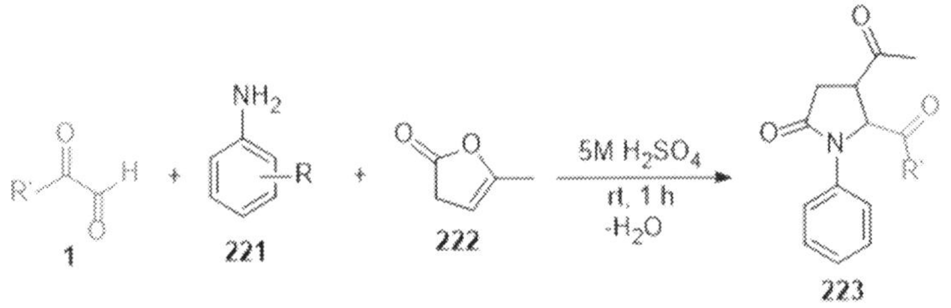

FIGURE 3.76   Synthesis of pyrrolidones.

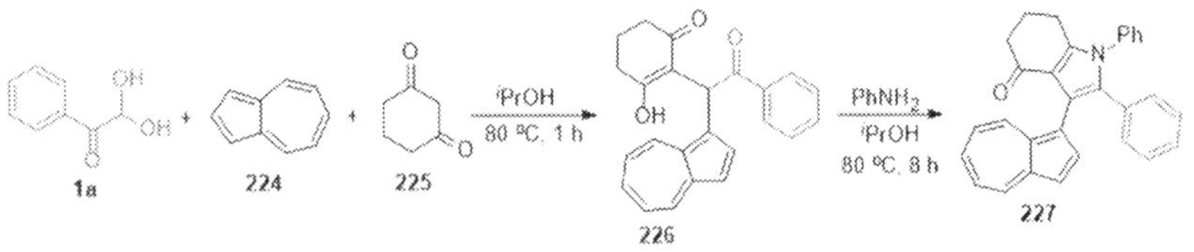

FIGURE 3.77   Synthesis of azulene indole adduct.

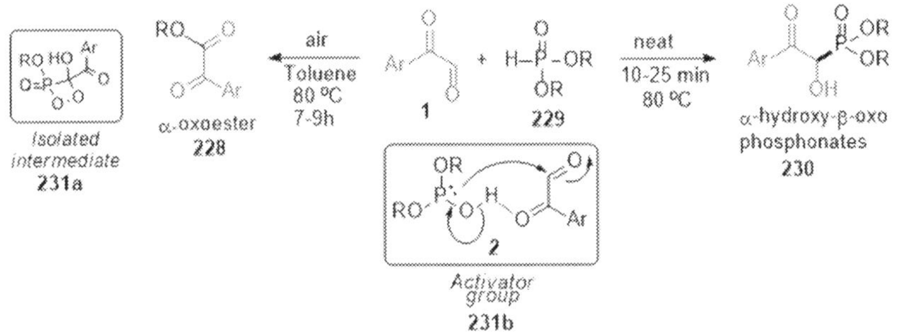

FIGURE 3.78   Synthesis of α-oxoester and α-hydroxy β-oxophosphonates.

form azulene indole adduct **226**. These azulene derivatives can be targeted with benzylic and aromatic amines to access heterocycles containing azulenes **227** (Fig. 3.77).[164]

3.2.29   Synthesis of α-hydroxy-β-oxophosphonates and α-oxoesters

The reactivity of 2-oxoaldehyde with phosphoesters **229** in two different environments presents a simple and direct method for the synthesis of α-oxoester **228** and α-hydroxy β-oxophosphonates **230** which explains the crucial role of the α-oxo group in promoting the reaction via H-bonding. The base or catalyst-free reaction in open air leads to the formation of a-oxoester **228** through an isolated intermediate **231a**. Other reaction forms α-hydroxy-β-oxophosphonates **230** in very neat conditions (Fig. 3.78). The reaction is promoted by the activator group **231b** generated as a result of H-bonding between α-carbonyl of the 2-oxoaldehyde and O atom in the phosphoester group, which makes phosphoryl group attack on activated -CHO group feasible.[165]

3.2.30 Synthesis of dicarbonyl compounds

AGs **1** can also promote *ortho*-functionalization of simple phenols **232** (Fig. 3.79). The reaction is catalyzed by copper acetate to form 2-hydroxyphenyl-1, 2-diones **233**, which are found in many natural products that exhibit various biological activities and also block the enzymatic activity of carboxylesterases.[166]

Phenylglyoxals and nitroalkanes react in an aldol type reaction called nitro aldol or Henry reaction to form 1, 2-diketones **236** under both heterogeneous and MW conditions. The overall reaction is a single step and uses amberlyst A21 (0.5 g mmol^{-1}) as a base in 2-MeTHF at room temperature for 5 h (Fig. 3.80). The compound **235** is converted to the desired product **236** at 110°C in 1 h under MW conditions in 75% yield.[167]

Another method that shows the reactivity of arylglyoxal is their reaction with thiols to form α-ketothioesters **238**. The reaction proceeds via iminium ion **237**, which is formed when pyrrolidine reacts with arylglyoxal in toluene in an open flask and without any metal reagent or catalyst with 76% yields (Fig. 3.81).[168]

Recently, a new visible light that promoted regioselective dicarbonylation of C—H bond was reported by Hua Cao and coworkers. The reaction takes place between 2-oxoaldehydes **239** by utilizing different indolizines **240** in the presence of *Rose Bengal* **241** as a photosensitizer to produce **242** in open air (as oxidant) conditions without the need of any metal catalysts. Some important 1,2-dicarbonyl indolizine structures **243** and **244** were obtained as a result of direct dicarbonylation of the planar C-H bond in 50%−91%

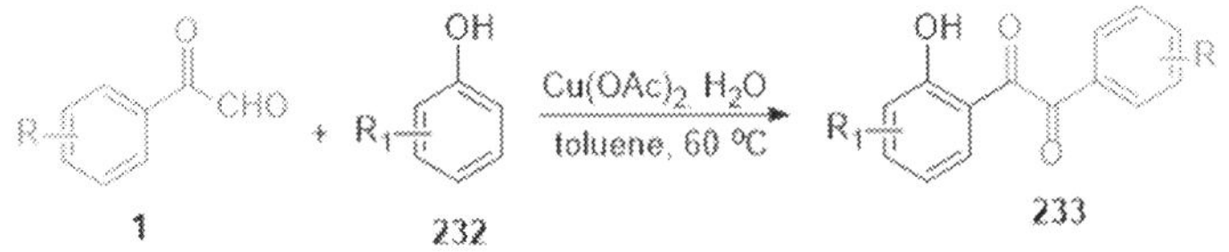

FIGURE 3.79   *Ortho*-functionalization of simple phenols.

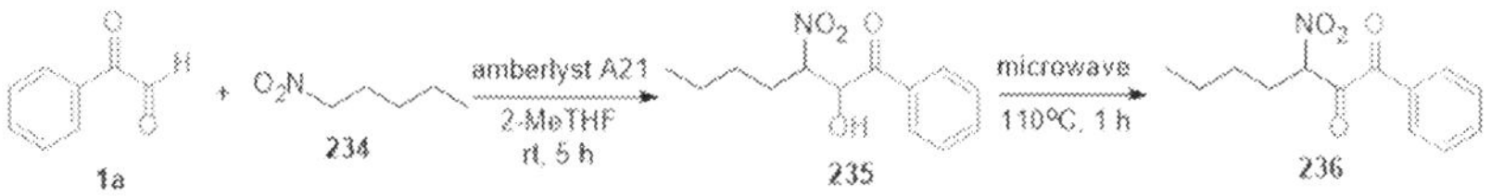

FIGURE 3.80   Synthesis of NO$_2$-containing 1,2-diketones.

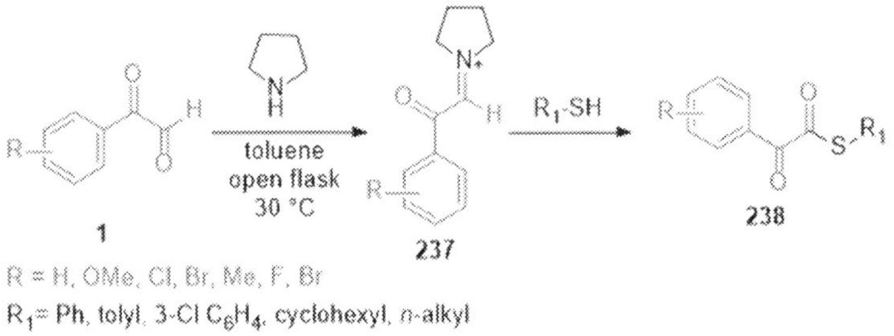

R = H, OMe, Cl, Br, Me, F, Br

R$_1$ = Ph, tolyl, 3-Cl C$_6$H$_4$, cyclohexyl, n-alkyl

FIGURE 3.81 Synthesis of α-ketothioesters.

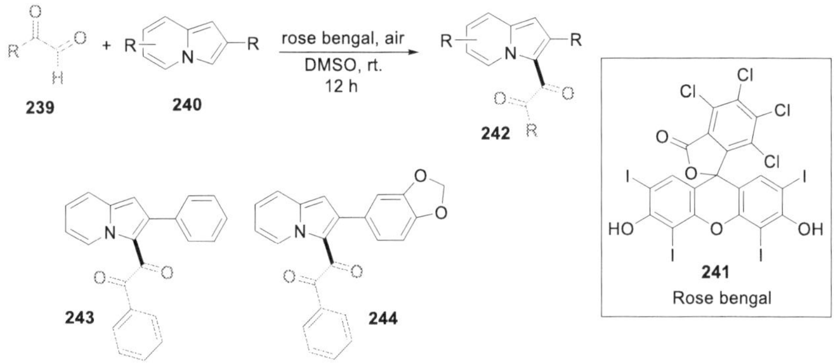

FIGURE 3.82 Regioselective dicarbonylation of 2-oxoaldehydes.

yields. The indolizine analoges thus formed could exhibit fluorescent properties, which could bear application in sensors in luminescent materials. In this context, **243** showed strong absorption intensity in DCM and highest fluorescent emission in DMSO solvent (Fig. 3.82).[169]

3.2.31 Synthesis of 2-oxoesters and 2-oxoacids

2-Oxaldehyde generates different substituted 2-oxoesters and 2-oxoacids **246** in a cross-coupling reaction with hydroperoxides **245** or iminium ions driven by 2-oxo unit at room temperature (Fig. 3.83). The reaction proceeds by using 10 mol% of pyrrolidine through direct coupling or amine-catalyzed pathway without any additional catalyst or oxidant. This reaction reveals the tendency of self-decomposition in hydroperoxides induced as a result of hydrogen bonding for the generation of 2-oxoesters and 2-oxoacids.[170]

3.2.32 Synthesis of acetamides

2-Oxaldehydes show application toward the Ritter reaction by selectively forming mono and di-acetamides **247** and **249**.[171] The mono-Ritter product

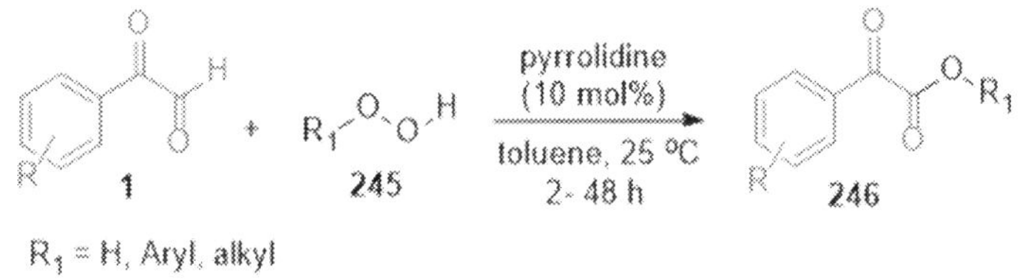

R_1 = H, Aryl, alkyl

FIGURE 3.83 Synthesis of 2-oxoesters and acids.

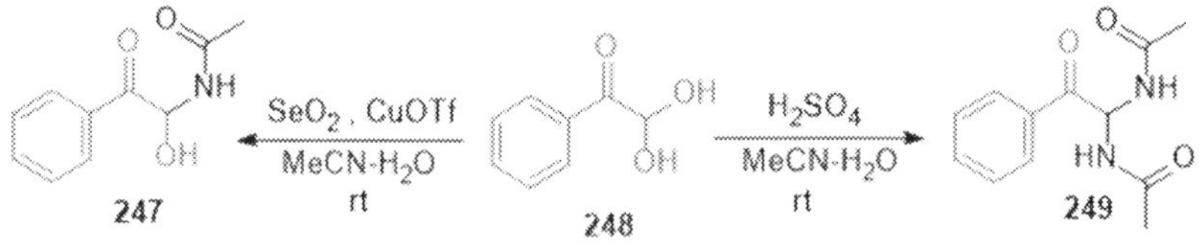

FIGURE 3.84 Regioselective dicarbonylation of 2-oxoaldehydes.

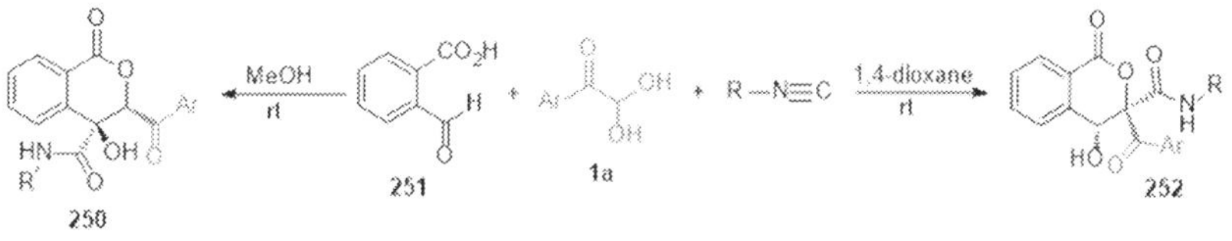

FIGURE 3.85 Synthesis of mono and di-acetamides.

247 was formed in presence of 50 mol% SeO_2 and 20 mol% Cu(OTf) as a catalyst while a di-Ritter product **249** was obtained in 20 mol% H_2SO_4 (Fig. 3.84).

3.2.33   Synthesis of isocoumarins

A three-component annulation reaction between AG-hydrates **1a**, 2-acetylbenzoic acid **251**, and isonitriles is easy to access toward the synthesis of substituted isocoumarins **250** and **252**. The product contains two chiral centers. The Passerini three-component condensations occur in a first step followed by cyclization through aldol condensation (Fig. 3.85).[172]

3.2.34   Synthesis of 2-oxoacetamidines

Similarly, a special reactivity toward AGs as compared with common aldehydes is shown by amines. AG combines in a domino fashion with anilines and secondary amines to yield 2-oxoacetamidines **253** through iminium ion formation (Fig. 3.86). This reaction does not require any additives or dehydrogenative α, α-diamination modes, which are important steps in many 2-oxoiminium ions promoted 2-oxoacetamidines yielding reactions.[173]

3.2.35 Synthesis of esters

Esterification of AGs **1** in the presence of different alcohols **254** and cata-lytic amounts of oxone ($2KHSO_5 \cdot KHSO_4 \cdot K_2SO_4$) is a very simple and eco-nomical way of obtaining esters **255** (Fig. 3.87). The aldehyde carbonyl function in 2-oxoaldehyde undergoes CO-C bond cleavage after which oxone promotes the esterification.[174]

3.2.36 Synthesis of 1,2-disubstituted benzimidazoles

The divergent reactivity of 2-oxoaldehydes is extendable to benzotriazoles **256** through the in situ generation of RCOCHN1N2 systems, generated by employ-ing amines **257** in the reaction. The formation of 6-aminophenanthridines **259** with amines takes place through decarbonylative cyclization with the loss of N_2 from benzotriazole motif **258** which is facilitated by the 2-oxo unit in the RCOCHN1N2 system (2.75). The same concept is applied in the reaction between 2-oxoaldehdyes **1**, benzotriazoles (BTZ) **256**, and 4-azidophenol **260** that leads to the formation of 1, 2-benzimidazoles **262**. The reaction is also facilitated by 4-azidophenol in its RCOCHN1N2 system by extruding N_2 through benzoquinone imine intermediate (Fig. 3.88).[175]

3.2.37 Synthesis of azolated derivatives

The reaction between malonate half esters **263** and AGs **1** forms an aldol intermediate **264** that carries a simple base controlled selective transforma-tion toward α-hydroxy carbonyl derivative **265** and α, β-unsaturated esters **266** using different C−C bond-forming sequences (Fig. 3.89).

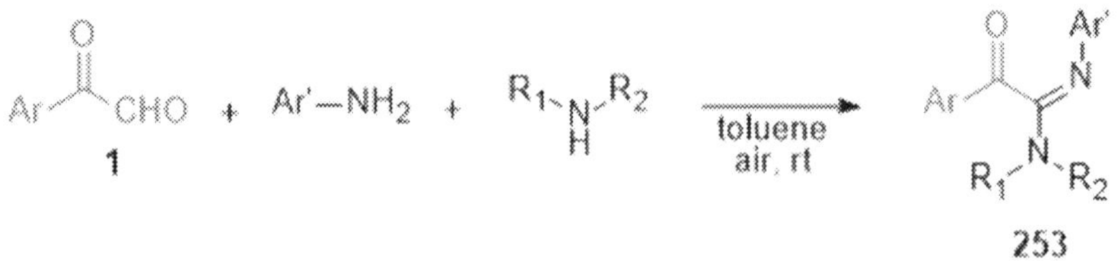

FIGURE 3.86 Synthesis of substituted isocoumarins.

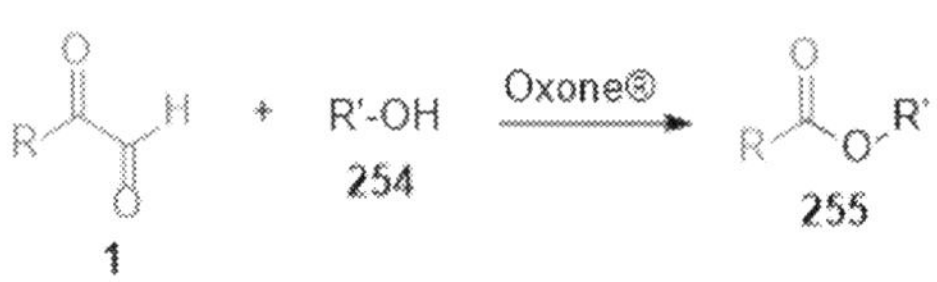

FIGURE 3.87 Synthesis of 2-oxoacetamidines.

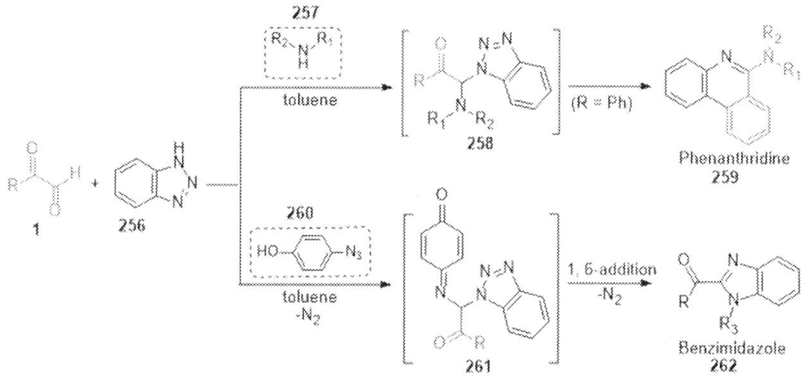

FIGURE 3.88 Synthesis of substituted esters.

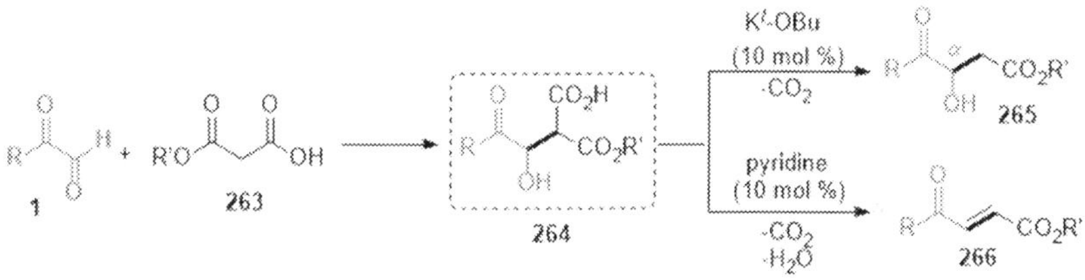

FIGURE 3.89 Synthesis of 1,2-disubstituted benzimidazoles.

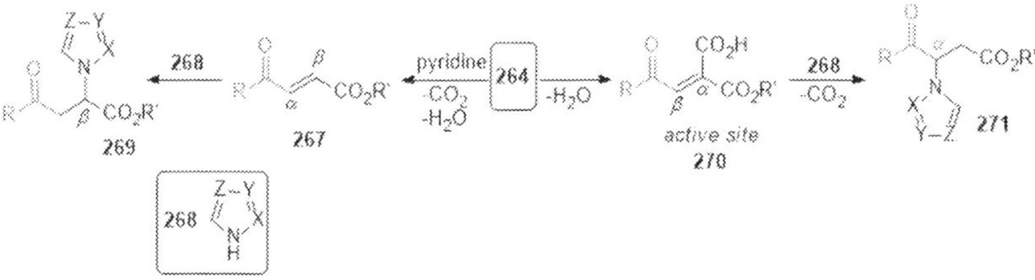

FIGURE 3.90 Synthesis of α-hydroxy carbonyl derivative and α, β-unsaturated esters.

Azoles are widely used as antifungal arylglyoxalents like clotrimazole and fluconazole, however, their synthesis of α- and β-azolated products are usually met with certain challenges related to atom economy and expensive materials.

These issues can overcome by the use of 2-oxoaldehyde generated intermediate **264** that leads to the formation of β-azolated products **269** in a one-pot two-step process by using pyridine (10 mol%) before the formation of (E)-α, β-unsaturated ester **267** in the presence of azole **268**. Another reaction occurs through α, β-unsaturated ester **270** in one pot to form α-azolated product **271**. Both reactions are accompanied by the release of CO_2 and H_2O molecules (Fig. 3.90).[176]

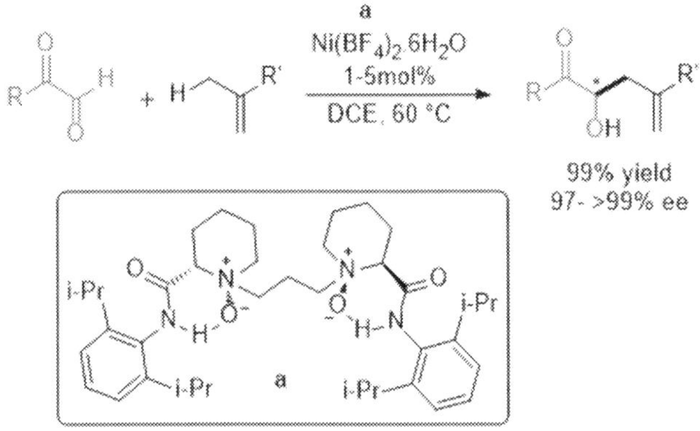

FIGURE 3.91 Synthesis of α and β-azolated products.

3.2.38 α-Hydroxy carbonyl compounds

AG and glyoxylate derivatives react with various alkenes in a carbonyl-ene protocol catalyzed by novel nickel (II)-N, N'-dioxide ligand to give α-hydroxy carbonyl moiety of biological significance under normal reaction conditions (Fig. 3.91). The reaction is highly enantioselective and nickel (II) ligand is necessary for facilitating the asymmetric synthesis of the product. The reaction has a remarkably wide reaction scope by using different substituted starting materials. Different aliphatic, aromatic, as well as glyoxylate substrates, work very well in the reaction system, and the product is obtained with high enantioselectivity (97%−99% ee) and up to 99% yield. The enantioselectivity is still preserved with high capacity even after catalyst loading was decreased up to 1 mol%.[177]

3.3 Reactions on keto group

3.3.1 Synthesis of cyclic ketals

Apart from the more reactive aldehyde part 2-oxo unit (keto group) in arylglyoxals can also participate in the reaction. 2-oxoaldehydes react with optically active substrate **272** to form diastereomeric cyclic ketal **273**. This transformation takes place with p-TsOH in refluxing benzene while azeotropic mixture formed during the reaction is removed by the Dean−Stark apparatus (Fig. 3.92). The product **273** was obtained in lower quantities while using $BF_3 \cdot OEt_2$ over a polymeric material.[178]

3.3.2 Synthesis of *R*- and *S*-diphenylphosphinoyl hydroxy aldehydes

Stuart Warren and his coworkers reported a multistep reaction for the stereo-controlled synthesis of *R*- and *S*-diphenylphosphinoyl hydroxy

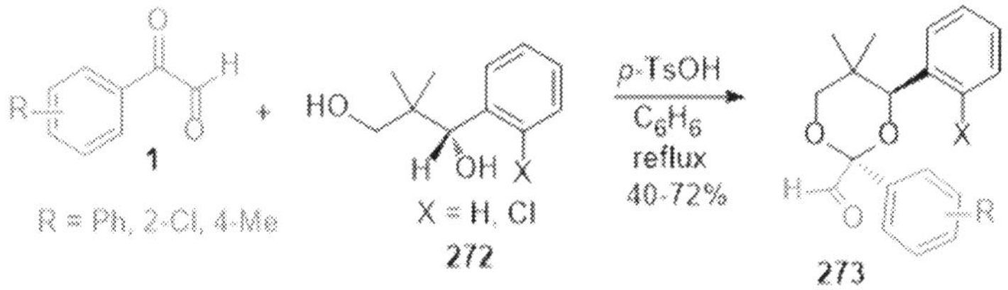

FIGURE 3.92 Synthesis of α-hydroxy carbonyl moiety.

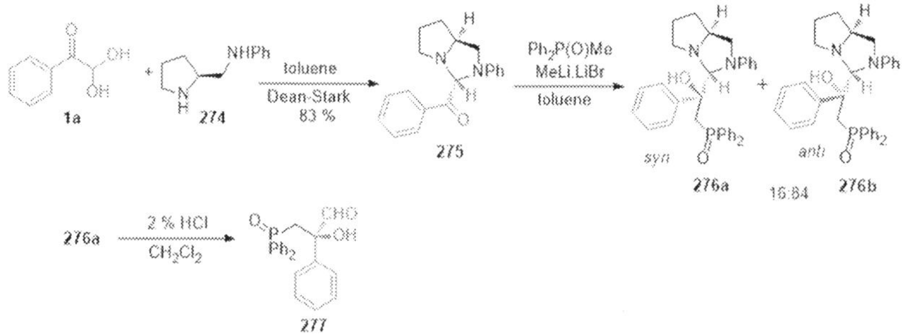

FIGURE 3.93 Synthesis of diastereomeric cyclic ketals.

aldehydes **277** by conducting a reaction between AGs **1a** and **274** in toluene. Initially, compound **275** was obtained, which was further reacted with methyldiphenylphosphine oxides and MeLi.LiBr complex in toluene to give *syn* **276a** and *anti* **276b** products in16:84 ratios. The *syn* product **276a** was further converted into diphenylphosphinoyl hydroxy aldehydes **277** in the presence of 2% HCl in CH$_2$Cl$_2$ solvent. The anainals diamine **274** was synthesized in four steps from (S)-(-)-proline (Fig. 3.93).[179]

3.3.3 Synthesis of trifluoromethylated 1,2-diols

Pedrosa et al. established an efficient protocol for the synthesis of enantiomerically enriched trifluoromethylated 1,2-diols **282** from a multistep reaction. AG **1a** reacts with **278** in benzene under refluxing condition to give 2-acyl-1,3-perhydrobenzoxazines **279**, which was further reacted with TMSCF$_3$ in the presence of TBAF or CsF to form a mixture of diastereoisomers **280a** and **280b**. The isomer **280a** is converted into trifluoromethylated hydroxyl aldehyde **281** in the presence of 2% HCl in EtOH solvent under the refluxing condition for 1 h followed by reduction with NaBH$_4$ to obtain trifluoromethylated 1,2-diols **282** (Fig. 3.94).[180]

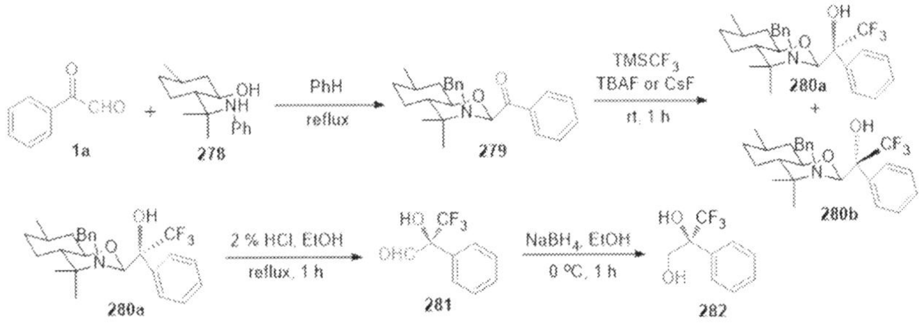

FIGURE 3.94 Synthesis of R- and S-diphenylphosphinoyl hydroxy aldehydes.

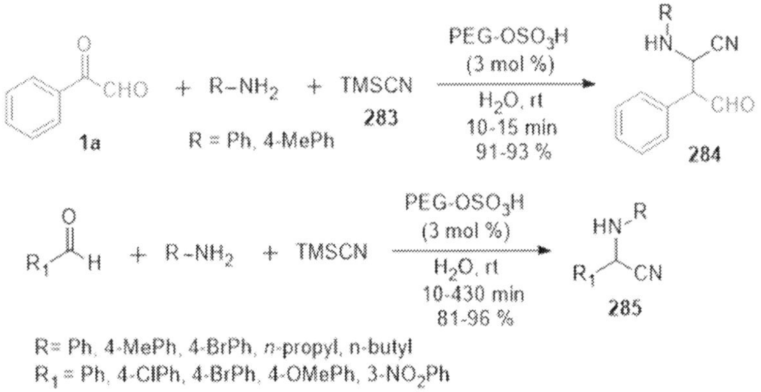

FIGURE 3.95 Synthesis of enantiomerically enriched trifluoromethylated 1,2-diols.

3.3.4 Synthesis of α-aminonitriles

A one-pot, three-component reaction for the synthesis of α-aminonitriles **284** in water by using sulfuric acid-modified polyethylene glycol 6000 (PEG-OSO$_3$H) was established by Shekouhy et al.[181] On conducting a reaction between AGs **1** or simple aldehydes with primary amines (aliphatic and aromatic) and trimethylsilyl cyanide (TMSCN) **283** in the presence of PEG-OSO$_3$H in the water at room temperature to obtain α-aminonitriles **284** and **285**, respectively, in excellent yields (Fig. 3.95).

3.3.5 Multifluorination reactions

Singh et al. reported[182] a new approach for the fluorination of glyoxal hydrates **1** with deoxofluor [(CH$_3$OCH$_2$CH$_2$)$_2$NSF$_3$] **286**. On the treatment of arylglyoxal-hydrate with deoxofluor in 3 mL of CH$_2$Cl$_2$ at room temperature gave a mixture of four products (**287**, **288**, **289**, and **290**) but the dilution of reaction from 3 to 200 mL of CH$_2$Cl$_2$ selectively gave two products (**289** and **290**) (Fig. 3.96).

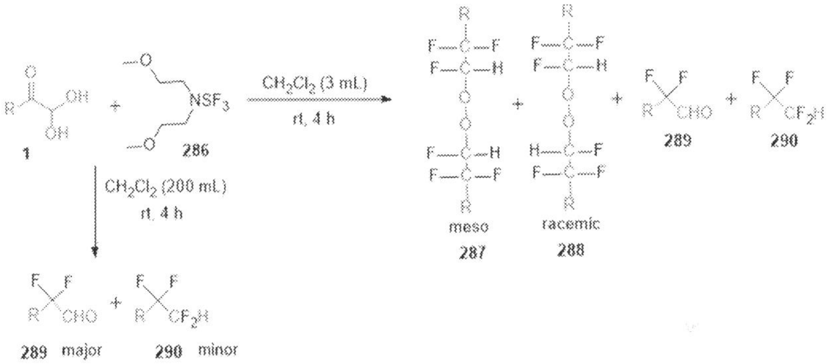

FIGURE 3.96 Synthesis of α-aminonitriles.

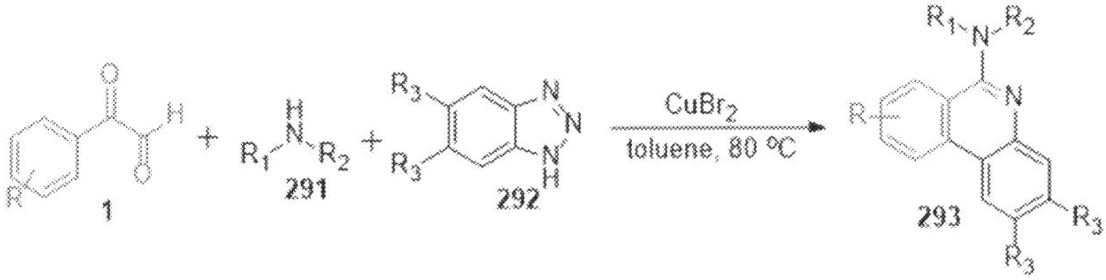

FIGURE 3.97 Fluorination of glyoxal hydrates.

3.4 Reaction from the participation of both keto and aldehyde groups

3.4.1 Synthesis of 6-aminophenanthridines

An efficient α-oxo group strategy has been used for the generation of 6-aminophenanthridines (6AP) **293** via an in situ generated novel system "CO − CH(N1N2)" from α-oxoaldehydes **1**, secondary amine **291**, and benzotriazole **292**. This reaction goes through N_2 extrusion in benzotriazoles followed by decarbonylative cyclization to give 6AP (Fig. 3.97). This protocol explains the new concept of benzotriazoles based on ring-opening chemistry and successfully applied for the synthesis of different secondary amine substituted phenanthridines (6AP).[183]

3.4.2 Synthesis of pyrrolotriazines

AGs **1** is also used for the synthesis of pyrrolo[1,2-b]-1,2,4-triazines by treating 1-NH-*boc*-protected 1,2-diaminopyrroles **294** with α-oxoaldehydes and concentrated HCl in THF at 0°C. After the addition of concentrated HCl at 0°C, two regioisomers of highly substituted pyrrolo[1,2-b]-1,2,4-triazines (**294a** and **294b**) were obtained in 11% − 44% and 8% − 75% yields (Fig. 3.98).[184]

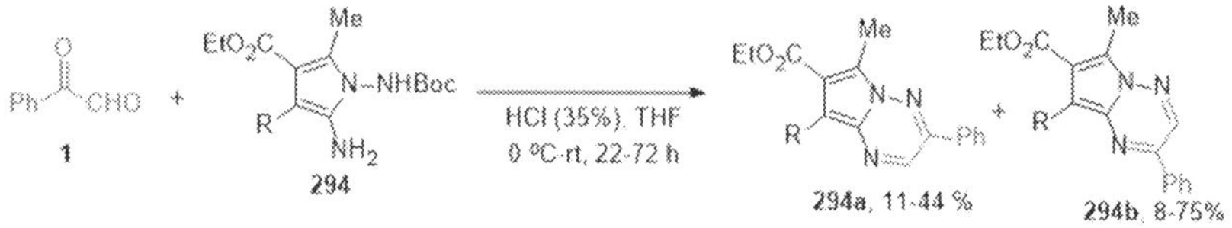

FIGURE 3.98 Synthesis of 6-aminophenanthridines.

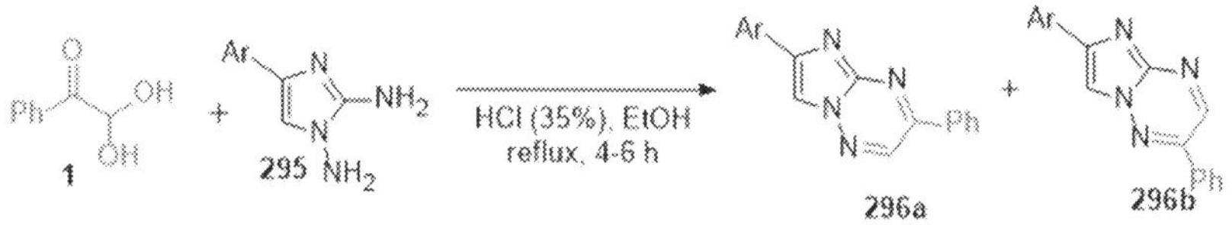

FIGURE 3.99 Synthesis of pyrrolo[1,2-b]-1,2,4-triazines.

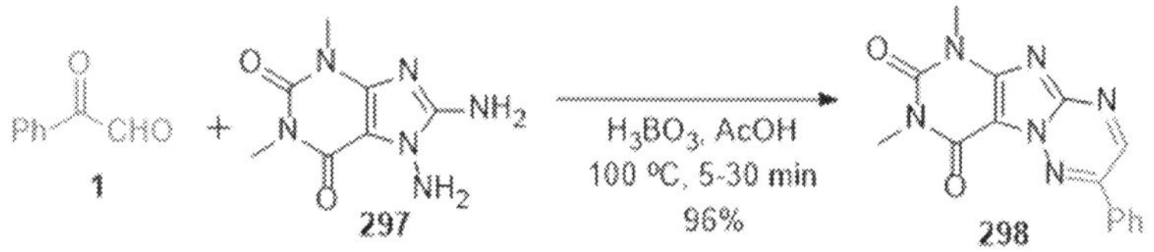

FIGURE 3.100 Synthesis of imidazo[1,2-b]-1,2,4-triazines.

3.4.3 Synthesis of imidazotriazines

The synthesis of imidazo[1,2-b]-1,2,4-triazines **296** was successfully achieved by Lalezari and his group from AGs (PG-hydrate) **1** and 4-substituted 1,2-diaminoimidazoles **295** in good yields (Fig. 3.99).[185] These reactions take place in a solution of PG-hydrate in EtOH when refluxed for for 1 h. Then add HCl and reflux for an additional 4 h to give the product **296** in 40% − 83% yields. The addition of HCl at the beginning gives a mixture of two regioisomers (**296a** and **296b**).

3.4.4 Synthesis of triazinopurinedione

A similar reaction was also done by Murata and his group by reacting AG **1** with 7,8-diamino-1,3-dimethylxanthine **297** in the presence of boric acid in AcOH at 100°C and the desired product 1,3-dimethyl-7-phenyl-[1,2,4]triazino [2,3-f]purine 2,4(1H,3H)-dione **298** was obtained in 96% yield (Fig. 3.100).[186]

3.4.5 Substituted pyrrolidines synthesis

Pyrrolidines are an important class of nitrogen-based heterocycles with different biological functions and are present in various natural products,[187]

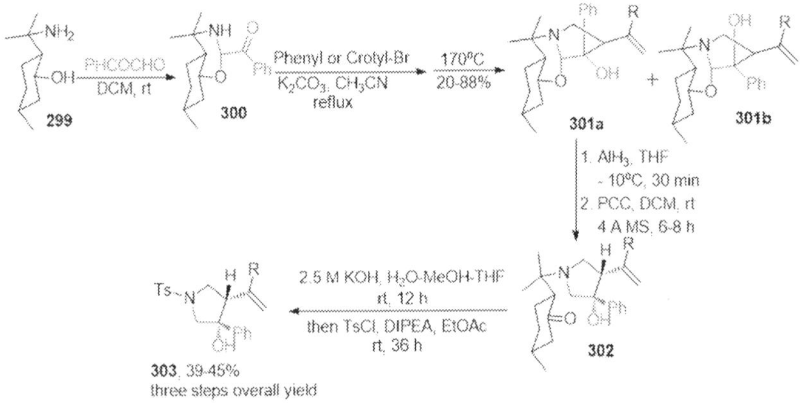

FIGURE 3.101   Synthesis of 1,3-dimethyl-7-phenyl-[1,2,4]triazino [2,3-f]purine 2,4(1H,3H)-dione.

pharmaceuticals,[188] and biologically active molecules.[189] Pyrrolidines also play an important role in organic synthesis and widely used as an organocatalyst,[190] chiral auxiliaries,[191] and ligands[192] for asymmetric synthesis. There are various literature reports available for the synthesis of biologically active pyrrolidines and their derivatives.

3.4.5.1   Synthesis of pyrrolidin-3-ols

A six step protocol for the synthesis of 3-phenyl-1-tosyl-4-vinylpyrrolidin-3-ols **303** was established by Pedrosa and coworkers (Fig. 3.101) from phenyl glyoxal and (−)-8-aminomenthol **299**.[141] The first step is the condensation reaction between phenyl glyoxal and (−)-8-aminomenthol in DCM at room temperature to afford 2-benzoyl-1,3-oxazine **300**. The next step involves the formation of diastereomeric 3-hydroxypyrrolidines **301a** and **301b** by the reaction of 2-benzoyl-1,3-oxazine with prenyl or crotyl bromide in the presence of K$_2$CO$_3$ in acetonitrile at refluxing conditions, followed by carbonyl-ene reaction under different thermal conditions. Later, mentone derivatives were achieved by reductive ring opening of **301a** in presence of in situ generated AlH$_3$, prepared from AlCl$_3$ and LiAlH$_4$, in THF at −10°C for 10 min followed by oxidation with PCC in DCM at room temperature for 6 − 8 h to obtain **302**. The Mentone derivatives can be used for the final step without purification. The final product pyrrolidine **303** was obtained by elimination of N-protection with KOH in H$_2$O − MeOH − THF (1:1:2) at room temperature followed by tosylation with TsCl in the presence of DIPEA in EtOAc for 36 h.

Bossio et al. reported[193] Ugi-reaction for the synthesis of N-substituted 4-cyano-2,5-dihydro-5-oxopyrrole-2-carboxamides **306** from aryl glyoxals (AGs) **1**, anilines, isocyanides, and cyanoacetic acid. The reactions proceed through the in situ formation of AG-imines **304** from AGs and anilines in toluene at heating conditions in a Dean − Stark apparatus to remove water, followed by reaction with isocyanides and cyanoacetic acid in Et_2O at room temperature for 6 days to afford Ugi adducts **305**. The obtained product was then reacted with Et_3N in EtOH to give the desired product **306** in 40%− 60% yields (Fig. 3.102).

3.4.6 Substituted pyrroles synthesis

AGs are also used for the synthesis of different biologically active pyrroles. They are an important class of heterocycles and are present in various natural products,[194] pharmaceuticals,[195] and are also used in material science.[196]

Feliciano et al.[197] reported the synthesis of 3-hydroxypyrroles **309** from enamino esters **307** with AG-hydrate **1** in MeOH at refluxing conditions to obtain the desired product. The reaction proceeds through the nucleophilic addition of enamino esters at the aldehyde group of phenyl glyoxal, followed by condensation reaction along with the removal of a water molecule to give the corresponding pyrroles **309**. The hydroxyl group of pyrroles was acetylated by using Ac_2O in pyridine at room temperature (Fig. 3.103).

A one-pot, three-component strategy was developed by Khalili and coworkers[198] for the synthesis of substituted 2-alkyl-5-aryl-(1H)-pyrrole-4-ols **312**. This approach proceeds through the reaction between 1,3-dicarbonyl compounds **311** and AGs **1** in the presence of an excess amount of NH_4OAc in H_2O at room temperature (Fig. 3.104).

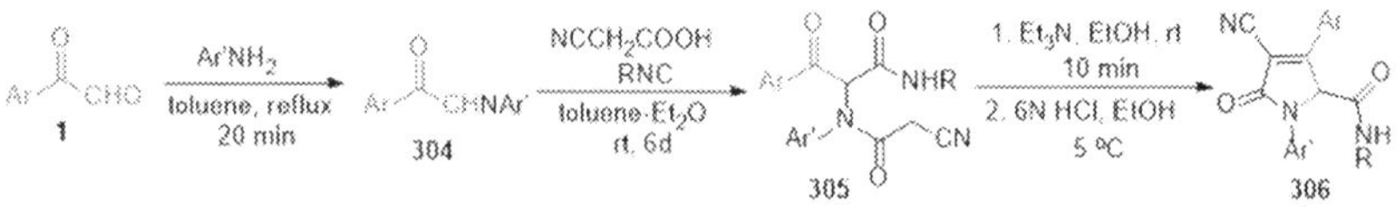

FIGURE 3.102 Synthesis of 3-phenyl-1-tosyl-4-vinylpyrrolidin-3-ols.

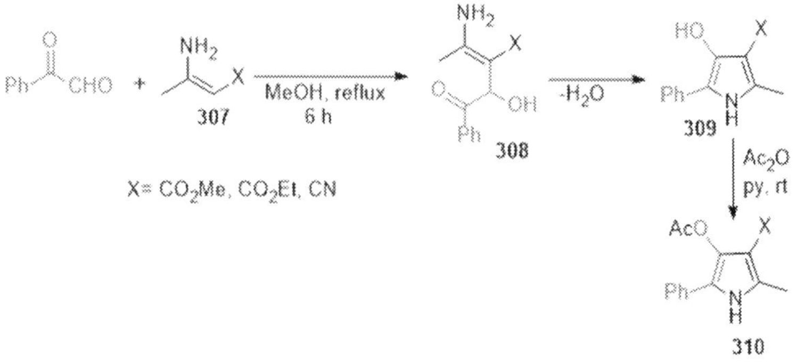

FIGURE 3.103 Synthesis of N-substituted 4-cyano-2,5-dihydro-5-oxopyrrole-2-carboxamide.

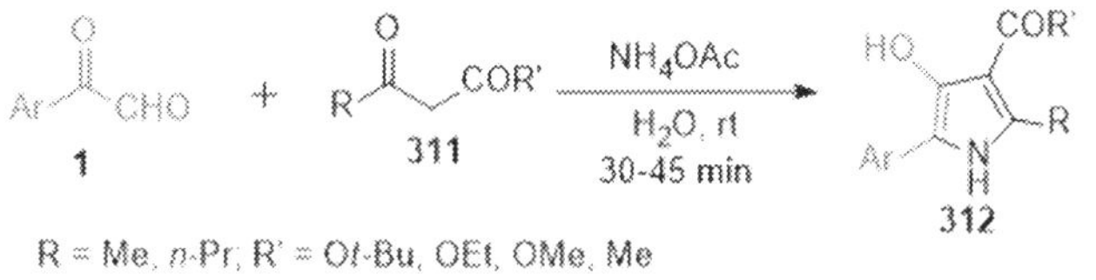

FIGURE 3.104   Synthesis of 3-hydroxypyrroles.

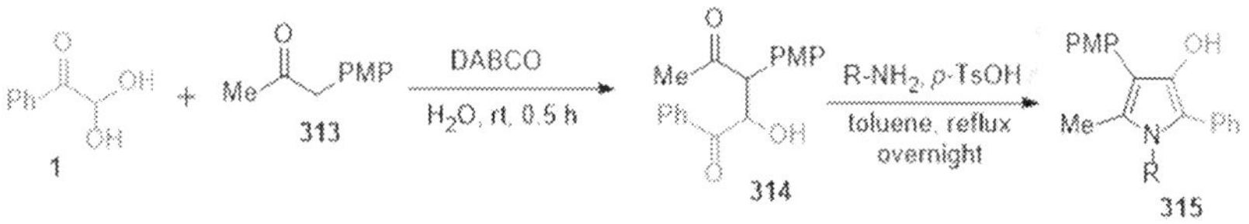

FIGURE 3.105   Synthesis of substituted 2-alkyl-5-aryl-(1H)-pyrrole-4-ols.

Recently, a new synthetic approach was established for the synthesis of N-alkyl(aryl)-2,4-diaryl-2-methyl-1H-pyrrole-3-ol derivatives **315** in 61% − 82% yields and proceeds through an aldol-Paal − Knorr reaction sequence. The first step involves an aldol reaction between 1-(p-methoxyphenyl) propan-2-one **313** and PG-hydrate **1** to generate 1,4-dicarbonyl compound **314** in the presence of DABCO in water as a solvent at room temperature. The next step involves the formation of various highly substituted pyrroles **315** by refluxing a solution of primary amines and 1,4-dicarbonyl compound in toluene in the presence of a catalytic amount of p-TsOH (Fig. 3.105).[199]

A mild and efficient strategy for the construction of 3-methylthio-substituted pyrroles **320** was established by Yin et al.[200] This approach was started from acetophenones **316** and the plausible mechanism shows the in situ formation of AGs **1** via DMSO-promoted oxidation of α-iodoacetophenones **317** in the presence of CuO, and I_2 followed by the loss of MeI and H_2O to form a mixture of E and Z isomer of 2-(methylthio)-1,4-diaryl-2-butene-1,4-diones **319** in good to excellent yields. Both isomers were then reacted with KI and concentrated HCl in acetone at room temperature followed by ammonium formate in AcOH under reflux conditions for 2 − 4 h to afford the corresponding pyrroles **320** in 80% − 92% yields (Fig. 3.106).

Furthermore, an efficient, catalyst-free, novel four-component domino reaction was reported by Shi et al.[201] for the synthesis of polysubstituted pyrroles **324** from an arylglyoxal monohydrate **1**, aniline **321**, malononitrile **322**, and dialkyl but-2-ynedioate **323** in ethanol under reflux conditions for 30 min. Different polysubstituted pyrroles **324** were synthesized from this transformation without any catalyst (Fig. 3.107).

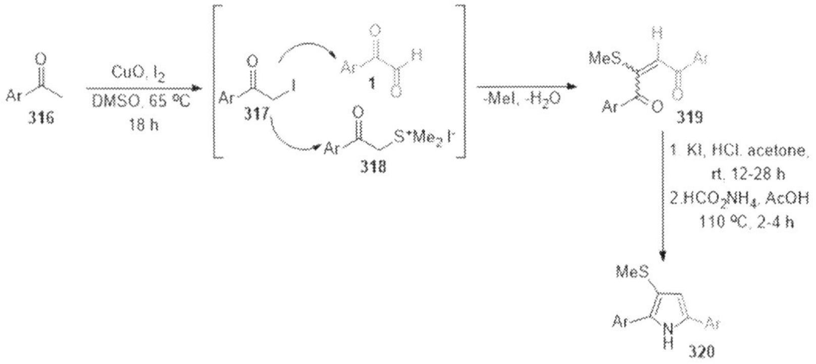

FIGURE 3.106 Synthesis of N-alkyl(aryl)-2,4-diaryl-2-methyl-1H-pyrrole-3-ol derivatives.

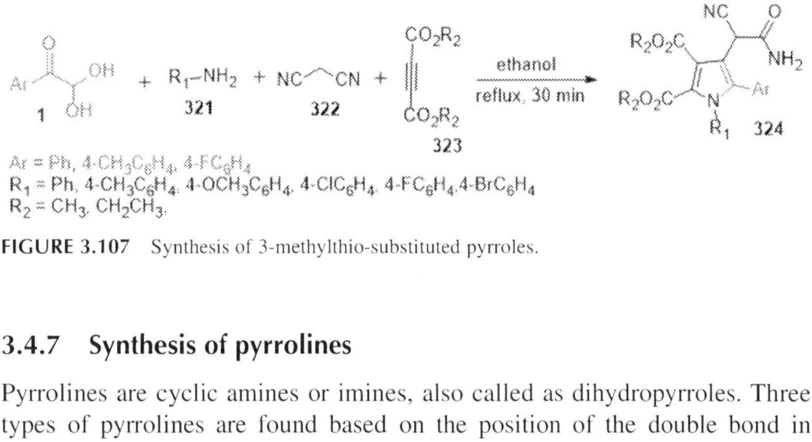

Ar = Ph, 4-CH$_3$C$_6$H$_4$, 4-FC$_6$H$_4$
R$_1$ = Ph, 4-CH$_3$C$_6$H$_4$, 4-OCH$_3$C$_6$H$_4$, 4-ClC$_6$H$_4$, 4-FC$_6$H$_4$, 4-BrC$_6$H$_4$
R$_2$ = CH$_3$, CH$_2$CH$_3$.

FIGURE 3.107 Synthesis of 3-methylthio-substituted pyrroles.

3.4.7 Synthesis of pyrrolines

Pyrrolines are cyclic amines or imines, also called as dihydropyrroles. Three types of pyrrolines are found based on the position of the double bond in five-membered ring.

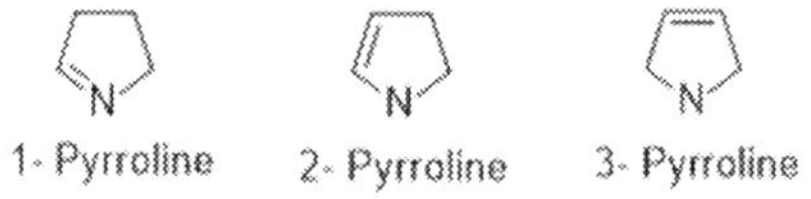

1- Pyrroline   2- Pyrroline   3- Pyrroline

Since pyrrolines are found in many biologically active molecules, natural products and certain drug molecules like thienamycin are resistant to bacterial β-lactamase enzymes and show excellent activity in Gram-negative and Gram-positive bacteria (Fig. 3.108).[202] They are also beneficial as synthetic intermediates like pyrroles and pyrrolidines which bear biological significance.[203] Therefore various strategies for the synthesis of pyrroline and their analogs have been developed.

Arylglyoxal **1** reacts with different esters like *N*-hydroxyalkyl and *N*-tolyl enamino esters to form pyrrolinones, 3-hydroxypyrols, and pyrrole in good yields. The stability of pyrrolinones **329** is due to H-bonding between OH

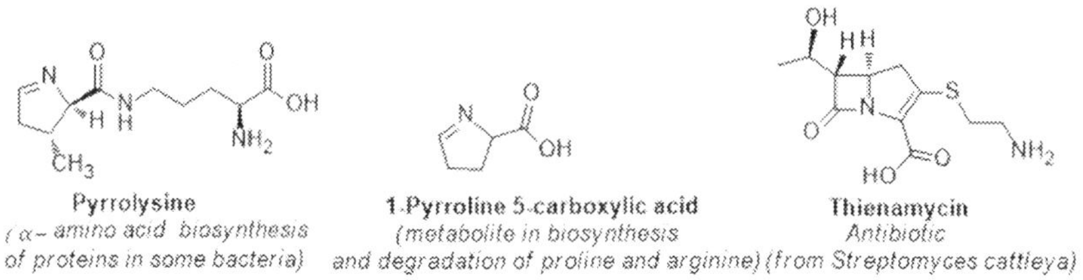

Pyrrolysine
(α-amino acid biosynthesis
of proteins in some bacteria)

1-Pyrroline 5-carboxylic acid
(metabolite in biosynthesis
and degradation of proline and arginine)

Thienamycin
Antibiotic
(from Streptomyces cattleya)

FIGURE 3.108 Synthesis of polysubstituted pyrroles.

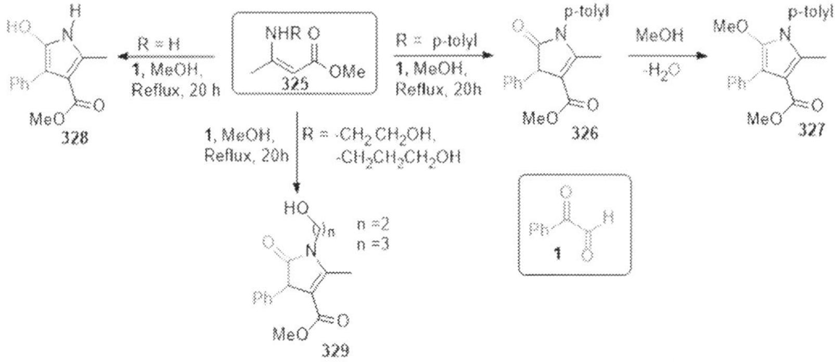

FIGURE 3.109 Biologically active molecules containing pyrrolines moiety.

and carbonyl group which is in close spatial proximity. Unsubstituted (R = H) enamino ester only yields **328** which proves the stability of pyrrolinone **328**. With *p*-Tolylaminocrotonate **327** was formed through **326** as an intermediate (Fig. 3.109).[204]

3.4.8 Synthesis of fused pyrroles and dibenzo[b,e][1,4]diazepin-1-ones

Li and his group[205] described two different domino approaches for the selective synthesis of fused pyrroles **334** and dibenzo[b,e][1,4]diazepin-1-ones **332**. Tricyclic fused pyrrole derivatives **334** were obtained by reacting arylglyoxal monohydrate **1** with N-amino acid **333** in HOAc under MW irradiation. On the other hand, N-amino acids were replaced by benzene-1,2-diamine-derived enaminones **331** to get dibenzo[b,e][1,4]diazepin-1-ones **332**. The application of these approaches is the controlled pathways to obtain selectively fused pyrroles and dibenzo[b,e][1,4]diazepin-1-ones (Fig. 3.110).

3.4.9 Synthesis of pyrazoles

Pyrazoles are one of the important heterocyclic compounds and are present in various small molecules of biological importance.[206] These molecules have a wide range of pharmaceutical[207] and agricultural activities such as

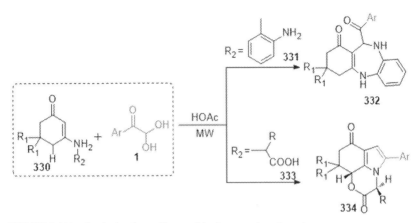

FIGURE 3.110 Synthesis of pyrrolinones, 3-hydroxypyrols and pyrole.

herbicidal,[208] insecticidal,[209] fungicidal,[210] antiinflammatory,[211] and analgesic properties.[212] Furthermore, they also show the applications in polymer[213] and supramolecular chemistry[214] and acts as ligands[215] for metal-catalyzed reactions. Various conventional approaches are available for the synthesis of substituted pyrazoles, which involves the condensation reaction between hydrazines and 1,3-dicarbonyl compounds or by 1,3-dipolar cycloaddition reactions.

An efficient approach for the synthesis of dihydropyrazoles was established by Del Buttero et al.[216] via 1,3-dipolar cycloaddition of 3(R)-phenyl-4 (S)-(4-benzoyl-E,E-1,3-butadienyl)-2-azetidinone **336** with in situ generated nitrilimines **337** by performing a reaction between hydrazonoyl chloride **342** with AgOAc in dioxane at room temperature for 24 h in dark to form a mixture of four isomeric dihydropyrazoles (**338**, **339**, **340**, and **341**). The initial step is a witting reaction between PG and (triphenylphosphoranylidenec)-acetaldehyde **335** followed by the addition of p-anisidine at room temperature in ethanol solvent and stir the reaction for 5 min to afford the imine derivative in good yield. The imine derivative was then reacted with phenyl acetyl chloride in the presence of Et₃N in DCM to form 2-azetidinone derivatives (Fig. 3.111).

Begtrup et al.[217] developed a new approach for the construction of 4-hydroxypyrazoles **344** by conducting reactions between aldehyde hydrazones **343** and phenyl glyoxal in the presence of n-BuOAc, MgSO₄, and AcOH in perfectly dry conditions at 110°C for 1 h (Fig. 3.112). When the same reaction was planned in water, cyclo condensation of PG with PG-hydrazones **345** takes place which was generated in situ by transhydrazonation of PG **1** with aldehyde hydrazones to form 3-benzoyl-4-hydroxy-5-phenylpyrazoles **346**. To confirm this, a reaction was performed between separately prepared oxohydrazones **345** and PG in the same conditions to form the corresponding benzoylpyrazoles **346**.

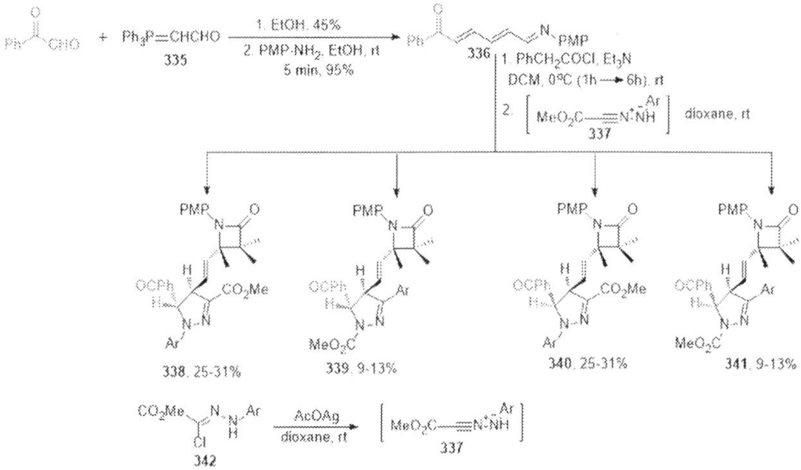

FIGURE 3.111   Synthesis of fused pyrroles and dibenzo[b,e][1,4]diazepin-1-ones.

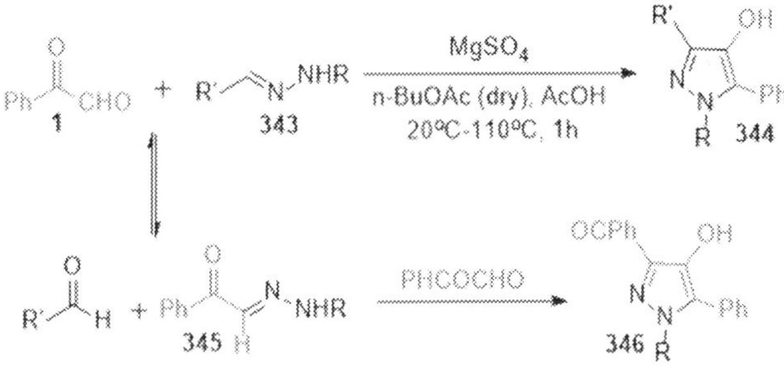

FIGURE 3.112   Synthesis of dihydropyrazoles.

A two-step strategy was used for the synthesis of 3-benzoyl-4-phenyl-1-methyl pyrazole **350** by condensation reaction of phenyl glyoxal with protected 1-methyl-1-phenacylhydrazine **347** in the presence of AcOH in EtOH at room temperature for 20 min to obtain **348** followed by addition of 75% H_2SO_4 and stirring the reaction for two weeks. The first step is the hydrazone formation from **347** with PG **1** and then deketalization to **349** in presence of 75% H_2SO_4, followed by the intramolecular condensation to obtain the desired product **350a** in 56% yield. Formation of 5-benzoyl-4-phenyl-1-methyl pyrazole **350b** from the aldol type condensation of the methylene and carbonyl group of phenyl glyoxal was not successful (Fig. 3.113).[217]

AGs reacts with 2-nitromethylenpyrrolidine **351** to form substrate **352** which later on can be utilized for the synthesis of many cyclized products **353, 354, 355** by optimizing different conditions in a very short duration (Fig. 3.114).[68]

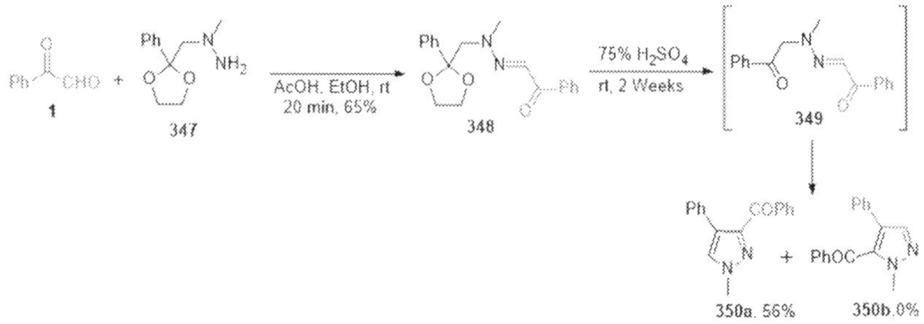

FIGURE 3.113   Synthesis of 4-hydroxypyrazoles.

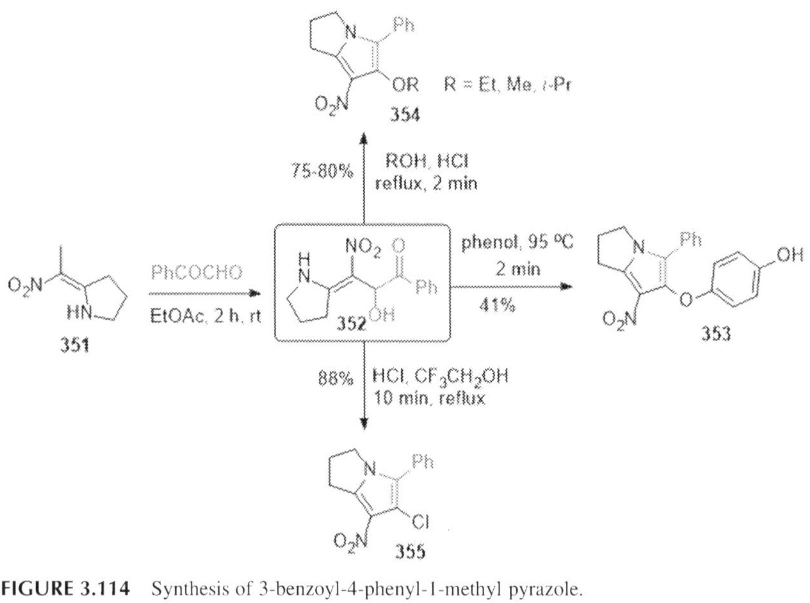

FIGURE 3.114   Synthesis of 3-benzoyl-4-phenyl-1-methyl pyrazole.

3.4.10   Synthesis of hydantoins and thiohydantoins

Imidazolin-2-one, imidazolidin-2,4-dione (hydantoin) and their derivatives are of great importance because of their attractive biological activities, such as antiinflammatory,[218] herbicidal,[219] antitumor,[220] antioxidant,[221] cardiotonic,[222] antitubercular,[223] antiarrhythmics.[224] In the literature, there are several methods reported for the synthesis of biologically important imidazolin-2-ones and hydantoins.

An efficient method for the synthesis of 1,5-disubstituted hydantoins and thiohydantoins **357** was established by Paul et al.[225] in MW conditions in 80% − 95% yields. The reaction proceeds through the condensation reaction of AGs and phenyl urea or phenylthiourea **356** in presence of polyphosphoric ester (PPE), which acts as a reaction mediator. This reaction was also reported in presence of PPE under neat conditions at 120°C for 3 min but the yields of the final products **357** are low as compared to MW conditions (Fig. 3.115).

In a similar way, dihydroxyimidazolidin-2-one **361** was synthesized from the condensation reaction of PG-hydrate and 5-*t*-butyl-1,3,4-thiadiazol-2-yl substituted urea **358** by using aqueous NaOH in EtOH at room temperature. The dihydroxyimidazolidin-2-one **359** was then transformed into the corresponding hydantoin **360** in the presence of *p*-TsOH at the refluxing condition in CH$_3$CN in an 87% yield. Hydantoin **360** was then converted into 4-hydroxyimidazolidin-2-one **361** by using NaBH$_4$ in EtOH in 76% yield as the only *trans*-isomer (Fig. 3.116).[226]

The condensation reaction of AG-hydrate with *N*-hydroxyurea **362** was reported by Shtamburg et al.[228] in water for the synthesis of 5-aryl-3-hydroxyhydantoins **366**. 3,4,5-Trihydroxy-5-arylimidazolidin-2-ones **364** was formed as a reaction intermediate, which undergo intramolecular proton transfer resulting in the formation of zwitterion **365**, followed by the loss of water molecule by hydride rearrangement to afford 5-aryl-3-hydroxyhydantoins **366** in 46% − 77% yields (Fig. 3.117).

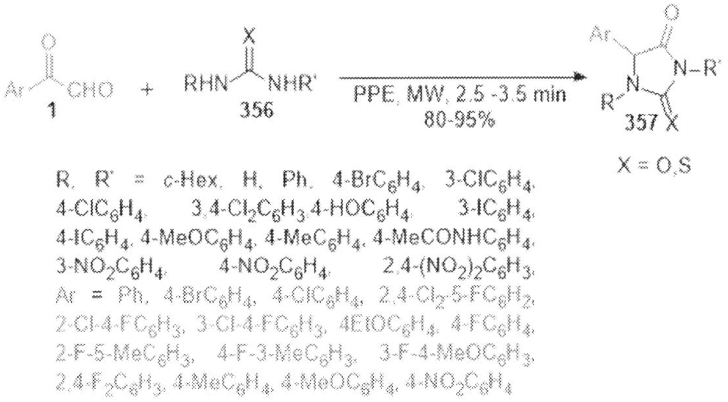

FIGURE 3.115 Synthesis of pyrrole based cyclized products.

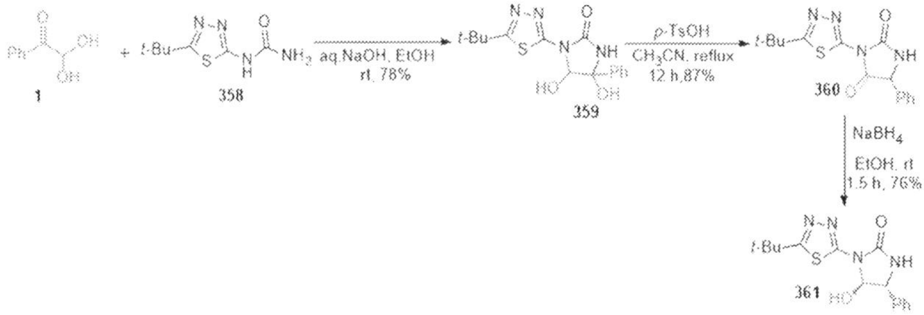

FIGURE 3.116 Synthesis of 1,5-disubstituted hydantoins and thiohydantoins.

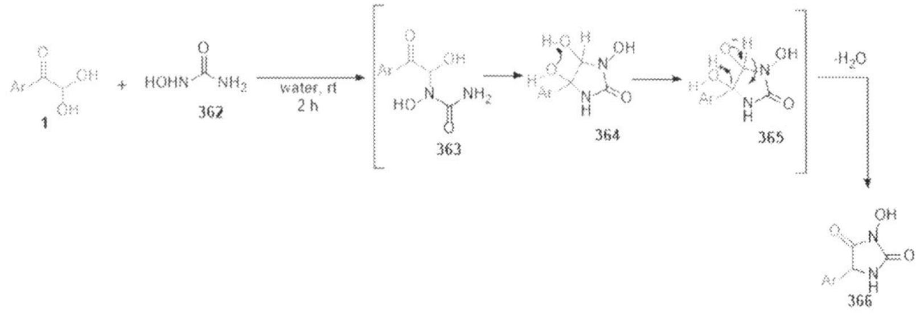

FIGURE 3.117 Synthesis of dihydroxyimidazolidin-2-one.

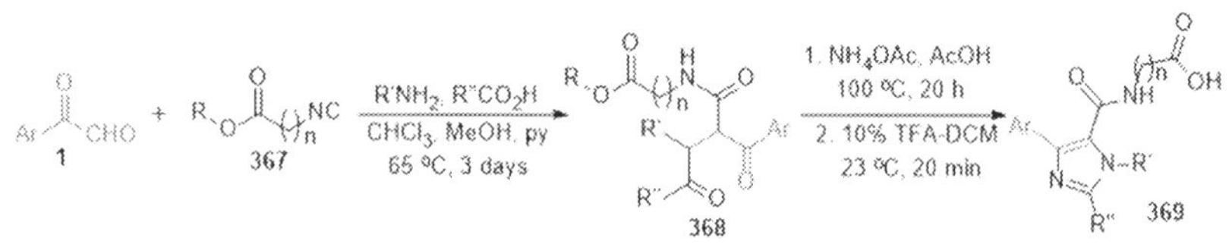

FIGURE 3.118 Synthesis of 5-aryl-3-hydroxyhydantoins.

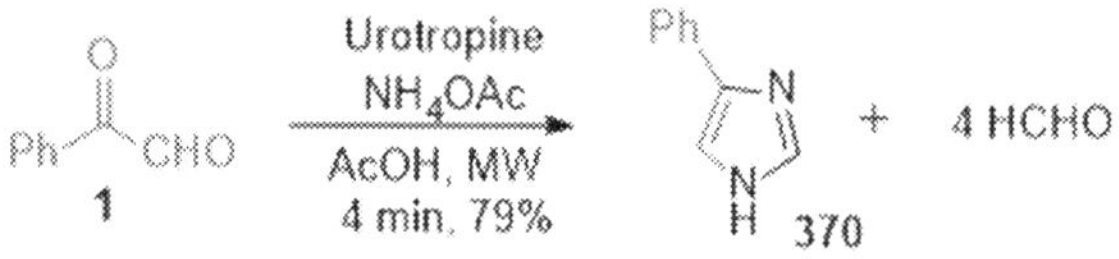

FIGURE 3.119 Synthesis of substituted imidazoles.

3.4.11 Synthesis of imidazoles

Imidazoles are an important class of heterocyclic compounds and are present in various natural products,[229] essential amino acid histidine, histamine, and the pilocarpine alkaloids.[230] Imidazole scaffolds are also present in various biologically active synthetic compounds such as losartan,[231] cimetidine, fungicides, and herbicides.[232] Because of their biological,[233] synthetics,[234] and industrial applications,[235] there are various literature reports available for the synthesis of different imidazoles.

Substituted imidazoles **369** are synthesized from Ugi-type four-component reactions of AGs, carboxylic acids, primary amines, and isocyanides **367**. Ketoamides **368** were initially obtained, which undergoes cyclization by treatment with NH$_4$OAc in AcOH at 100°C for 20 h, followed by the addition of 10% TFA-DCM to obtain **369** in 16% − 56% yields (Fig. 3.118).[236]

Synthesis of 4-phenylimidazole **370** was reported by Bratulescu by conducting a reaction in MW condition between PG **1** and urotropine in the presence of NH$_4$OAc and a few drops of AcOH. Urotropine acts as a synthon of formaldehyde and the reaction was conducted without solvent to obtain the desired product **72** in 79% yield (Fig. 3.119).[236]

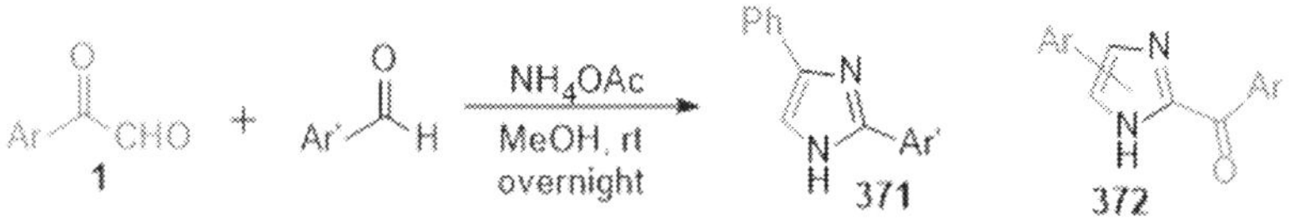

FIGURE 3.120   Synthesis of 4-phenylimidazole.

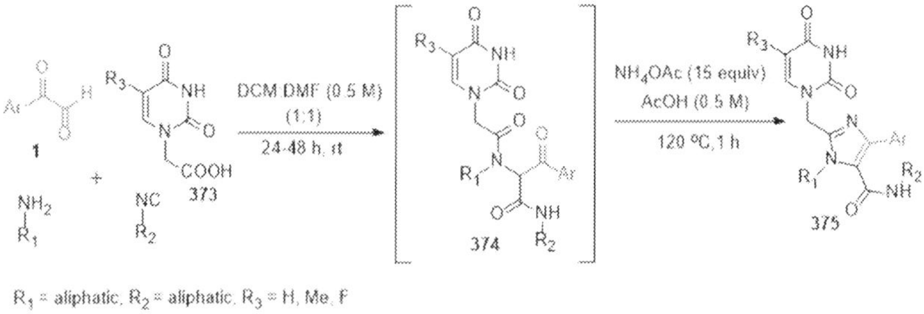

R_1 = aliphatic, R_2 = aliphatic, R_3 = H, Me, F

FIGURE 3.121   Synthesis of 2,4(5)-diarylimidazoles.

Zuliani et al.[237] reported the selective synthesis of 2,4(5)-diarylimidazoles **371** from a reaction between AG **1** and aldehydes in the presence of NH_4OAc in methanol at room temperature in $52\% - 83\%$ yields. The unwanted side product 2-aroyl-4(5)-arylimidazoles **372** is not formed in this reaction. The same reaction was conducted in different conditions and it was observed that the selectivity was affected by using different solvents, MeOH at room temperature gives the best yield of 83% (Fig. 3.120).

Recently, Alanexder Domling and his coworker[238] established Ugi-type four-component reaction for the synthesis of polysubstituted imidazole derivatives **375** from phenylglyoxal **1**, amine, isocyanide, and uracil derived acetic acid **373** in methanol. After a detailed screening for obtaining the best condition, it was observed that a combination of DCM: DMF (1:1) was appropriate for this Ugi-reaction to obtain the product **374** in 90% yields within 48 h. The Ugi product was further converted into corresponding tetra-substituted imidazoles **375** in presence of an excess of NH_4OAc (15 equivalent) in CH_3COOH at $120°C$ for 1 h. The desired product was obtained in 75% yield (Fig. 3.121).

An efficient multicomponent one-pot two-step approach for the synthesis of biologically active imidazo-[1,5-a]quinoxalines derivatives **378** from phenylglyoxaldehyde **1**, ortho-N-Boc-phenylenediamine **376**, and unsubstituted TOSMIC **377** in the presence of K_2CO_3 under MW conditions was successfully established. The first step involves the imine formation between phenylglyoxaldehyde and *ortho*-N-Boc-phenylenediamine and the other step involves the attack of TOSMIC in presence of K_2CO_3 to get the desired product **378** in good yields (Fig. 3.122).[239]

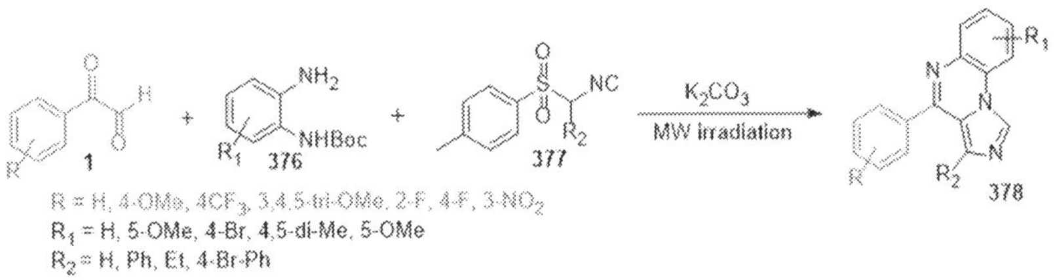

FIGURE 3.122   Synthesis of polysubstituted imidazole derivatives.

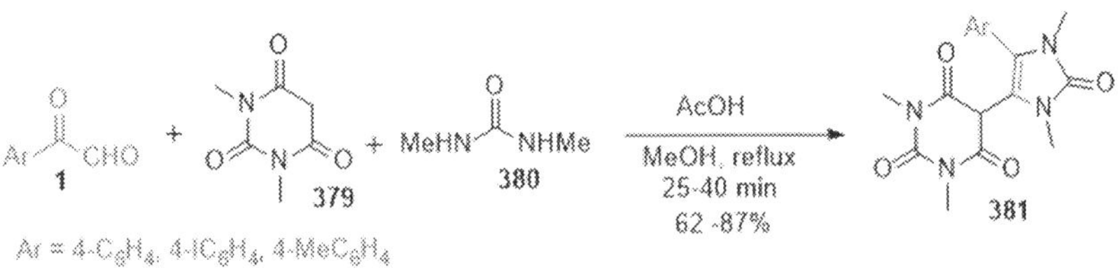

FIGURE 3.123   Synthesis of biologically active imidazo-[1,5-a]quinoxalines derivatives.

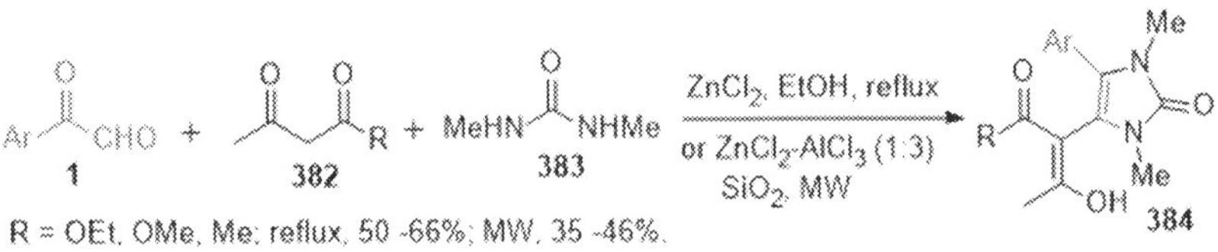

FIGURE 3.124   Synthesis of imidazol-4-yl-pyrimidine-2,4,6-triones.

3.4.12   Synthesis of imidazol-2-ones and imidazolin-4-one

Gozalishvili et al.[240] reported an efficient method for the synthesis of imidazol-4-yl-pyrimidine-2,4,6-triones **381** from 1,3-dimethylbarbituric acid **379**, AGs **1**, and N, N'-dimethyl urea **380** in MeOH in the presence AcOH under reflux conditions for 25 − 40 min in 62% − 87% yields (Fig. 3.123).

Another similar approach was reported by using 1,3-dicarbonyl compounds **382**, PG **1** and N, N'-dimethyl urea **383** in the catalytic amount of $ZnCl_2$ in heating conditions or in the combination of $ZnCl_2/AlCl_3$ (1:3) on silica gel in the MW conditions to obtain the multisubstituted imidazolin-2-one derivatives **384** in 50% − 66% or 35% − 46% yields, respectively (Fig. 3.124).[116]

Waugh et al.[241] reported a condensation reaction between PG-hydrate and benzamidine **385** to obtain the hydroxy-phenacyl benzamidine **386** in H_2O in the presence of KOH at room temperature, followed by cyclization in heating condition to afford the imidazole derivative **387** in 64% yield (Fig. 3.125).

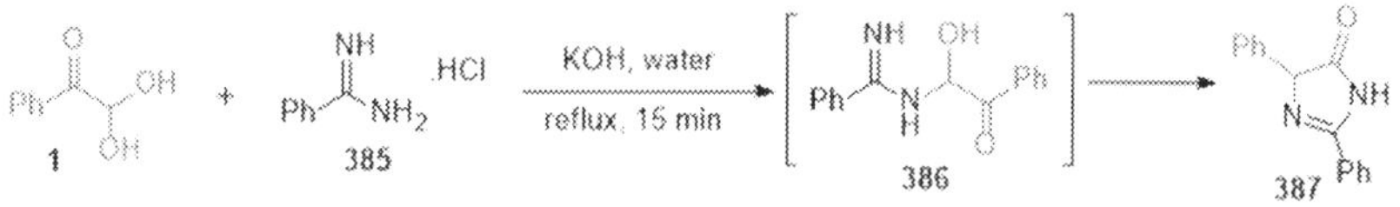

FIGURE 3.125   Synthesis of multisubstituted imidazolin-2-one derivatives.

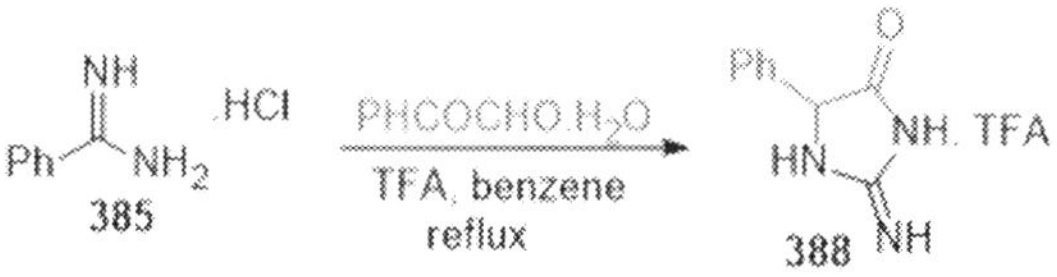

FIGURE 3.126   Synthesis of hydroxy-phenacyl benzamidine.

A reaction between guanidine hydrochloride 385, PG-hydrate and TFA in heating benzene undergoes azeotropic removal of water by using Dean – Stark apparatus to get the trifluoroacetate salt of 2-imino-5-phenylimidazolidin-4-one 388 (Fig. 3.126).[242]

3.4.13   Synthesis of imidazothiazoles and imidazopyridines

Fused heterocyclic compounds are of great importance and are found in various types of synthetic and naturally occurring biologically important compounds.[243] Imidazothiazoles and imidazopyridines possess a variety of applications such as dyes,[244] pesticides, fungicides,[245] antitumor,[246] antiinflammatory, and analgesic activity.[247] In most of the cases, oxidative condensation – cyclization or Pictet – Spengler strategy can be used for the synthesis of these types of fused heterocycles.

Drach et al.[248] reported a two-step reaction for the synthesis of 5-acylamido-6-phenylimidazo[2,1-b]-thiazoles 393 and 2-phenyl-3-acylamidoimidazo[1,2-a]-pyridines 394 in good to moderate yields. The first step involves the condensation reaction between PG and amides 389, followed by the addition of thionyl chloride or phosphorus pentachloride to obtain α-chloro-α-acylaminoacetophenones 390. In the next step, 2-aminothiazole 391 and 2-aminopyridine 392 were added to the reaction and kept as such for 24 h and the resulting residue was refluxed in MeOH for 1 h (Fig. 3.127).

A MW-assisted, three-component reaction for the synthesis of a novel class of imidazo[1,2-a]azine derivatives 397 was developed by Nikolay Yu. Gorobets and his coworkers[249] from 2-aminoazine 395, AG 1, and cyclic 1,3-dicarbonyl compound 396. Three reaction conditions were optimized in order to increase the substrate scope of the reaction that results in the generation of a small library of imidazo[1,2-a]azine derivatives 397 (Fig. 3.128).

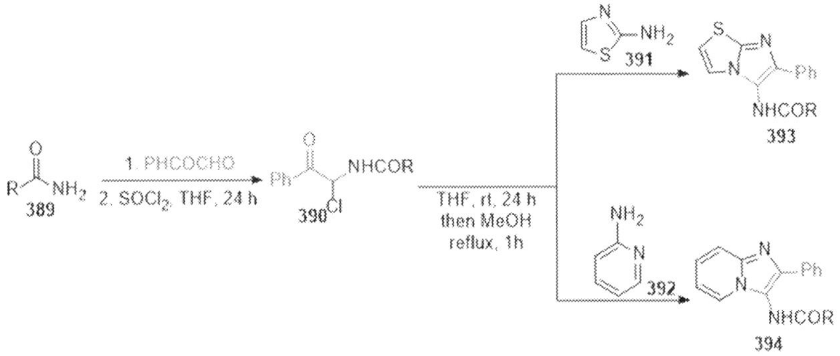

FIGURE 3.127 Synthesis of trifluoroacetate salt of 2-imino-5-phenylimidazolidin-4-one.

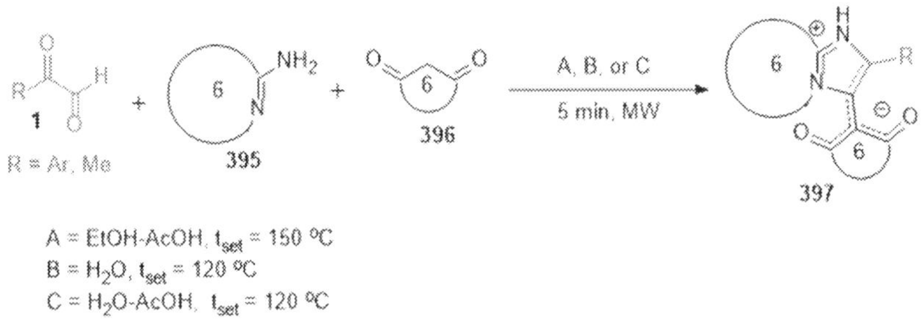

A = EtOH-AcOH, t_{set} = 150 °C
B = H$_2$O, t_{set} = 120 °C
C = H$_2$O-AcOH, t_{set} = 120 °C

FIGURE 3.128 Synthesis of 5-acylamido-6-phenylimidazo[2,1-b]-thiazoles and 2-phenyl-3-acylamidoimidazo[1,2-a]-pyridines.

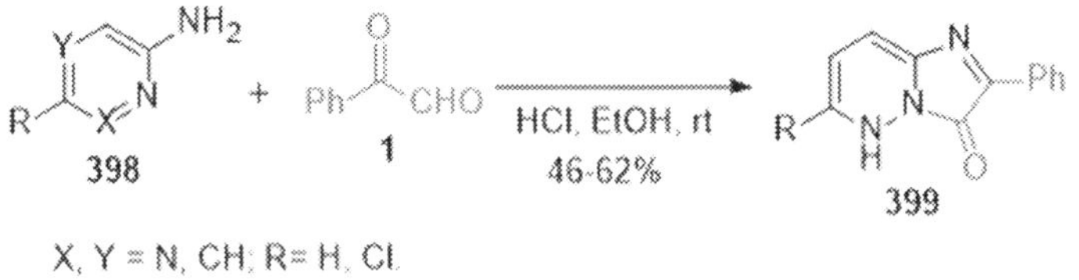

X, Y = N, CH; R= H, Cl

FIGURE 3.129 Synthesis of imidazo[1,2-a]azine derivatives.

3.4.14 Synthesis of imidazopyridazines and imidazopyrazinones

Barlin et al.[250] reported a condensation reaction method for the synthesis of substituted imidazopyridazines **399** from PG and **398** in the presence of an ethanolic solution of concentrated HCl at room temperature to obtain the desired product **399** in 46% − 62% yields (Fig. 3.129).

The synthesis of substituted imidazopyrazinones **402** from the condensation reaction of 2-aminopyrazines **400** and AGs or AG acetals **401** in the presence of HCl in heating EtOH under argon atmosphere for 4 h was reported by Devillers and coworkers.[251] These substituted imidazopyrazinones **401** shows antioxidant activity and also behave as quenchers of superoxide anion (Fig. 3.130).

3.4.15   Synthesis of 1,2,3-triazoles and tetrazoles

1,2,3-Triazole is an entity that has become one of the most important heterocycles in recent chemistry research from the past few years because of its importance in industries, agrochemicals,[252] biological science,[253] material chemistry,[254] and medicinal chemistry.[255] There are various reported methods for the synthesis of biologically active 1,2,3-triazoles and tetrazoles in different conditions. The most efficient, easy, and attractive approach to synthesize 1,2,3-triazoles is the thermal 1,3-dipolar cycloaddition of azides and alkynes in the presence of metal and the tetrazoles synthesis from a cycloaddition reaction between a nitrile and an azide. In recent years, tetrazole chemistry is gaining attention owing to its significance in a variety of synthetic and industrial processes. 1,2,3-Triazoles and tetrazoles are also act as efficient ligands in coordination chemistry and can also be used as precursors to generate other important heterocycles.[256]

Tang and Hu reported[257] the synthesis of 2,4-diaryl-1,2,3-triazoles **406** by conducting a reaction between α-hydroxy acetophenone **403** and phenylhydrazines in the presence of $CuCl_2$ in refluxing glacial AcOH. The reaction goes through $CuCl_2$-catalyzed oxidative cyclization reaction between α-hydroxy acetophenone and phenylhydrazine that is similar to the formation of sugar osazone to obtain PG bisphenylhydrazone **404**, which underwent $CuCl_2$-catalyzed oxidative cyclization to give intermediates **405**. Finally, intermediates **405** undergo the loss of aryl nitrene molecule to form the desired product **406** in 52% − 86% yields under thermal conditions (Fig. 3.131).

Yates et al.[258] reported the synthesis of benzoyl tetrazole **412** by using PG-mono hydrazone **409** and α-diazo acetophenone **407** in the presence of a base in MeOH at room temperature for 1 h (Fig. 3.132). The PG-mono hydrazone **409** was prepared in situ by the reduction of α-diazo acetophenone by using methanolic NaOMe. The PG-mono hydrazone reacts with α-diazo acetophenone in methanolic NaOMe to obtain intermediate **410**,

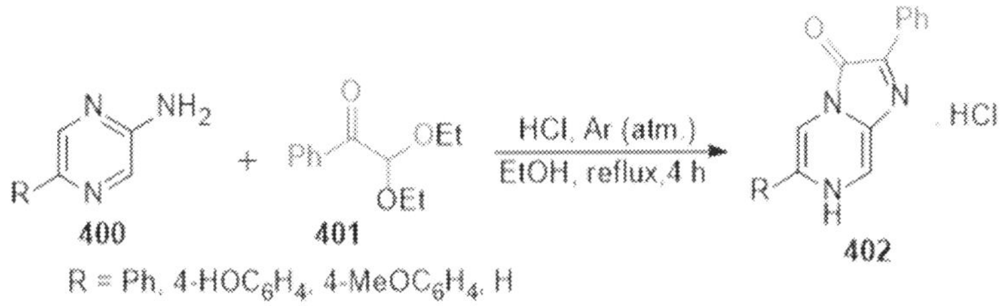

R = Ph, 4-HOC$_6$H$_4$, 4-MeOC$_6$H$_4$, H

FIGURE 3.130   Synthesis of substituted imidazopyridazines.

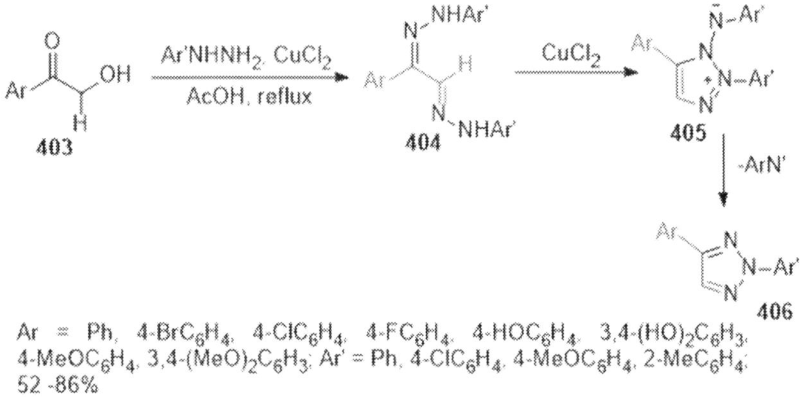
Ar'NHNH2, CuCl2
AcOH, reflux
CuCl2
-ArN'
403
404
405
406
Ar = Ph, 4-BrC6H4, 4-ClC6H4, 4-FC6H4, 4-HOC6H4, 3,4-(HO)2C6H3, 4-MeOC6H4, 3,4-(MeO)2C6H3; Ar' = Ph, 4-ClC6H4, 4-MeOC6H4, 2-MeC6H4; 52 -86%

FIGURE 3.131 Synthesis of substituted imidazopyrazinones.

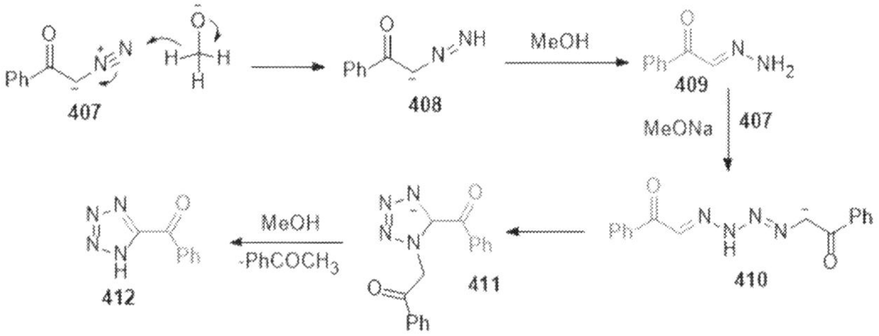
MeOH
MeONa
407
MeOH
-PhCOCH3
407
408
409
410
411
412

FIGURE 3.132 Synthesis of 2,4-diaryl-1,2,3-triazoles.

which underwent cyclization to obtain **411** followed by elimination of acetophenone to obtain the desired product **412**.

3.4.16 Synthesis of furan

Furan is a five-member heterocyclic aromatic organic compound consisting of four carbon and one oxygen atoms. Furans and their derivatives are very important synthons, which are used for the synthesis of different biologically important compounds and are also found in various bioactive natural products,[259] such as cembranolides, kallolides, furan fatty acids, calicogorgins, gersolanes, rosefuran, pseudopteranes, agassizin, furodysin, and α-clausenan. Even though there are numerous reports for the synthesis of furan especially the Paal – Knorr reaction, the construction of new and efficient strategies is of substantial interest.

The $BF_3 \cdot OEt_2$-catalyzed reaction of 3-acetyl-1-aryl-2-pentene-1,4-diones **413** was studied by Onitsuka and coworker[260] for the synthesis of bis-furans

417 and a small amount of **416** in the presence of water in THF under reflux conditions. Stable furfuryl carbocation (**414** and **415**) was formed as an intermediate, which underwent electrophilic substitution and then reduction to afford *bis*-furans **417** in 21%−79% yields and intramolecular cyclization reaction followed by reduction to give **416** in 3%−10% yields (Fig. 3.133).

In addition, a reaction was conducted between **413** which was prepared by Knoevenagel condensation of AGs with acetylacetone in CH_3CN and an electron-rich furan derivative **418** in the catalytic amount of $BF_3 \cdot OEt_2$ in the presence of H_2O in THF at 23°C for 30 min and dimeric furan **419** was obtained in 49% yield. When **413** was treated with Ph_3P in $CHCl_3$ for 30 min in heating conditions, deoxygenation occurs to form **420** (Ar = Ph) in 88% yield. In presence of concentrated HCl in THF at 23°C, pentenedione was converted into chloromethylfurane **421** in 87% yield, which was further converted into 2-ethoxymethylfuran **422** in ethanol after heating for 30 min or furfuryl alcohol by hydrolysis in water − THF when heated for 3 h. When Diels − Alder reaction was carried out between pentenedione and cyclopentadiene in EtOAc at 23°C for 2 h, 2-oxabicyclo [3.3.0]octene **423** was obtained along with the corresponding Diels − Alder cycloadduct **424** in 35:65 ratio (Fig. 3.134).

Furthermore, the reaction between **413** and ethyl acetoacetate or acetylacetone **425** in the presence of a catalytic amount of $BF_3 \cdot OEt_2$ in THF at room temperature for 15 min lead to the formation of **426** in 50% yields along with the isomeric **427** or corresponding furan carboxylate **428** in 31%, respectively (Fig. 3.135).[261]

Yin et al.[200] reported the synthesis of different substituted furan derivatives from 1,4-diones in different conditions. The 1,4-diones **429** were transformed into 3-methylthio-substituted furans **430** in heating conditions by using $SnCl_2$ in a mixture of concentrated HCl and AcOH (4/6) in 69% − 90% yields, followed by the selective removal of methylthio group by using Raney-Ni in EtOH in refluxing conditions for 1.5 h to obtain

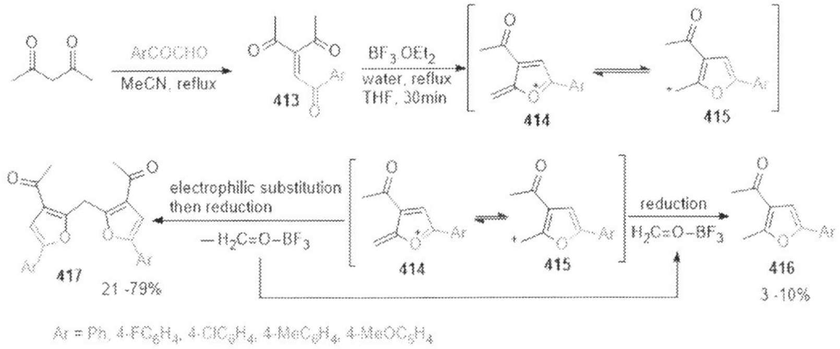

FIGURE 3.133 Synthesis of benzoyl tetrazole.

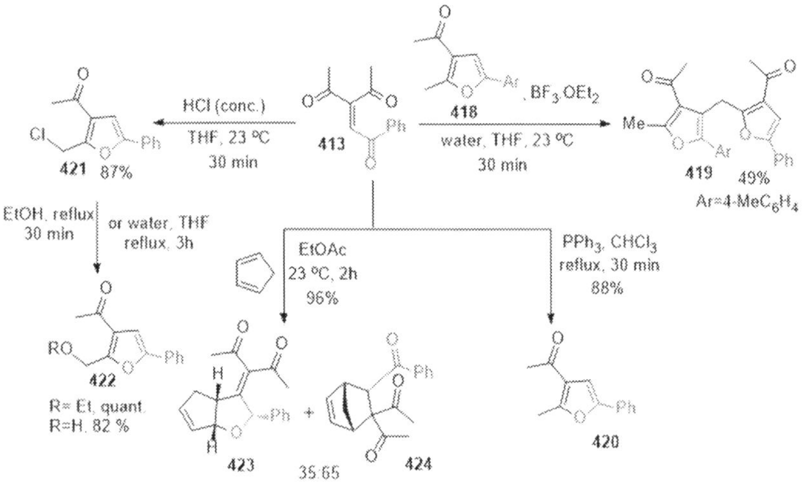

FIGURE 3.134   Synthesis of bis-furans.

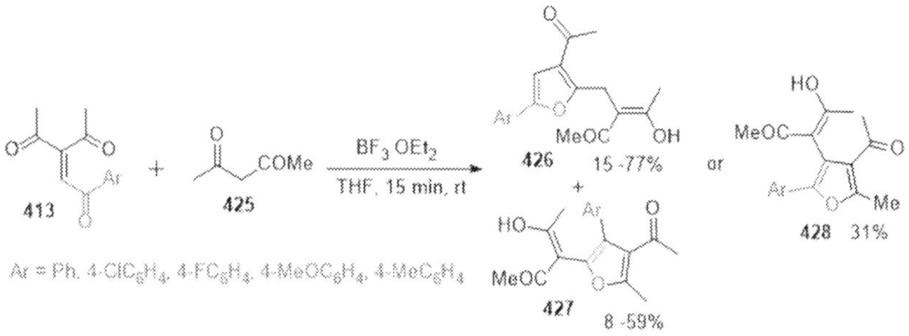

FIGURE 3.135   Multidivergent reactions of AGs with acetylacetone.

2,5-diaryl furans **431** in 85% − 91% yields. Also the reaction of 1,4-diones **429** with a 30% solution of HBr in AcOH and one drop of H_2SO_4 at 0°C in $CHCl_3$ solvent afford 2,5-diaryl furans **432** in 58% − 92% yields (Fig. 3.136).

Indole − furan conjugates (**435** and **436**) were synthesized by Yang and coworkers[262] from the reaction of aryl methyl ketones, indoles, and 1,3-dicarbonyl compounds. Aryl methyl ketones undergo self-condensation or condensation with 1,3-dicarbonyl compounds to form **433** and **429** via AG intermediates by using I_2, CuO, and DMSO. Indoles **434** when reacted with **429** and **433** underwent domino Friedel − Crafts alkylation/Paal − Knorr cyclization to get 3-(furan-3-yl)indole derivatives **435** and **436** in good yields in presence of Lewis and Brønsted acids. Different solvents, as well as acids, were screened for the synthesis of the furan skeleton. MeOH in CH_3CN under refluxing conditions was selected as the best condition for the

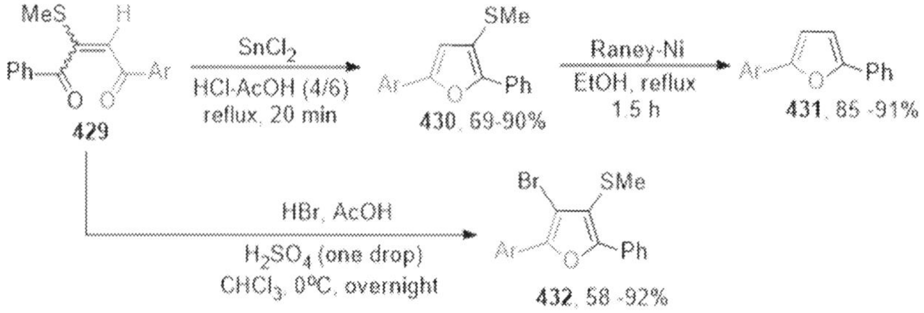

Ar = Ph, 4-BrC_6H_4, 4-ClC_6H_4, 4-$EtOC_6H_4$, 4-$MeOC_6H_4$, 4-MeC_6H_4, 2-benzofuryl, 1-naphthyl, 2-naphthyl, 2-thienyl, 3-thienyl

FIGURE 3.136 Synthesis of isomeric furans.

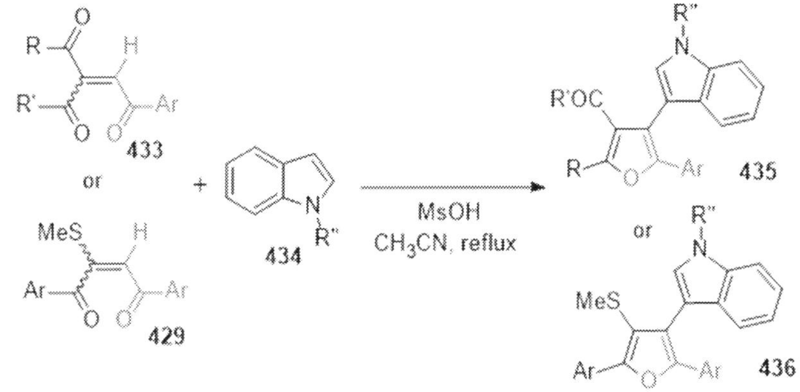

A: R = Ph, 4-ClC_6H_4, 4-FC_6H_4, 3,4,5-$(MeO)_3C_6H_2$, 4-MeC_6H_4, 3-$NO_2C_6H_4$, 4-$NO_2C_6H_4$, 2-furyl; R' = OEt, OMe, Me, Ph; R" = H, Me; Ar = Ph, 4-BrC_6H_4, 4-ClC_6H_4, 3,4-$Cl_2C_6H_3$, 4-FC_6H_4, 4-OHC_6H_4, 4-$MeOC_6H_4$, 4-MeC_6H_4, 4-$NO_2C_6H_4$, 2-benzofuryl, 1-naphthyl, 2-naphthyl, 3-thienyl; reflux, 8 -10 h, 24-99%.
B: Ar = Ph, 4-BrC_6H_4, 4-ClC_6H_4, 4-FC_6H_4, 4-$MeOC_6H_4$, 4-MeC_6H_4, 4-$NO_2C_6H_4$, 2-naphthyl, 3-thienyl; reflux, 5 h, 47 -97%.

FIGURE 3.137 Synthesis of different substituted furan derivatives.

synthesis of the desired indole – furan conjugates in 24% – 99% and 47% – 97% yields, respectively (Fig. 3.137).

Recently, Ming et al.[263] reported mild synthetic protocol for the synthesis of a substituted furans **438** from β-ketothioamides **437** and arylglyoxals **1** in the presence of catalytic amount $Yb(OTf)_3$ in CH_3CN at room temperature for 1.5 h. This synthesis proceeds through a tandem sequence of reactions that involves aldol condensation, followed by N-cyclization, then ring-opening, O-cyclization, S-cyclization, and finally Eschenmoser sulfide contraction (Fig. 3.138).

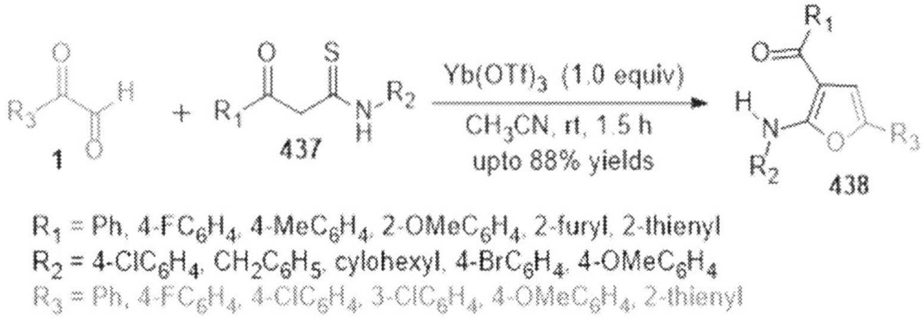
1
437
Yb(OTf)3 (1.0 equiv)
CH3CN, rt, 1.5 h
upto 88% yields
438
R1 = Ph, 4-FC6H4, 4-MeC6H4, 2-OMeC6H4, 2-furyl, 2-thienyl
R2 = 4-ClC6H4, CH2C6H5, cylohexyl, 4-BrC6H4, 4-OMeC6H4
R3 = Ph, 4-FC6H4, 4-ClC6H4, 3-ClC6H4, 4-OMeC6H4, 2-thienyl

FIGURE 3.138 Synthesis of indole – furan conjugates.

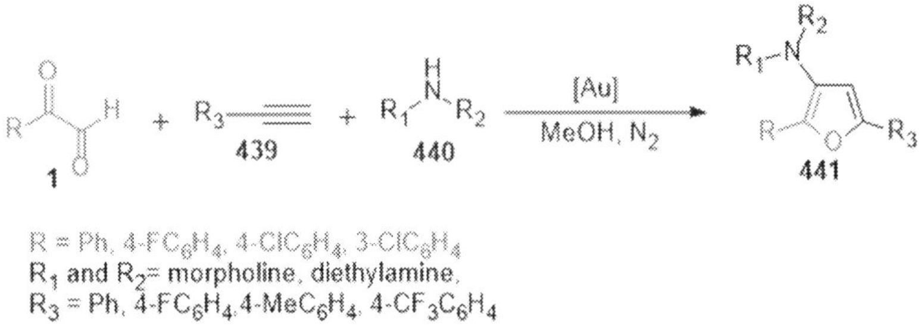
1
439
440
[Au]
MeOH, N2
441
R = Ph, 4-FC6H4, 4-ClC6H4, 3-ClC6H4
R1 and R2= morpholine, diethylamine,
R3 = Ph, 4-FC6H4,4-MeC6H4, 4-CF3C6H4

FIGURE 3.139 Synthesis of a substituted furans.

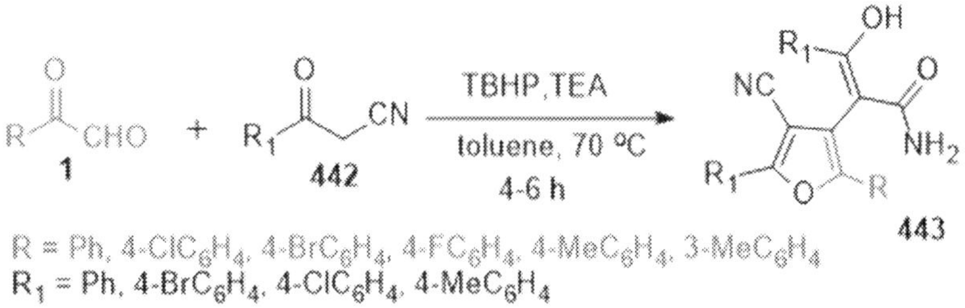
1
442
TBHP,TEA
toluene, 70 °C
4-6 h
443
R = Ph, 4-ClC6H4, 4-BrC6H4, 4-FC6H4, 4-MeC6H4, 3-MeC6H4
R1 = Ph, 4-BrC6H4, 4-ClC6H4, 4-MeC6H4

FIGURE 3.140 Synthesis of different derivatives of furan.

An efficient three-component coupling reaction for the generation of different derivatives of furan **441** was developed by Liu et al. from arylglyoxal **1**, terminal alkynes **439**, and secondary amines **440** in presence of the catalytic amount of Au in MeOH under N_2 atmosphere (Fig. 3.139).[264]

Ahmed and his group[265] develop a new approach for the synthesis of 3-hydroxy-2-furanyl-acrylamides **443** from AGs **1** and aroylacetonitriles **442** in the presence of TBHP and triethylamine (TEA) in toluene at 70°C for 3–4 h. This approach undergoes different reactions sequentially in one-pot-like Knoevenagel condensation, Michael addition, selective amidation, and Paal – Knorr cyclization to obtain the desired product **149** in good yields (Fig. 3.140).

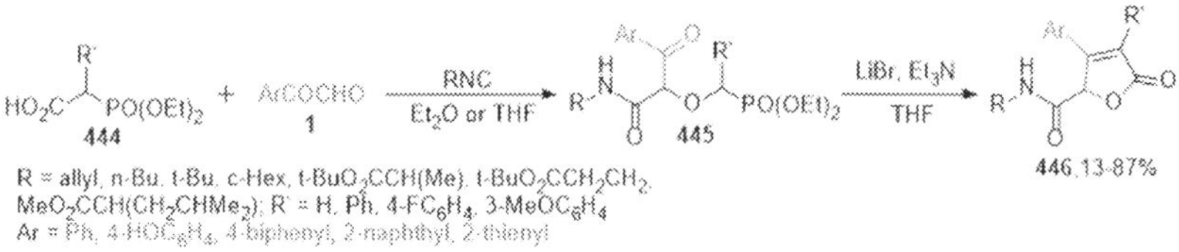

FIGURE 3.141   Synthesis of 3-hydroxy-2-furanyl-acrylamides.

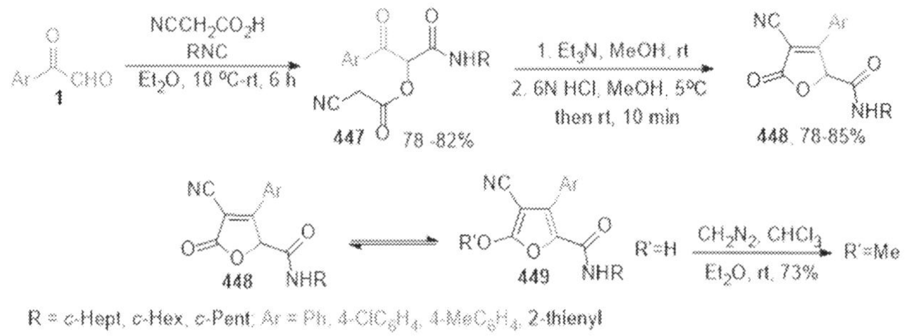

FIGURE 3.142   Synthesis of 5-acylamino butenolides.

3.4.17   Synthesis of tetrahydrofuran and dihydrofuran

Tetrahydrofuran and dihydrofuran are repeatedly found in various biologically active compounds,[266] natural products,[267] and also serve as valuable intermediates for the synthesis of different active scaffolds.[268] Therefore more interest has been shown in recent years for the development of new synthetic routes for the synthesis of different substituted tetrahydrofuran and dihydrofurans. This includes [3 + 2] cycloaddition reactions, cyclization of alkenols, and oxidative cyclization reactions.

Beck et al. reported[269] the MCR for the synthesis of 5-acylamino butenolides **446** from AGs **1**, isocyanides, and **444** in Et$_2$O or THF to obtain Passerini adducts **445**, which undergo intramolecular Horner — Wadsworth — Emmons reaction in presence of LiBr and Et$_3$N in THF to afford the desired product **446** in 13% — 87% yields (Fig. 3.141).

Bossio et al. reported[270] the synthesis of substituted dihydrofuran-2-ones **448** from AGs, cyanoacetic acid, and isocyanides to get N-substituted 3-aryl-2-cyano acetoxy-3-oxo propionamides **447**, which undergoes cyclization to obtain the desired product **448** in the presence of Et$_3$N. The reaction was started by treating AGs with isocyanides and cyanoacetic acid in diethyl ether at room temperature, followed by Et$_3$N or piperidine addition in MeOH at room temperature for 10 min. The reaction mixture was then acidified with 6N HCl, to furnish the dihydrofuran-2-ones **448** in 78% — 85% yields. The dihydrofuran-2-ones is converted into 5-methoxy furan **449** via hydroxyl furan by the addition of an excess amount of CH$_2$N$_2$ in Et$_2$O — CHCl$_3$ at room temperature for 6 h (Fig. 3.142).

An efficient protocol for the synthesis of 5-hydroxy dihydrofurans (**453** and **454**) was established by Anary-Abbasinejad et al. by conducting a reaction between AG-hydrates **1**, **450**, and Ph_3P. The reaction between **450**, Ph_3P, and AG-hydrates gives DAAD − PPh_3 zwitterions **451** in DCM at room temperature. The DAAD − PPh_3 zwitterions **451** undergo protonation and conjugate addition with AGs to afford **452**, which underwent intramolecular Wittig reaction to give the desired product (**453** and **454**) in 71% − 85% yields. The product was obtained as two diastereomers and their *cis/trans* ratio (10/90 to 62/38) was determined from ^{1}H-NMR spectroscopy (Fig. 3.143).[271]

3.4.18 Synthesis of furmoins

The benzoin condensation reaction of AGs **1** with potassium cyanide was established by Peter et al.[272] for the synthesis of furmoins **455** in aqueous EtOH. A solution of AGs in EtOH was added to the 50% aqueous ice-cold solution of KCN in EtOH. After some time, the reaction was solidified and the products **455** were obtained in 27% − 71% yields (Fig. 3.144).

The same concept was applied for the synthesis of formoin diacetate **456** by refluxing PG in Ac_2O in the presence of pyridine in good yield (Fig. 3.145). The reaction proceeds through the proton removal from the diacetyl derivative **457** by pyridine to form an anion which then reacts with another molecule of PG to generate an intermediate **458**, which undergoes cyclization to form the corresponding product **459**.[273]

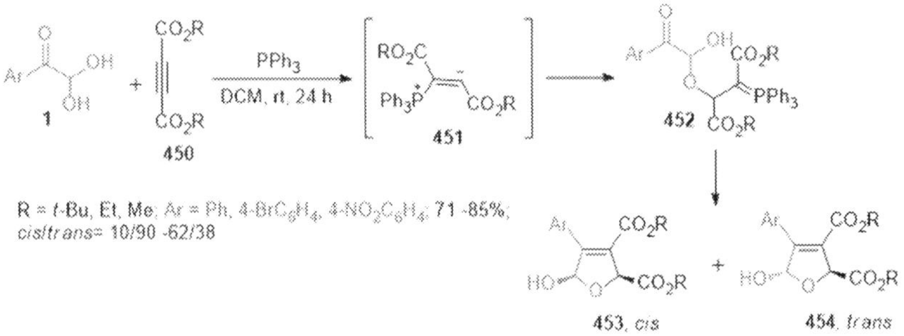
PPh3
DCM, rt, 24 h
1
450
451
452
R = t-Bu, Et, Me; Ar = Ph, 4-BrC6H4, 4-NO2C6H4; 71 -85%;
cis/trans= 10/90 -62/38
453, cis
454, trans

FIGURE 3.143 Synthesis of substituted dihydrofuran-2-ones.

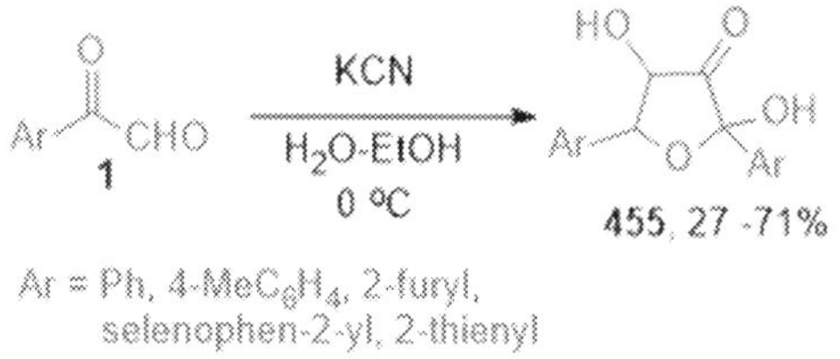
KCN
H2O-EtOH
0 °C
1
455, 27 -71%
Ar = Ph, 4-MeC6H4, 2-furyl,
selenophen-2-yl, 2-thienyl

FIGURE 3.144 Synthesis of 5-hydroxy dihydrofurans.

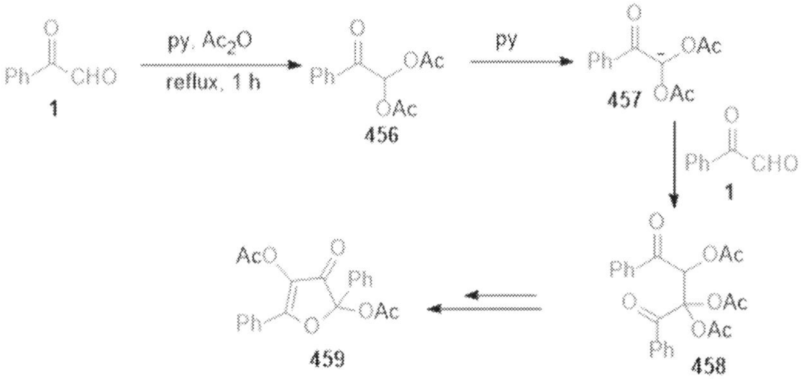

FIGURE 3.145   Synthesis of furmoins.

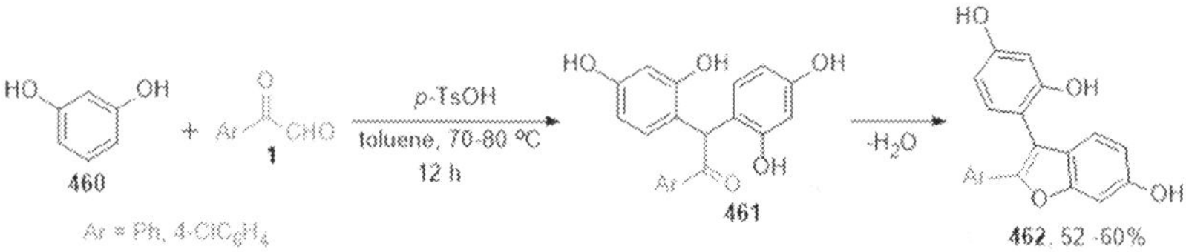

FIGURE 3.146   Synthesis of formoin diacetate.

3.4.19   Synthesis of benzofurans and furofurans

Benzofuran is one of the most significant oxygen-containing heterocycles consisting of fused benzene and furan ring, which shows potent biological properties such as antifungal,[274] antibacterial,[275] antiviral,[276] antiinflammatory,[277] and are widely presented in various naturally occurring and synthetically active compounds. Substituted benzofurans find applications such as fluorescent sensors,[278] oxidants,[279] in drug discovery, and in another field of chemistry and agriculture.[280] Based on the applications, a diversity of synthetic approaches can be applied in literature for the synthesis of benzofurans and substituted benzofurans. AGs are also used as the starting materials for the synthesis of different substituted benzofurans in presence of different metals and in metal-free conditions. These synthetic approaches are discussed here.

On the treatment of resorcinol **460**, with AGs **1** in acidic conditions, substituted benzofurans **462** depending on the type of phenolic compounds were synthesized by Talinli et al. A reaction was conducted between AGs **1** with resorcinol (2 equivalent) **460** in the presence of p-TsOH in toluene at 70°C − 80°C, the desired product 2-phenyl-3-(2,4-dihydroxy)-6-hydroxy-benzo[b]furans **462** were obtained in 52% − 60% yields via an intermediate **461**. This reaction proceeds through the sequence of Friedel − Crafts semiacetalization reactions followed by the loss of water molecule (Fig. 3.146).[91]

However, benzo[b]naphtho[2,1-f]oxepin-13-ones **464** were obtained by a reaction of AGs **1** with 2-naphthol (2 equivalent) **463**. This reaction prevents the semi-acetalization step due to the bulkiness of the 2-naphthol ring and the reaction was carried out in presence of the catalytic amount of H_2SO_4 in AcOH at 50°C for 12 h. The final products **464** were obtained in 35% − 60% yields (Fig. 3.147).

The three-component reaction of AG-hydrates **1**, phenols **465**, and *p*-$TsNH_2$ in presence of the catalytic $InCl_3$ was established by Chen et al. for the synthesis of 2-aryl-3-amino benzofuran derivatives **466** in good yields. This reaction was optimized by using different Lewis acids but the best yields were obtained in the presence of $InCl_3$. AG-hydrate, *p*-$TsNH_2$, and the catalytic amount of $InCl_3$ are dissolved in DCM after which phenol is added to form the product **466** in 45% − 94% yields. The electronic nature of the substitutions affects the yields of the desired product, electron-donating groups present on phenols gives the corresponding product in high yields, whereas electron-withdrawing groups give the product in moderate yields (Fig. 3.148).[281]

1H-pyrano[2,3-c]isoquinoline-2-carbonitrile **470** and furo[2′,3′:2,3]furo[5,4 c]isoquino-line-10-carbonitrile **471** were synthesized via Michael reaction of 2-methyl-4 phencyclidine-1,3-(2H,4H)-isoquinolinedione **468** with malononitrile and Et_2NH at 60°C (Fig. 3.149). The reaction was conducted in a mixture of the benzene − EtOH solvent system to form **469**. When reaction proceeds through **route a**, 1H-pyrano[2,3-c]isoquinoline-2-carbonitrile **470** was formed and when it goes through **route b**, furo[2′,3′:2,3]furo[5,4 c]isoquinoline-10-carbonitrile **471** was formed. The first step involves the condensation reaction of 2-methyl-1,3-(2H,4H)-isoquinolinedione **467** with PG to form **468**.[282]

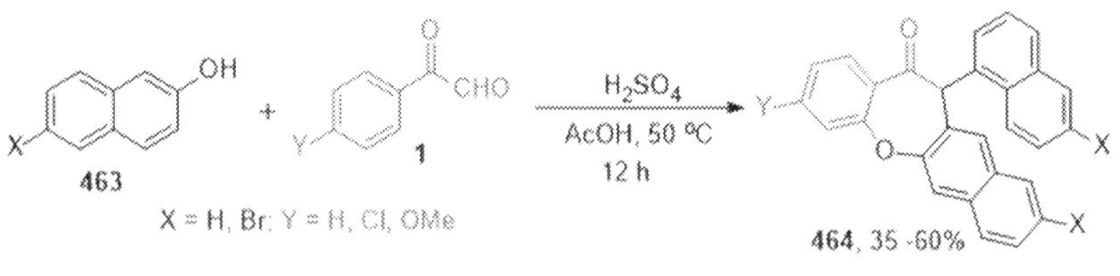
OH
463
CHO
1
H2SO4
AcOH, 50 °C
12 h
X = H, Br; Y = H, Cl, OMe
464, 35 -60%

FIGURE 3.147 Synthesis of substituted benzofurans.

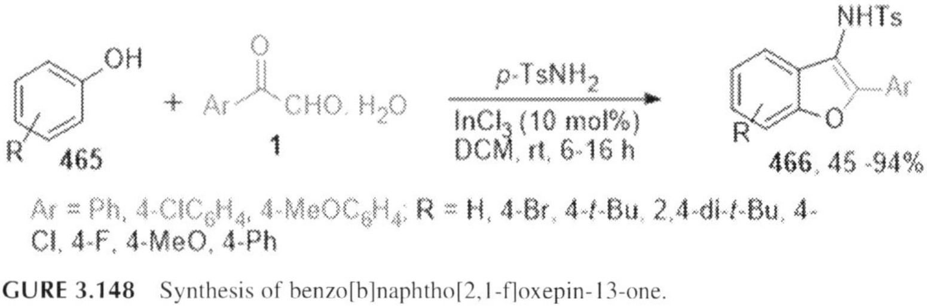
OH
R
465
Ar
CHO. H2O
1
p-TsNH2
InCl3 (10 mol%)
DCM, rt, 6-16 h
NHTs
Ar
466, 45 -94%
Ar = Ph, 4-ClC6H4, 4-MeOC6H4; R = H, 4-Br, 4-t-Bu, 2,4-di-t-Bu, 4-Cl, 4-F, 4-MeO, 4-Ph

FIGURE 3.148 Synthesis of benzo[b]naphtho[2,1-f]oxepin-13-one.

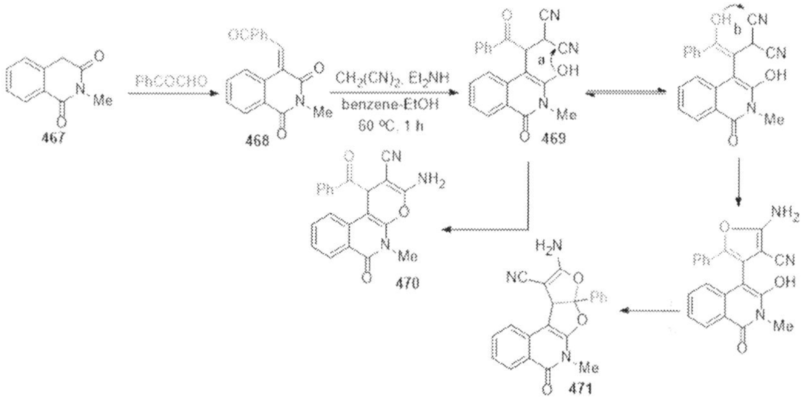

FIGURE 3.149   Synthesis of 2-aryl-3-amino benzofuran derivatives.

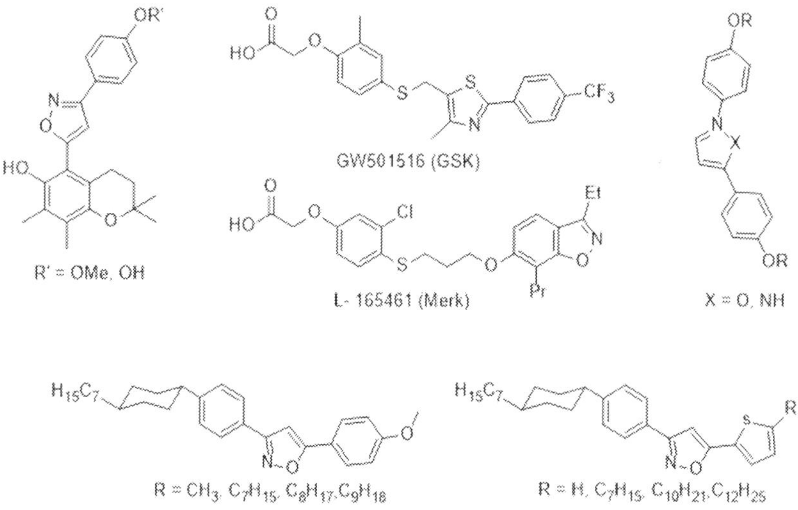

FIGURE 3.150   Synthesis of 1H-pyrano[2,3-c]isoquinoline-2-carbonitrile and furo[2',3':2,3] furo[5,4 c]isoquino-line-10-carbonitrile.

3.4.20   Synthesis of isoxazoles

Isoxazoles are an important class of heterocyclic compounds composed of nitrogen and oxygen atom. It is present in various natural products and synthetic compounds of biological importance like antibacterial,[283] antagonists,[284] antiinflammatory,[285] analgesics,[286] and also show the applications in functional materials.[287] Isoxazole derivatives are used as starting material for the synthesis of different biologically active molecules as shown in Fig. 3.150. Various synthetic methods have been reported for the synthesis of different isoxazoles such as 1,3-dipolar cycloaddition of nitrile oxides, condensation of hydroxylamine with 1,3-dicarbonyl compounds. Various derivatives of isoxazoles have been synthesized by taking arylglyoxal as the starting material.

A two-step reaction for the synthesis of 5-substituted-3-acyl isoxazoles **474** was reported by Juhasz-Toth et al.[288] from 2-azido-3-hydroxy-1,4-diketones **473** in 39% − 63% yields. On the treatment of α-azido acetophenones **472** with AG-hydrates **1** in dry THF in the presence of DBU at 0°C to form 2-azido-3-hydroxy-1,4-diketones **473**. The next step involves the addition of excess amount of MsCl and Et₃N in dry DCM at −15°C for 1.5−5 h which results in the formation of 5-substituted-3-acyl isoxazoles product **474** in good yield (Fig. 3.151).

3.4.21 Synthesis of oxazolidines

Oxazolidines are commonly used as metal ligands in asymmetric catalysis,[289] synthetic intermediates in organic synthesis,[290] and also used as the protecting groups.[291] This moiety is present in various medicinally important natural products and biologically important synthetic compounds.[292] In the literature, numerous methods are available for the synthesis of oxazolidine derivatives by using AGs as the starting material.

Choi and coworkers reported[293] the stereoselective synthesis of α-hydroxy aldehydes (**479** and **480**) by using (R)-Piperidin-3-ol **475** as a chiral auxiliary. PG-hydrate when treated with (R)-Piperidin-3-ol **475** undergoes condensation reaction in the presence of 4 A° molecular sieves in anhydrous DCM to afforded (5R,7R)-7-Benzoyl-6-oxa-1-azabicyclo-[3.2.1]-octane **476** in 85% yield (Fig. 3.152). The treatment of **476** with Grignard reagent afforded carbinols **477** and **478** with excellent yields. This is due to the chelation of the metal between the oxygen of the carbonyl group and the ether oxygen. The stereoselectivity of carbinols was low in the case of RMgCl, while RMgBr gives better stereoselectivity. Hydrolysis of the carbinols A and B in the presence of silica gel in DCM gives α-hydroxy aldehydes (**479** and **480**).

N-Boc-2-benzoyloxazolidine **483** was used as a chiral auxiliary for the synthesis of enantiopure 1,2-diols **484** and was reported by Agami et al.

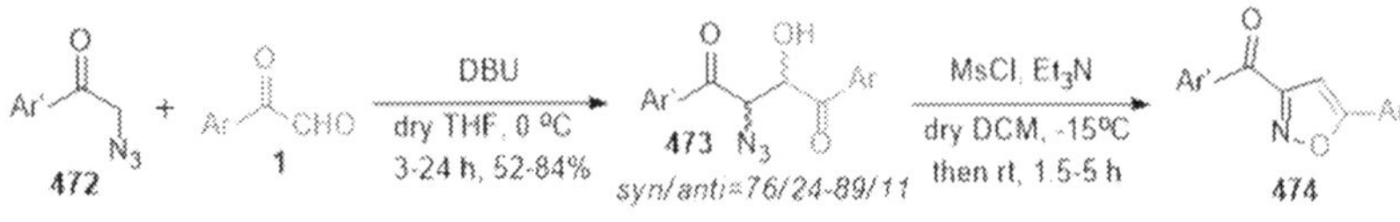

Ar = Ph, 4-ClC₆H₄, 4-MeOC₆H₄; Ar' = Ph, 4-ClC₆H₄, 4-FC₆H₄, 4-MeOC6H4; 39 -63%.

FIGURE 3.151 Biologically active molecules containing isoxazole scaffolds.

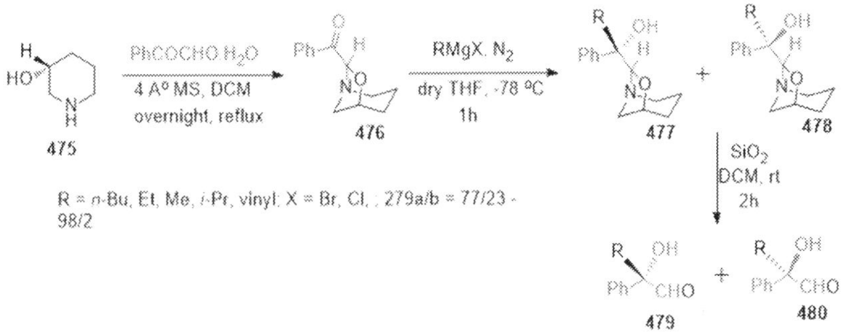

FIGURE 3.152 Synthesis of 5-substituted-3-acyl isoxazoles.

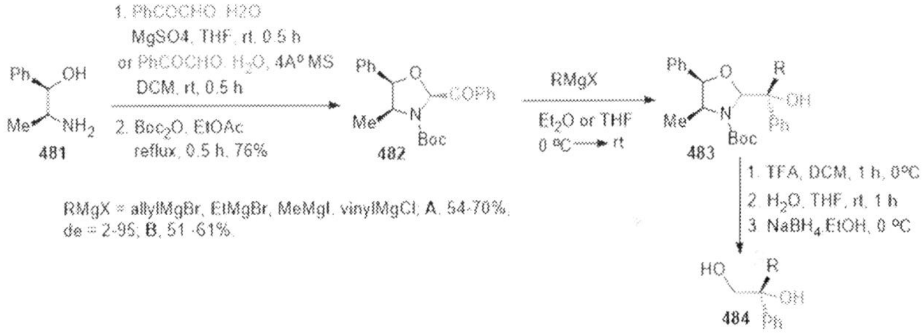

FIGURE 3.153 Synthesis of α-hydroxy aldehydes.

(Fig. 3.153). The reaction of (1R,2S)-norephedrine **481** with PG-hydrate in presence of anhydrous $MgSO_4$ in THF for 0.5 h at room temperature or in the presence of 4 Å Ms in DCM at room temperature for 0.5 h, followed by addition of Boc_2O in EtOAc under refluxing conditions to obtained oxazolidine **482** in good yield. The next step involves the Grignard reaction of oxazolidine in THF or Et_2O at 0°C to get compound **483**. Hydrolysis of compound **483** with TFA in DCM at 0°C for 1 h, followed by reduction with $NaBH_4$ in EtOH at 0°C to furnish enantiopure 1,2-diols **484**.[294]

On the treatment of PG-hydrate with (S)-prolinol **485** in the presence of 4 A° Ms in CH_2Cl_2 undergoes condensation reaction to obtain the chiral 2-benzoyl-3-oxa-1 azabicyclo[3.3.0]octane **486**. The Grignard reaction was carried out with chiral 2-benzoyl-3-oxa-1 azabicyclo[3.3.0]octane **486** in different solvent systems like Et_2O, THF − HMPA, or THF to obtained the desired carbinols (**487** and **488**) in 80% yields (Fig. 3.154).[295]

3.4.22 Synthesis of oxazoles and furocoumarins

Oxazoles are important heterocyclic compounds and are present in various compounds of biological importance such as antiviral,[296] antibacterial,[297]

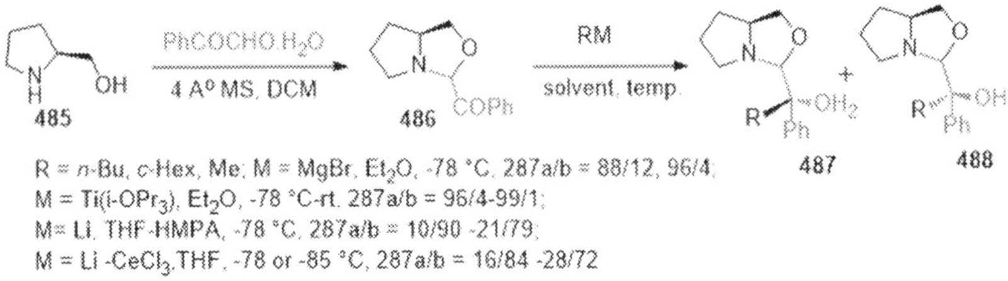
PhCOCHO.H2O
4 A° MS, DCM
RM
solvent, temp.
485
486
487
488
R = n-Bu, c-Hex, Me; M = MgBr, Et2O, -78 °C, 287a/b = 88/12, 96/4;
M = Ti(i-OPr3), Et2O, -78 °C-rt. 287a/b = 96/4-99/1;
M= Li, THF-HMPA, -78 °C, 287a/b = 10/90 -21/79;
M = Li -CeCl3,THF, -78 or -85 °C, 287a/b = 16/84 -28/72

FIGURE 3.154 Synthesis of enantiopure 1,2-diols.

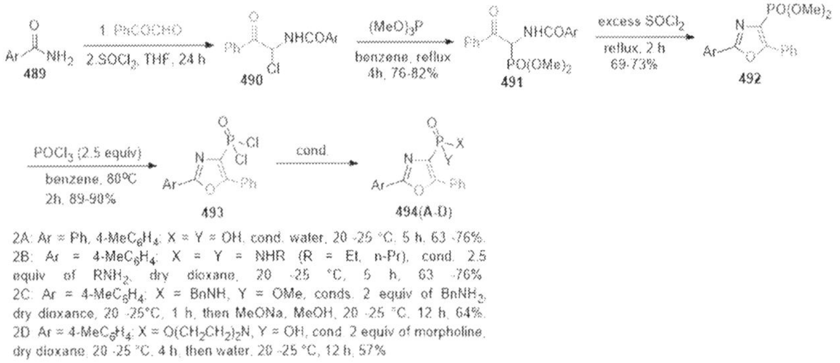
1. PhCOCHO
2.SOCl2, THF, 24 h
(MeO)3P
benzene, reflux
4h, 76-82%
excess SOCl2
reflux, 2 h
69-73%
POCl3 (2.5 equiv)
benzene, 80°C
2h, 89-90%
cond.
489
490
491
492
493
494(A-D)
2A: Ar = Ph, 4-MeC6H4; X = Y = OH, cond. water, 20 -25 °C, 5 h, 63 -76%.
2B: Ar = 4-MeC6H4; X = Y = NHR (R = Et, n-Pr), cond. 2.5 equiv of RNH2, dry dioxane, 20 -25 °C, 5 h, 63 -76%
2C: Ar = 4-MeC6H4; X = BnNH, Y = OMe, conds. 2 equiv of BnNH2, dry dioxance, 20 -25°C, 1 h, then MeONa, MeOH, 20 -25 °C, 12 h, 64%.
2D: Ar = 4-MeC6H4; X = O(CH2CH2)2N, Y = OH, cond. 2 equiv of morpholine, dry dioxane, 20 -25 °C, 4 h, then water, 20 -25 °C, 12 h, 57%

FIGURE 3.155 Synthesis of carbinols.

antifungal,[298] antiinflammatory,[299] and antibiotic.[300] In the literature, there are synthetic methods available for the preparation of oxazoles and their derivatives such as cyclodehydration reaction of α-acylaminoketone and Robinson–Gabriel synthesis.

Belyuga et al. reported[301] a multistep protocol for the preparation of methyl (2-aryl-5-phenyl-1,3-oxazol-4-yl) phosphonates **494** from Arbuzov products in 69% – 73% yields in the presence of excess amounts of $SOCl_2$ under heating conditions for 2 h. The treatment of PG with amides **489** followed by the addition of $SOCl_2$ afforded α-Chloro-α-acylamino ketones **490**. The compound **490** was then treated with trimethyl phosphite in refluxing benzene for 4 h to obtain compound **491** in 76% – 82% yields. (1,3-oxazol-4-yl)phosphonates **492** were transformed into phosphonic acid **494A**, phosphonic dialkyl amides **494B**, methyl *N*-benzyl phosphonamidate **494C**, and phosphonic morpholide **494D** by using different reaction conditions via **493** (Fig. 3.155).

Babu et al. developed[302] a novel multicomponent approach for the synthesis of oxazoles **498** and furocoumarins **499** Treatment of AGs **1** with **495** forms *N*-acyliminium ion (NaI) intermediate without any catalyst and solvent at 80°C. The in situ generated **496** undergoes further transformations with cyanide and **497** in presence of TfOH to form oxazoles **498** and furocoumarins **499** one pot same pot, respectively (Fig. 3.156).

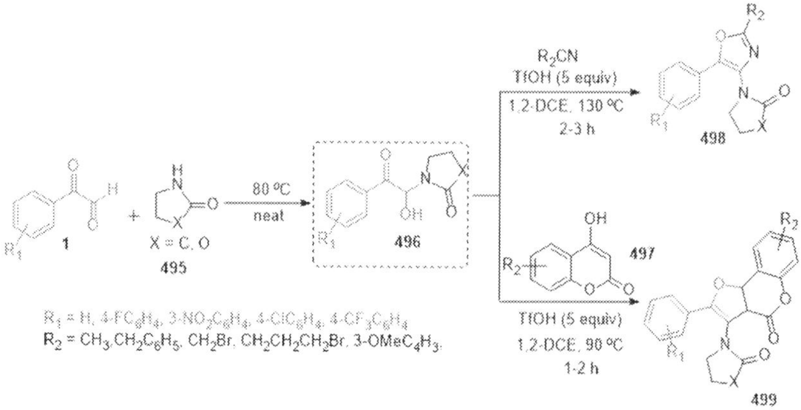

FIGURE 3.156 Synthesis of methyl (2-aryl-5-phenyl-1,3-oxazol-4-yl) phosphonates.

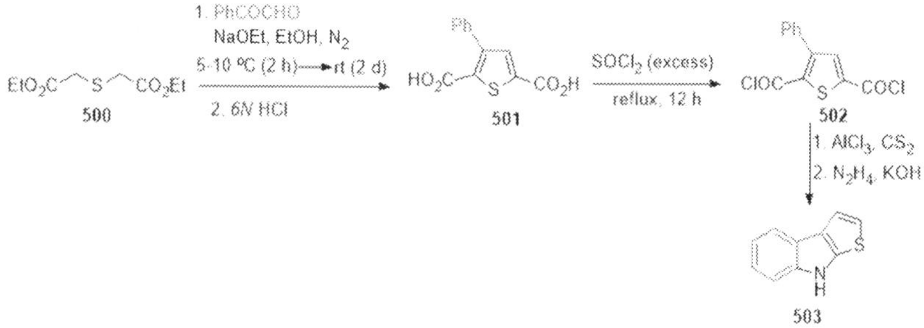

FIGURE 3.157 Synthesis of oxazoles and furocoumarins.

3.4.23 Synthesis of thiophenes

Thiophenes and their derivatives are known for showing a wide range of applications in electrochemical,[303] physical,[304] optical,[305] and biological properties.[306] In addition, thiophenes have shown remarkable applications in advanced materials, that is, organic conductors,[307] conjugated polymers,[308] light-emitting devices, and semiconductors.[309] Thiophenes containing drugs are also useful in the treatment of various diseases.[310] There are various methods reported such as Gewald and Paal−Knorr reactions for the synthesis of thiophene derivatives by using AGs.

Mac Dowell et al. described[311] a three-step reaction for the preparation of 8H-indeno[2,1-b]thiophene **503** in good yield. Treatment of PG with diethyl thiodiglycolate **500** undergoes Hinsberg − Stobbe type condensation to give 3-phenyl-2,5-thiophenedicarboxylic **501** in 57% yield in the presence of NaOEt followed by saponification. The compound **501** was converted into **502** by reacting with SOCl$_2$ under the refluxing condition. The **502** undergoes ring closure by using AlCl$_3$ and CS$_2$, followed by Wolff − Kishner reduction with N$_2$H$_4$ and KOH, which after decarboxylation and reduction lead to the formation of the product **503** in 64% yield (Fig. 3.157).

A comparable approach was reported for the preparation of 2,5-di-*p*-toluoyl-3-phenylthiophene **505** by conducting a reaction between diketo sulfide **504** and PG in the presence of NaOMe in MeOH under refluxing conditions (Fig. 3.158).[312]

Methylthio-substituted 1,4-diones **506** were converted into 3-methylthio-substituted thiophenes **507** in presence of KI and concentrated HCl in acetone at room temperature, followed by addition of Lawesson's reagent in refluxing toluene for 2 h in 75% − 88% yields (Fig. 3.159).[200]

3.4.24 Synthesis of thiazolidines and thiazolines

Thiazolidine and thiazoline are important heterocyclic scaffolds that are present in various natural products and synthetic compounds of biological importance like anticonvulsant,[313] antidepressant, sedative, antihypertensive, antiinflammatory, antihistaminic, and antiarthritic.[314] These compounds are also used as chiral auxiliaries for the synthesis of diastereoselective products in synthetic organic chemistry.[315] AGs are also used for the synthesis of thiazolidine and thiazoline rings.

Pinho e Melo et al. reported[89] the synthesis of a mixture of diastereoisomeric thiazolidines **509 a/b** and **510 a/b**, from 4-MeO − PG and PG with L-cysteine methyl ester **508**, respectively. Thiazolidine **509a** was converted into *N*-acylthiazolidine **513** by conducting a reaction between **509a** and prop-2-ynyloxyacetyl chloride **511** in presence of a base (K$_2$CO$_3$) in anhydrous DCM followed by addition of LiI (4 equivalent) in EtOAc and acidifying the resulting reaction mixture with aqueous HCl. The treatment of **512** with Ac$_2$O and chiral 3-benzoyl-1H,3H-pyrrolo[1,2-c]thiazole derivative **513** furnished through the intramolecular 1,3-dipolar cycloaddition of intermediate **514** in 42% yield. The same reaction when tried with **509b**, the other isomer of **513** was obtained. In contrast, the same thiazolidine was also obtained by the acylation reaction

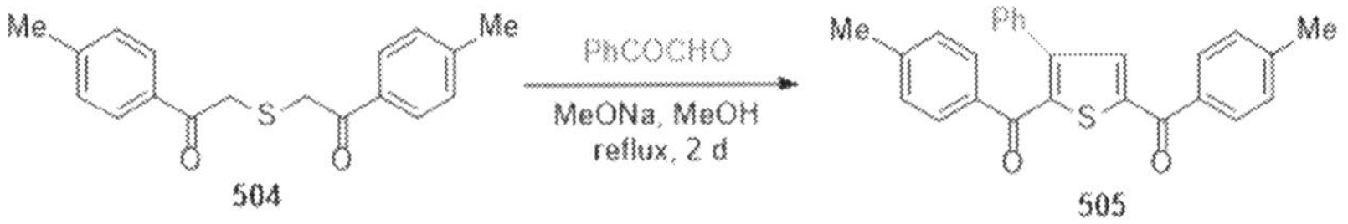

FIGURE 3.158 Synthesis of 8H-indeno[2,1-b]thiophene.

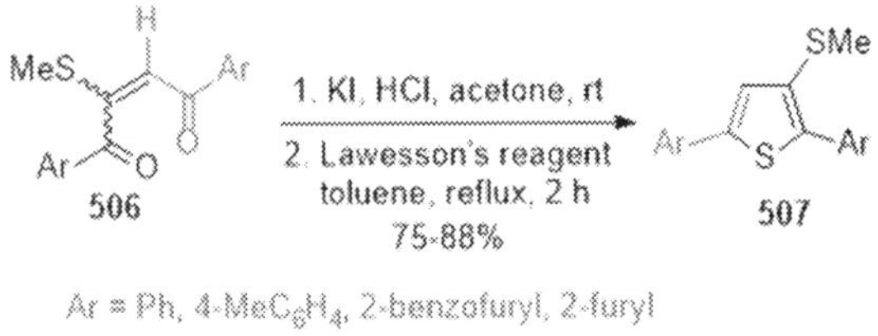

Ar = Ph, 4-MeC$_6$H$_4$, 2-benzofuryl, 2-furyl

FIGURE 3.159 Synthesis of 2,5-di-p-toluoyl-3-phenylthiophene.

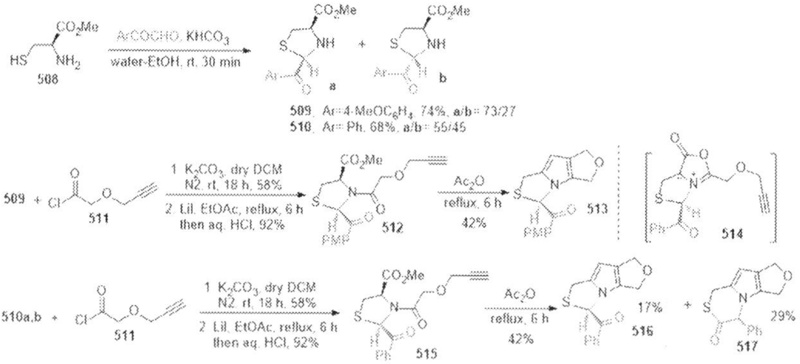

FIGURE 3.160   Synthesis of 3-methylthio-substituted thiophenes.

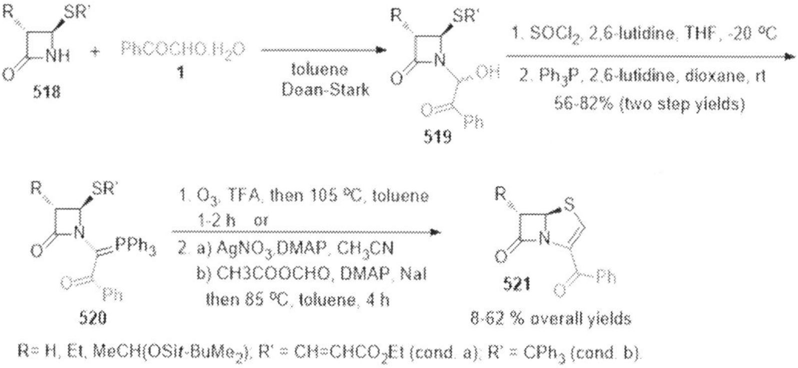

FIGURE 3.161   Synthesis of thiazolidine based scaffolds.

of **510a,b**, followed by reaction with LiI in EtOAc to furnish thiazolidine **515**. Two chiral products, that is, *1H,3H*-pyrrolo[1,2-c]thiazole **516** and pyrrolo[1,2-c][1,4]thiazine **517** in 17% and 29% were obtained when product **515** was treated with Ac$_2$O under reflux conditions for 6 h (Fig. 3.160).

A four-step protocol was reported by Pearson et al.[60] for the synthesis of 6-substituted-3-benzoyl penems **521**. The initial step involves the condensation of PG with azetidinones **518** in toluene to obtain **519**, followed by transformation of **239** to chloro derivatives. The addition of triphenylphosphine and 2,6-lutidine in dioxane at room temperature afforded the phosphorane derivatives **520** in good yields. The phosphorane derivatives were converted into penems **521** by ozonolysis (treated with O$_3$) in the presence of TFA and then heating the reaction at 105°C in toluene. This transformation of **520** to **521** was also done by using AgNO$_3$ in the presence of DMAP in CH$_3$CN, followed by formylation with acetic formic anhydride in the presence of DMAP and NaI, and then heating the reaction at 85°C in toluene (Fig. 3.161).

3.4.25   Synthesis of dioxophospholanes

Treatment of PG with an excess amount of trimethyl phosphite (TMP) in DCM at $0°C$ afforded unsaturated dioxophospholane **523** under N_2 atmosphere and was reported by Ramirez et al.[316] Pentaoxyphosphorane **522** was obtained from phenyl glyoxal in 48% yield in the presence of trimethoxy phosphite, which was further reacted with PG to form dioxophospholane as two diastereoisomers **523a** and **523b** in 35/65 ratio. In addition, the reaction of PG with diacetyl-TMP 1:1 adduct **524** was conducted to obtain dioxophospholane as two diastereoisomers (**525a** and **525b**) in quantitative yield. The treatment of diacetyl with **522**, gives the same product **525a** and **525b** in different condition (Fig. 3.162).

3.4.26   Synthesis of pyridazines

The pyridazine derivatives are mostly present in biologically active compounds and are also present with different pharmacophores.[317] Pyridazines are also important because of their utility in synthetic organic chemistry and in physical organic chemistry.[318] There are reports available for the synthesis of pyridazine and their derivatives by using 2-oxoaldehyes.

An efficient, mild, three-component reaction was reported by Rimaz et al.[319] for the generation of pyridazine-4-carboxylates **527** in 70% − 97% yields. All three components, that is, hydrazine hydrate (5 mmol), AGs (1 mmol) **1**, and 1,3-dicarbonyl compounds (1 mmol) **526** were dissolved in water at room temperature and stirred for 30−60 min. This type of reaction was also performed by using ultrasound irradiation to obtain the pyridazines product **527** in good yields and in short reaction times (Fig. 3.163).

A two-step protocol was established by Ismail and coworkers[320] for the production of steroidal pyridazine derivatives **530** from 2-acetylestradiol **528** and AGs. 1,4-dicarbonyl compounds **529** were prepared from a reaction of **528** with AGs in refluxing Ac_2O, followed by the reaction with hydrazine

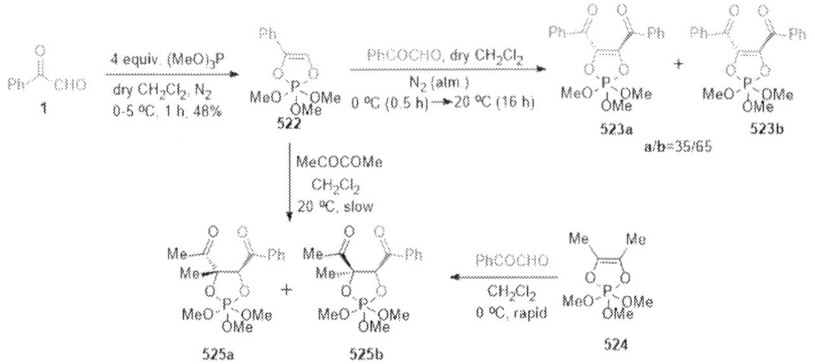

FIGURE 3.162   Synthesis of 6-substituted-3-benzoyl penems.

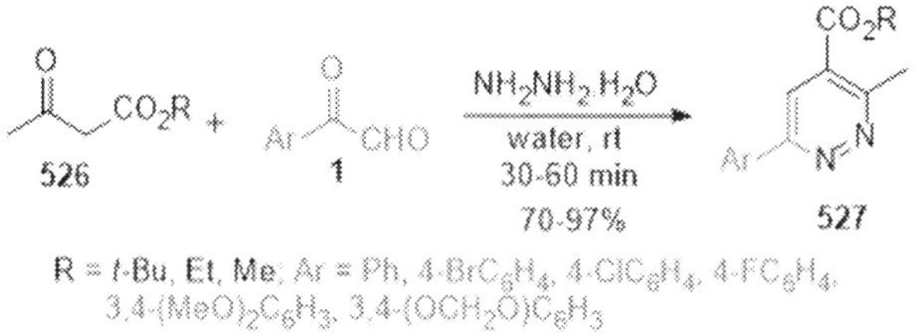

FIGURE 3.163 Synthesis of unsaturated dioxophospholane.

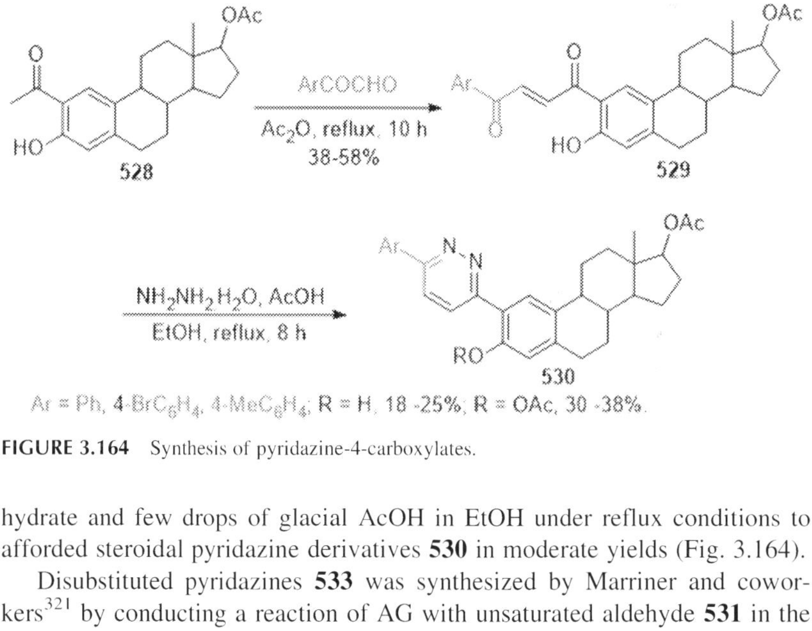

FIGURE 3.164 Synthesis of pyridazine-4-carboxylates.

hydrate and few drops of glacial AcOH in EtOH under reflux conditions to afforded steroidal pyridazine derivatives **530** in moderate yields (Fig. 3.164).

Disubstituted pyridazines **533** was synthesized by Marriner and coworkers[321] by conducting a reaction of AG with unsaturated aldehyde **531** in the presence of catalytic amount rhodium to afford β-hydroxy-γ-keto aldehydes **532**, which were transformed into disubstituted pyridazines **533** by reacting with an excess amount of hydrazine in MeOH at room temperature. A glass vessel was charged with 1 mmol of KOAc, AG (1 mmol), followed by addition of 5 mmol of **531**, Rh $(COD)_2OTf$ (1 mol%) and Ph_3P (2.4 mol%) in DCE to obtain **532**, the addition of excess hydrazine in MeOH at room temperature for 45 min afforded disubstituted pyridazines **533** in 30% − 62% yields (Fig. 3.165).

Furthermore, a different approach for the synthesis of pyridazine ring **536a** (R = morpholino) was established by conducting a reaction between 1,1-bis (thiomethyl)-2-nitroethylene **534** and 1-aminoethylmorpholine which undergoes the nucleophilic addition reaction along with hydrazine hydrate in refluxing EtOH, followed by addition of PG in 51% yield. Reduction of the nitro group of **536a** with H_2/Pd/C afforded aminopyridazine **538** in 75% yields.[322]

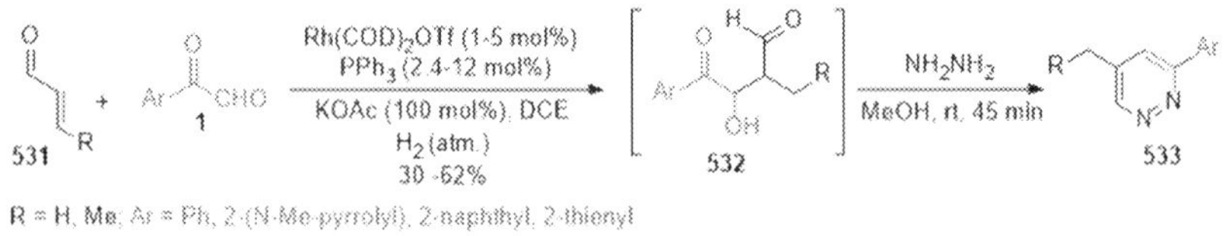

FIGURE 3.165 Synthesis of steroidal pyridazine derivatives.

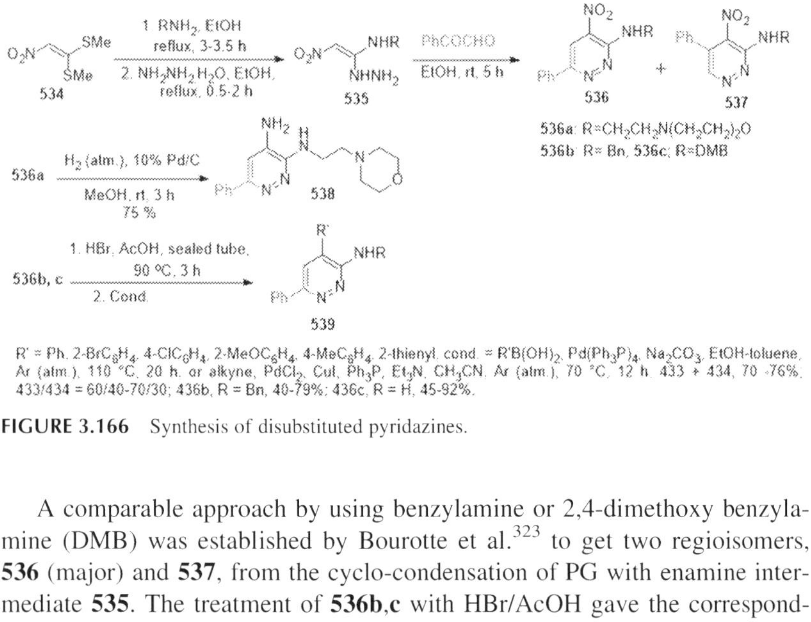

FIGURE 3.166 Synthesis of disubstituted pyridazines.

A comparable approach by using benzylamine or 2,4-dimethoxy benzylamine (DMB) was established by Bourotte et al.[323] to get two regioisomers, **536** (major) and **537**, from the cyclo-condensation of PG with enamine intermediate **535**. The treatment of **536b,c** with HBr/AcOH gave the corresponding 4-bromo derivatives **539** (Fig. 3.166). Different Suzuki and Sonogashira coupling reactions were carried out with the resulting 4-bromo pyridazines in different conditions to give their corresponding 3-amino pyridazines **539** products.

3.4.27 Synthesis of pyrazines

Pyrazine moiety is an important class of heterocyclic compounds and is showing different biological activities such as antibacterial, antidiabetic, antimycobacterial, and sedative agents.[324] Several derivatives of pyrazines have been famous as flavor components in foods.[325] Pyrazines are also showing applications as N, N'-bidentate ligands in metal coordination chemistry[326] and as versatile synthetic intermediates in organic chemistry.[327] There are various synthetic methods reported for the synthesis of different substituted pyrazines by using AGs.

The first step for the synthesis of botryllazine B **542**, a naturally occurring alkaloid involves the condensation reaction of *p*-methoxy phenylglyoxal with 2,3-diamino propionic acid hydrobromide **540** in methanolic NaOH solution under O_2 atmosphere at 20°C for 12 h. Pyrazine carboxylic acid **541** was obtained in 55% yields and further used for the synthesis of botryllazine B, which was first isolated from the red ascidian *Botryllus leachii* (Fig. 3.167).[328]

A similar condensation reaction between PG **1** and aminomalonamidamidine dihydrochloride **543** for the synthesis of 2-aminopyrazine-3-carboxamides **544** in the presence of dilute NH_4OH (pH = 8 − 9) at 0°C − 20°C was reported by Vogl and Taylor.[329] The product was obtained in 36.6% yield and the 5-phenyl-2-aminopyrazine regioisomer was isolated (Fig. 3.168).

A regioselective condensation reaction for the synthesis of 3-cyano-5-phenyl-3 phenylthiopyrazine **546a** from 2,3-diamino-3-phenylthioacrylonitrile **545** with PG acetal was described by Zhang et al.[330] Different reaction conditions were optimized to obtain better selectivity and better yields. When the reaction was carried out in *i*-PrOH and TFA, the isomer **546a** was obtained in high selectivity. Low yield and low selectivity were obtained when the reaction was conducted in the presence of mineral acids or weak acids in MeOH and EtOH. By using PG, the regioisomers **546a,b** were obtained without selectivity in 30% − 60% yields. The treatment of product **546a** with 3:1 mixture of AcOH and chloroacetic acid with 3.1 equivalent of sodium perborate at 50°C for 16 h, followed by the addition of 1.1 equivalent

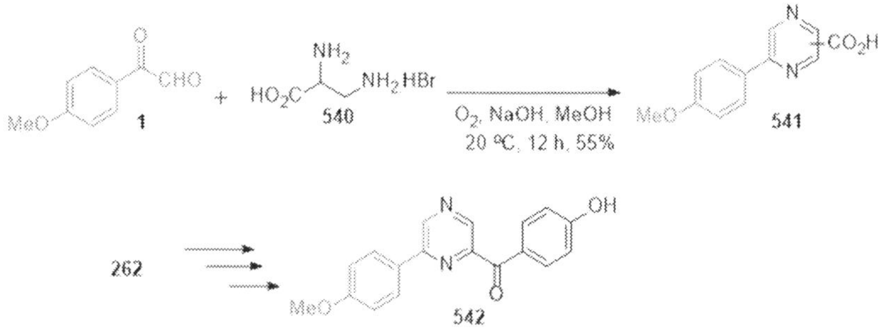

FIGURE 3.167   Synthesis of pyridazine ring and 3-amino pyridazines.

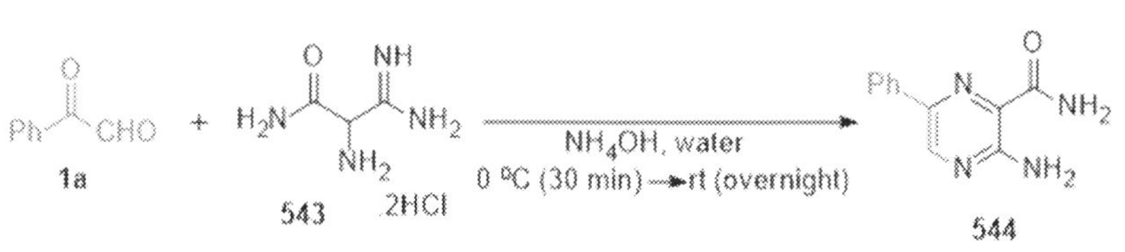

FIGURE 3.168   Synthesis of Botryllazine B.

of methyl thioglycolate in EtOH with Hunig's base in a crude mixture of **546a** produced thieno[2,3-b]pyrazine **547** in 93% yield (Fig. 3.169).

Haight et al. reported[331] a five-step protocol for the synthesis of thienopyrazine isocyanate **552**. The first step involves the condensation reaction of aminomalononitrile **548** with PG oxime **549** in *i*-PrOH at room temperature for 24 h to obtain 2-amino-3-cyano-5-phenylhydrazine *N*-oxide **550**. Compound **550** was converted into **551** in two steps by reacting with P(OEt)$_3$ in EtOH, followed by reaction with *t*-BuONO and CuBr$_2$ in DMF. Thienopyrazine isocyanate **552** was obtained from **551** in good yields (Fig. 3.170).

3.4.28 Synthesis of furo[2,3-b]pyrrole

Li et al.[332] described a multicomponent bicyclization approach for the synthesis of furo[2,3-b]pyrrole derivatives **555** from glyoxals **1**, β-keto thioamides **553**, and ethyl cyanoacetate **554** in the presence of a catalytic amount of *N*-ethyl diisopropylamine (DIPEA) in EtOH at room temperature for 1.5 h. Different bases were screened to obtain the selectivity in the products. The reaction proceeds through Knoevenagel condensation, Michael addition, and double cyclization sequence pathway that results in the formation of

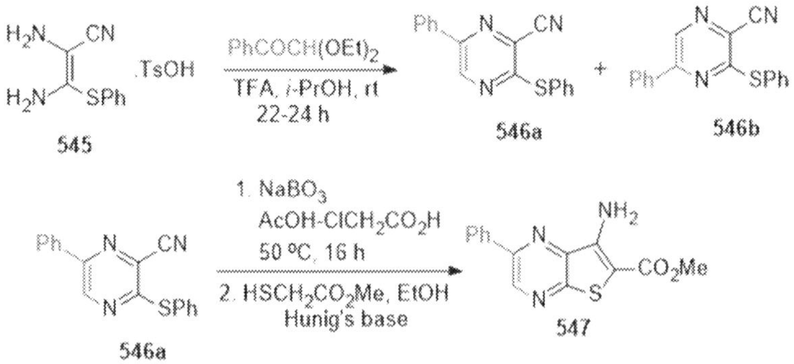

FIGURE 3.169 Synthesis of 2-aminopyrazine-3-carboxamides.

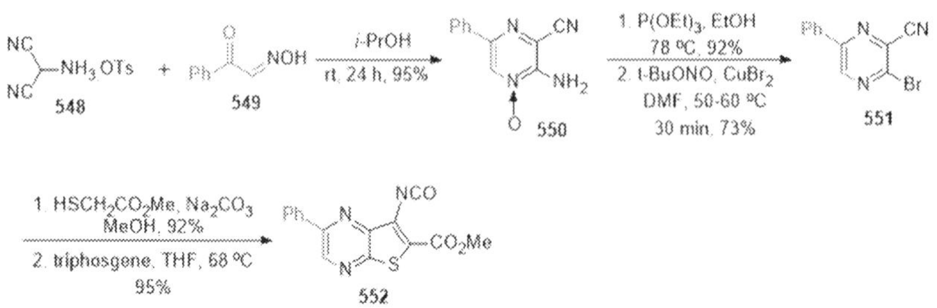

FIGURE 3.170 Synthesis of 3-cyano-5-phenyl-3 phenylthiopyrazine.

four chemical bonds, three stereogenic centers, and two five-membered cycles in a one-pot operation (Fig. 3.171).

3.4.29 Synthesis of 5,6-dihydropyridin-2(1*H*)-ones

A multicomponent synthetic approach was established by Ding et al.[333] for the synthesis of 5,6-dihydropyridin-2(1H)-ones **557** from arylglyoxals **1**, Baylis – Hillman phosphonium salts **556**, primary amines, and isocyanides, in presence of MeOH, followed by the addition of K_2CO_3. This reaction proceeds through a domino process involving Ugi, aldol, and hydrolysis reactions to obtain the desired product (Fig. 3.172).

3.4.30 Synthesis of polysubstituted carbazoles

Guo et al. reported[334] the synthesis of 2-indolyl substituted carbazoles **561** from arylglyoxal **1**, 3-vinyl indoles **558**, and indoles **559** in the presence of Brønsted acid catalyst in one pot. Different catalysts were screened to obtain the product in good yields and the simple phosphoric acid **560**, gave the polysubstituted carbazoles **561** in good to excellent yields. The **560** was derived from 2,2′-biphenol (Fig. 3.173).

3.4.31 Synthesis of pyrrolo[2,1-a]isoquinolines

An intermolecular [3 + 2] cycloaddition reaction for the synthesis of pyrrolo [2,1-a] isoquinolines **564** was established by An-Xin Wu and his group[335] from 1,2,3,4-tetrahydroisoquinoline (THIQ) **562**, phenylglyoxal monohydrate **1**, and (E)-(2-nitrovinyl)benzene **563** in the presence of a catalytic amount of

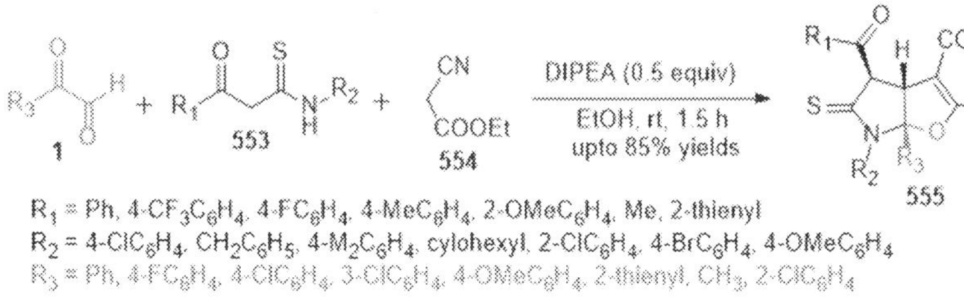
1
553
554
DIPEA (0.5 equiv)
EtOH, rt, 1.5 h
upto 85% yields
555
R1 = Ph, 4-CF3C6H4, 4-FC6H4, 4-MeC6H4, 2-OMeC6H4, Me, 2-thienyl
R2 = 4-ClC6H4, CH2C6H5, 4-M2C6H4, cylohexyl, 2-ClC6H4, 4-BrC6H4, 4-OMeC6H4
R3 = Ph, 4-FC6H4, 4-ClC6H4, 3-ClC6H4, 4-OMeC6H4, 2-thienyl, CH3, 2-ClC6H4

FIGURE 3.171 Synthesis of thienopyrazine isocyanate.

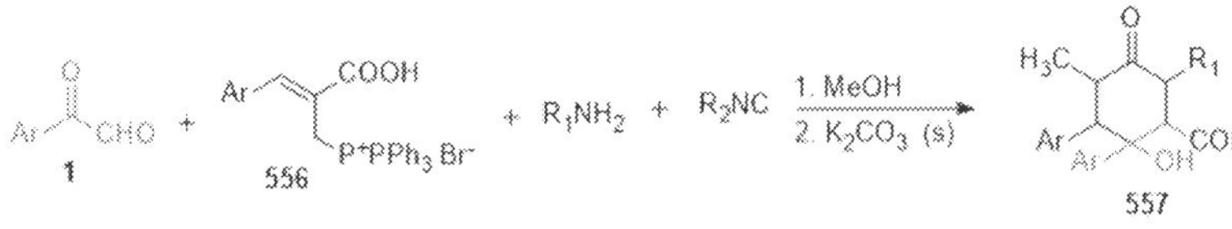
1
556
+ R1NH2 + R2NC
1. MeOH
2. K2CO3 (s)
557

FIGURE 3.172 Synthesis of furo[2,3-b]pyrrole derivatives.

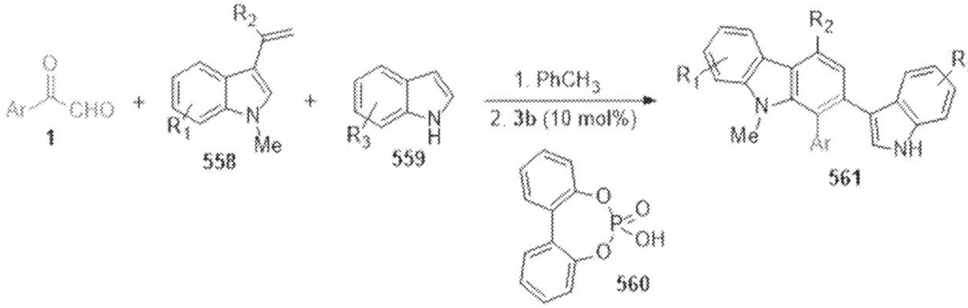

Ar = 4-FC$_6$H$_3$, 4-BrC$_6$H$_4$, 4-CF$_3$C$_6$H$_4$, 2-OMeC$_6$H$_4$, 2,4-F$_2$C$_6$H$_3$
R$_1$ = 5-Me, 5-Br, 6-Me;
R$_2$ = C$_6$H$_5$, 2-MeC$_6$H$_4$, 3-BrC$_6$H$_4$, 4-MeC$_6$H$_4$, 4-OMeC$_6$H$_4$, 4-BrC$_6$H$_4$;
R$_3$ = H, 5-Me, 5-OMe, 5-Cl, 5-Br, 5-I, 6-F, 6-Br, 7-Me,

FIGURE 3.173 Synthesis of 5,6-dihydropyridin-2(1H)-ones.

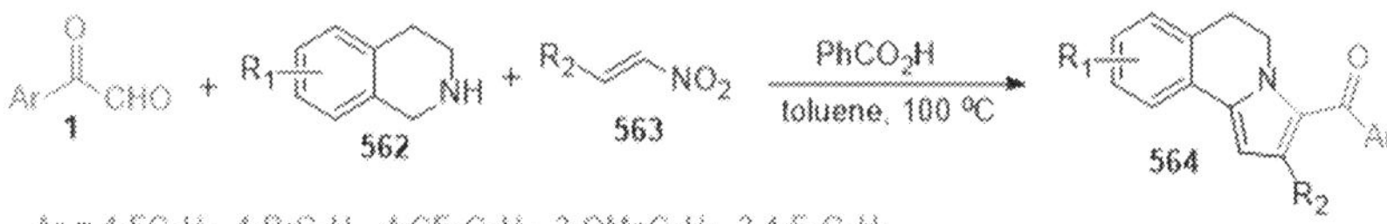

Ar = 4-FC$_6$H$_4$, 4-BrC$_6$H$_4$, 4-CF$_3$C$_6$H$_4$, 2-OMeC$_6$H$_4$, 2,4-F$_2$C$_6$H$_3$
R$_1$ = 5,6-diOMe
R$_2$ = C$_6$H$_5$, 4-MeC$_6$H$_4$, 4-CNC$_6$H$_4$, 4-FC$_6$H$_4$, 4-OMeC$_6$H$_4$, 4-BrC$_6$H$_4$;

FIGURE 3.174 Synthesis of 2-indolyl substituted carbazoles.

PhCO$_2$H in toluene at 100°C. This strategy was also used for the total synthesis of the lamellarin core and lamellarin G trimethyl ether (Fig. 3.174).

In addition, a multicomponent tandem cyclization was also developed by An-Xin Wu and his group[336] by replacing 1,2,3,4-tetrahydroisoquinoline **562** with tryptamines **565** to obtain nitro substituted dihydroindolizino[8,7-b] indoles **567** in the presence of CF$_3$SO$_3$H in CH$_3$CN at 80°C. The same concept was also applied for the synthesis of cyano substituted dihydroindolizino[8,7-b]indoles **566** by reacting arylglyoxal monohydrates **1**, tryptamines **565**, and malononitrile in the presence of TFA in CHCl$_3$ at 80°C (Fig. 3.175).

3.4.32 Synthesis of dihydropyrrolo[1,2-a]quinolines

A four-component synthetic protocol for the preparation of 4,5-dihydropyrrolo[1,2-a]quinolines **571** was established by Yaragorla et al.[337] from AG, amine **568**, alkene **569**, and alkyne **570**. This reaction takes place in the presence of calcium as a catalyst and forms *syn* diastereoselective product. The advantage of this approach is the formation of three new C − C bonds and two new C − N bonds (Fig. 3.176).

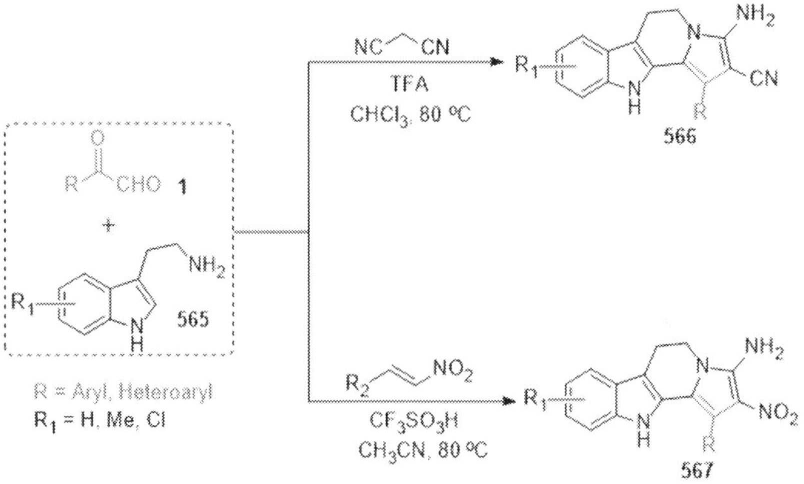

FIGURE 3.175   Synthesis of pyrrolo[2,1-a] isoquinolines.

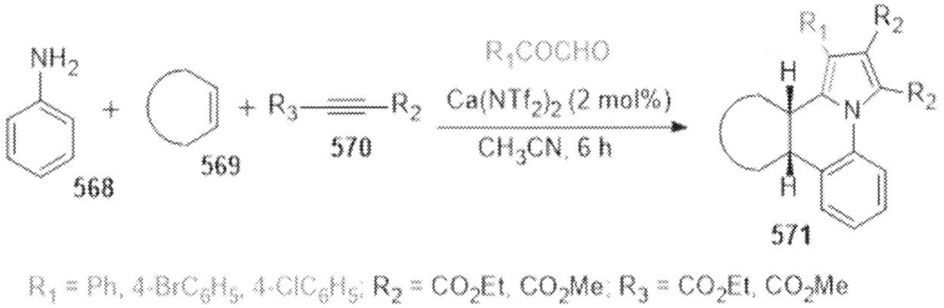

R_1 = Ph, 4-BrC$_6$H$_5$, 4-ClC$_6$H$_5$; R_2 = CO$_2$Et, CO$_2$Me, R_3 = CO$_2$Et, CO$_2$Me

FIGURE 3.176   Synthesis of nitro substituted dihydroindolizino[8,7-b]indoles.

3.4.33   Synthesis of α,β-epoxy-γ-lactams

Recently, Feng et al. reported[338] a new approach for the synthesis of chiral α,β-epoxy-γ-lactams **573** from glyoxals and α-bromo-β-esteramides or α-bromo-β-ketoamide **572** in presence of the catalytic amount of chiral N, N'-dioxide/Yb(III) complex in THF at 0°C−20°C. Substitution on glyoxal and α-bromo-β-esteramides or α-bromo-β-ketoamide does not affect this approach and a variety of chiral α, β-epoxy-γ-lactams **573** were synthesized in good yields with excellent selectivity (Fig. 3.177).

3.4.34   Synthesis of 3-functionalized indole

Da-Qing Shi and his group[339] described a MW-assisted, three-component domino reaction for the regioselective synthesis of 3-functionalized indole derivatives **576** from arylglyoxal monohydrates **1**, anilines **574**, and cyclic

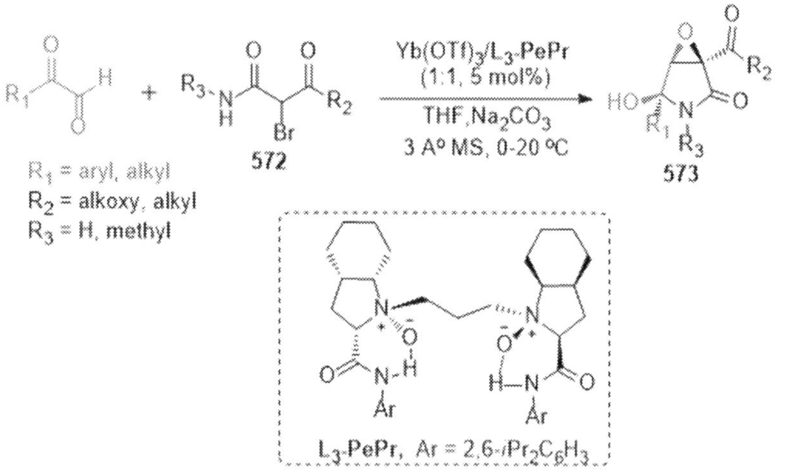

FIGURE 3.177 Synthesis of 4,5-dihydropyrrolo[1,2-a] quinolines.

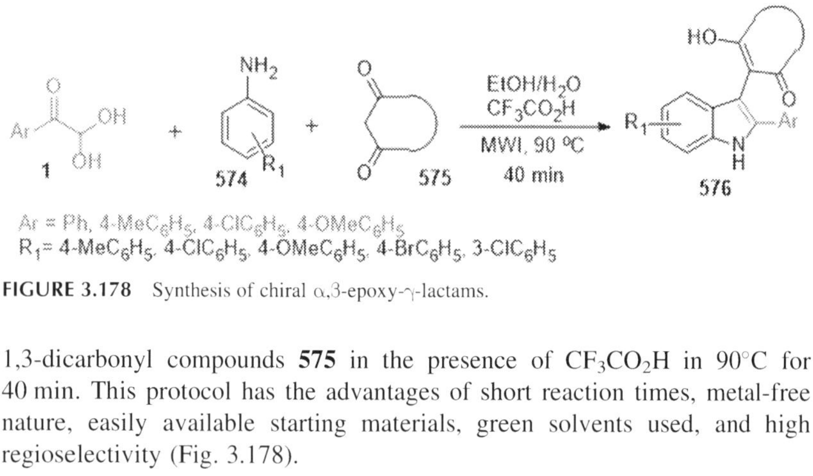

FIGURE 3.178 Synthesis of chiral α,3-epoxy-γ-lactams.

1,3-dicarbonyl compounds **575** in the presence of CF_3CO_2H in 90°C for 40 min. This protocol has the advantages of short reaction times, metal-free nature, easily available starting materials, green solvents used, and high regioselectivity (Fig. 3.178).

3.4.35 Synthesis of tetrahydrofuro[3,2- d]oxazoles and furanones

A MW-assisted, three-component reaction was developed by Ahmed and his group[340] for the synthesis of diastereoselective tetrahydrofuro[3,2-d]oxazole **579** and naphthofuranones **580**. Different reaction conditions were optimized for the selective synthesis of both products (Fig. 3.179). The synthesis of naphthofuranones was achieved by reacting arylglyoxal **1**, pyrrolidine **577**, and β-naphthol **578** in the presence of $Cu(OAc)_2$ in MW irradiation at 100°C for 15 min. On the other hand, tetrahydrofuro[3,2-d]oxazole **579** was

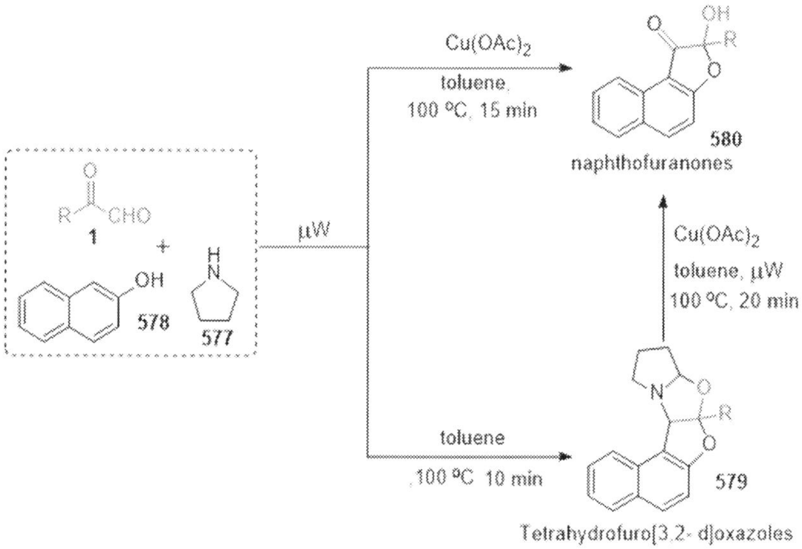

FIGURE 3.179 Synthesis of 3-functionalized indole derivatives.

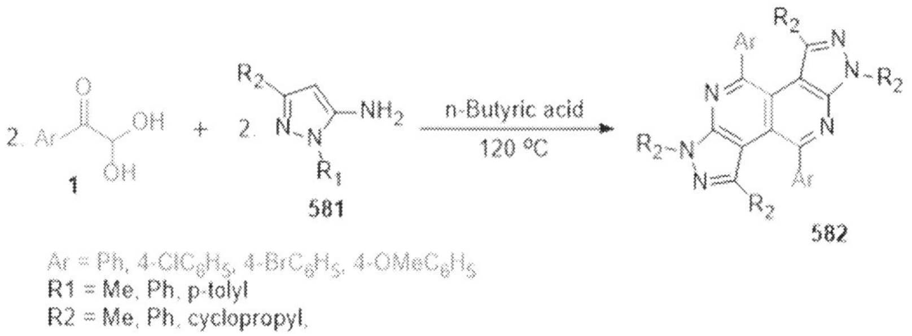

Ar = Ph, 4-ClC$_6$H$_5$, 4-BrC$_6$H$_5$, 4-OMeC$_6$H$_5$
R1 = Me, Ph, p-tolyl
R2 = Me, Ph, cyclopropyl.

FIGURE 3.180 Synthesis of diastereoselective tetrahydrofuro[3,2-d]oxazole and naphthofuranones.

obtained in metal-free conditions and is converted into naphthofuranones **580** by the addition of Cu(OAc)$_2$.

3.4.36   Synthesis of dipyrazolo-fused 2,6-naphthyridines

Tu et al. reported[341] the synthesis of tetracyclic dipyrazolo-fused 2,6-naphthyridines **582** from a four-component strategy of arylglyoxal monohydrate **1** and electron-rich pyrazol-5-amines **581**. This cascade reaction undergoes double [3 + 2 + 1] hetero annulation and involves an SN^2-type reaction/dipolymerization/6π-aza electrocyclization sequence. This multicomponent strategy shows the application for the construction of various target molecules of different biological activity and avoiding the use of transition metal catalysts (Fig. 3.180).

3.4.37 Synthesis of chiral 1,2-amino alcohols and morpholin-2-ones

Powell et al.[342] described a two-step protocol for the synthesis of chiral 1,2-amino alcohols **585** from arylglyoxals **1** and pseudoephedrine auxiliary **583**. Chiral 1,2-amino alcohols are important scaffolds with broad biological significances as drug candidates and are also used as chiral ligands for the synthesis of different chiral compounds. Arylglyoxals when treated with pseudoephedrine auxiliary **583** in the presence of a catalytic amount of trifluoroacetic acid (TFA) and MgSO$_4$ in toluene at 75°C to obtain morpholinone products **584** in high yields and selectivities. The morpholine ring was transformed into 1,2-amino alcohols **585** by reacting with red Al in toluene for 18 h and then SOCl$_2$ in THF for 1 h followed by the addition of KCN in MeCN: DMSO (1:1) for 20 h (Fig. 3.181).

3.4.38 Synthesis of fused five- and six-membered N-heterocycles

Choudhury et al. reported[343] different MCRs for the synthesis of different fused five- or six-membered N-heterocycles from arylglyoxal, 4-hydroxycoumarin **586**, and different cyclic 1,3-C,N-binucleophiles (Fig. 3.182). These reactions were performed in acetic acid under MW conditions. Five-membered binucleophiles, like amino-pyrazole/oxazole (**587/591**) generate fused six-membered dihydropyridines **588** or pyridine derivatives **592**. In addition, when the reaction was conducted by using six-membered binucleophiles such as aminouracil,

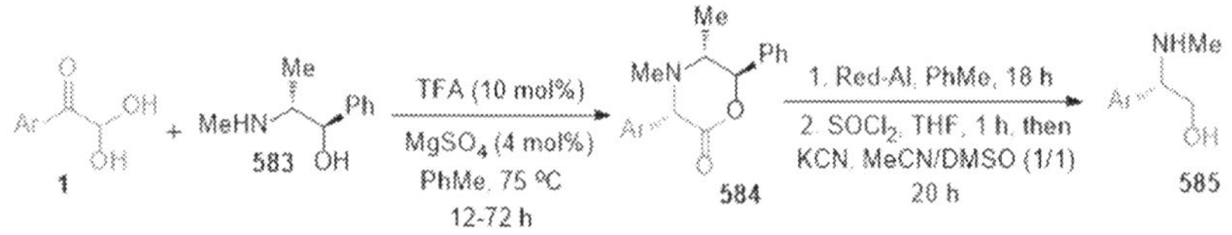

FIGURE 3.181 Synthesis of tetracyclic dipyrazolo-fused 2,6-naphthyridines.

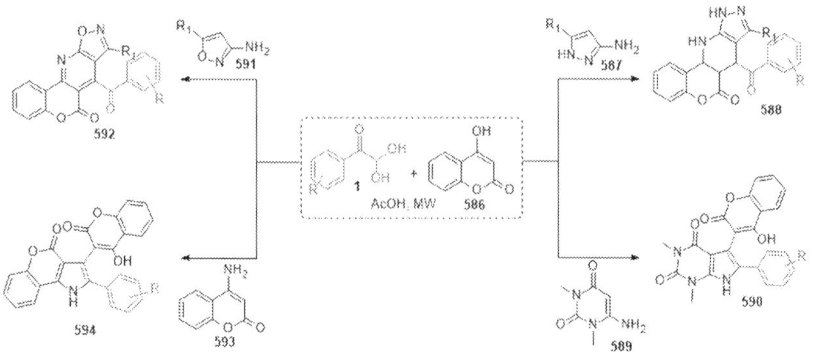

FIGURE 3.182 Synthesis of 1,2-amino alcohols and morpholin-2-ones.

cytosine **589**, and 4/3-aminocoumarines **593** corresponding five-membered fused pyrroles (**590** and **594**) were formed.

3.4.39 Synthesis of 2-acetamido-4-arylthiazol-5-yl derivatives

Alizadeh-Bami et al.[344] described an efficient approach for the construction of 2-acetamido-4-arylthiazol-5-yl derivatives (**597** and **599**) from a three-component reaction of arylglyoxals **1**, acetylthiourea **595**, and 1,3-dicarbonyl compounds such as Meldrum's acid **596** or barbituric acid **598** in the presence of triethylamine (Et$_3$N) in ethanol (Fig. 3.183).

3.4.40 Synthesis of diazepines and benzodiazepines

The 1,4-diazepine and 1,4-benzodiazepine are important scaffolds and are present in various biologically active compounds of different biological activities such as antimalarial,[345] antibiotic,[346] and anticancer agents.[347] In recent times, benzodiazepines are used as the prescribed minor tranquilizers that can act on the central nervous system. Different synthetic methods are present in the literature for the construction of these heterocycles by using AGs as one of the easily available starting materials in different optimized conditions.

Sañudo and coworkers reported[348] a two-step, multicomponent Ugi-reaction between AGs **1**, cyclohexyl isocyanide **600**, benzylamines **602**, and 2-azidobenzoic acid **601** for the synthesis of 5-oxobenzo[e][1,4]diazepine-3-carboxamides in MeOH at room temperature to obtain **602** in first step (45% − 90% yields) and in second step, the product **602** was treated with Ph$_3$P (1.5 equivalent) in toluene at room temperature under nitrogen atmosphere to obtain the desired product **603** in 59% − 99% yields (Fig. 3.184).

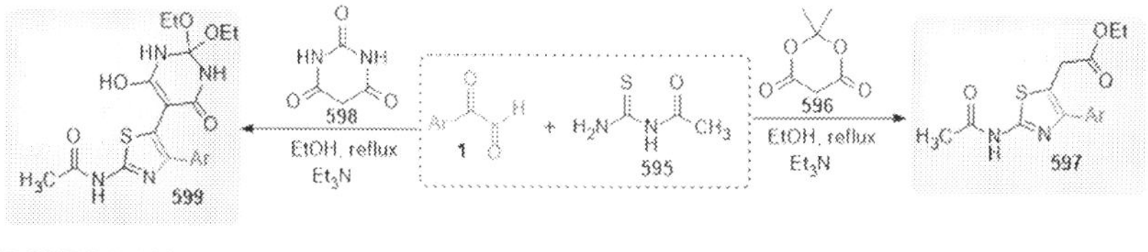

FIGURE 3.183 Synthesis of different fused five- or six-membered N-heterocycles.

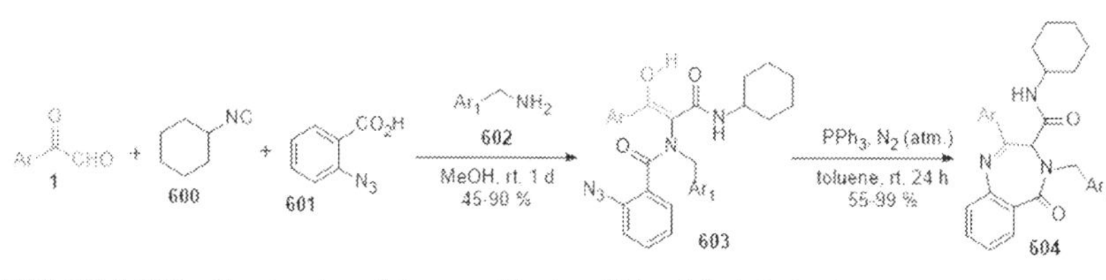

FIGURE 3.184 Construction of 2-acetamido-4-arylthiazol-5-yl derivatives.

Similar approach was established for the synthesis of enantiomerically pure tetrahydro-1,4-diazepine-3-carboxamides **607** by using 3-azido-(S)-2-(*N*-boc-amino)propanoic acid **605** in 61% − 98% yields via an intermediate **606** (Fig. 3.185).[349]

3.4.41 Synthesis of 1,4-oxazepane

1,4-Oxazepanes are showing a wide variety of biological activities and are present in different natural products.[350] Different synthetic procedures are available in the literature for the synthesis of oxazepane and their derivatives.

A reaction between **608** and PhSeCl, and SnCl$_4$ was conducted at −78°C for 2 h in which a mixture of two products **609** and **610** in the ratio of 93/7 was obtained. The product **610** was transformed into 1,4-oxazepane **611** by treating with AlCl$_3$ and LiAlH$_4$ under the nitrogen atmosphere at 0°C in 90% yield (Fig. 3.186).[351]

3.4.42 Synthesis of oxadiazines

Oxadiazines are biologically important heterocycles containing two nitrogen atoms and one oxygen atom in a ring. In some cases, Oxadiazines also behave as a synthetic intermediate for the synthesis of different biologically active heterocyclic compounds.[352]

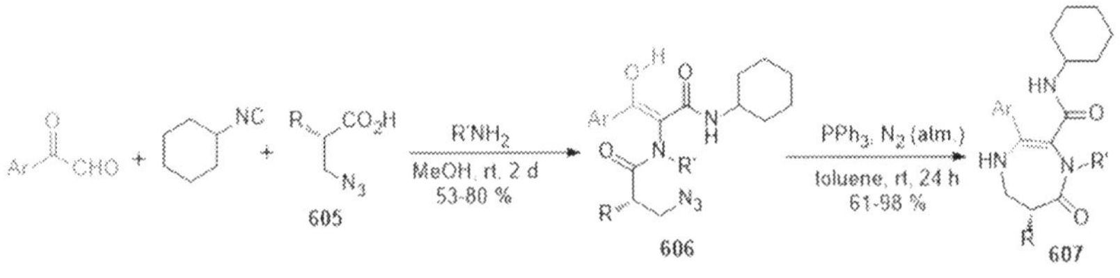

FIGURE 3.185 Synthesis of 5-oxobenzo[e][1,4]diazepine-3-carboxamides.

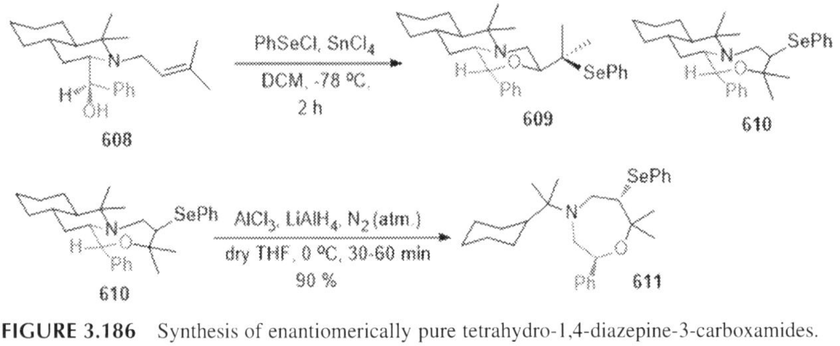

FIGURE 3.186 Synthesis of enantiomerically pure tetrahydro-1,4-diazepine-3-carboxamides.

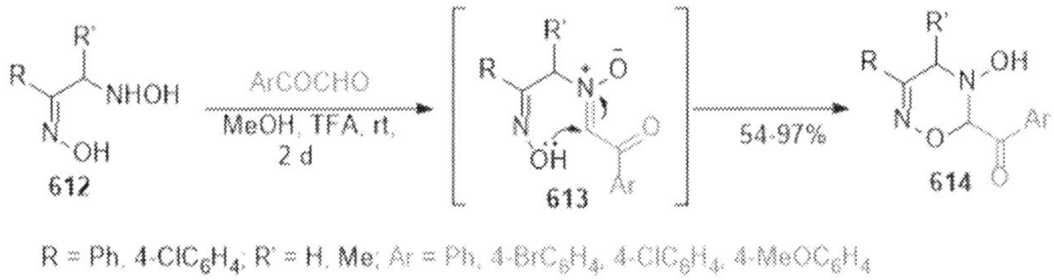

FIGURE 3.187   Synthesis of 1,4-oxazepane.

One of the simplest approaches for the synthesis of substituted oxadiazines **614** was established by Amitina and coworkers[144] by conducting a reaction between AGs and Z-isomers of hydroxylamino oximes **612** in the presence of TFA in MeOH at room temperature for 2 days. The product 6-aroyl-5-hydroxy-5,6-dihydro-4H-1,2,5-oxadiazines **614** was obtained via intermediate **613** in 54% − 97% yields (Fig. 3.187).

References

1.   Eftekhari-Sis, B.; Zirak, M.; Akbari, A. Arylglyoxals in synthesis of heterocyclic compounds. *Chem. Rev.* **2013,** *113,* 2958−3043.

2.   (a) Zhang, Q.; Xu, C.-M.; Chen, J.-X.; Xu, X.-L.; Ding, J.-C.; Wu, H.-Y. *Appl. Organomet. Chem.* **2009,** *23,* 524.
     (b) Blay, G.; Hernandez-Olmos, V.; Pedro, J. R. *Chem.-Eur. J.* **2011,** *17,* 3768.

3.   (a) Schmitt, E.; Schiffers, I.; Bolm, C. *Tetrahedron Lett.* **2009,** *50,* 3185.
     (b) Kobayashi, S.; Araki, M.; Yasuda, M. *Tetrahedron Lett.* **1995,** *36,* 5773.
     (c) Itoh, T.; Nagata, K.; Miyazaki, M.; Ishikawa, H.; Kurihara, A.; Ohsawa, A. *Tetrahedron* **2004,** *60,* 6649.

4.   Trimurtulu, G.; Ohtani, I.; Patterson, G. M. L.; Moore, R. E.; Corbett, T. H.; Valeriote, F. A.; Demchik, L. J. *Am. Chem. Soc.* **1994,** *116,* 4729.

5.   (a) Hodgkinson, T. J.; Shipman, M. *Tetrahedron* **2001,** *57,* 4467.
     (b) Coleman, R. S.; Perez, R. J.; Burk, C. H.; Navarro, A. *J. Am. Chem. Soc.* **2002,** *124,* 13008.
     (c) Narylglyoxaloaka, K.; Matsumoto, M.; Oono, J.; Yokoi, K.; Ishizeki, S.; Nakashima, T. *J. Antibiot.* **1986,** *39,* 1527.

6.   Kupchan, S. M.; Court, W. A.; Dailey, R. G., Jr; Gilmore, C. J.; Bryan, R. F. *J. Am. Chem. Soc.* **1972,** *94,* 7194.

7.   Hanada, M.; Sugawara, K.; Kaneta, K.; Toda, S.; Nishiyama, Y.; Tomita, K.; Yamamoto, H.; Konishi, M.; Oki, T. *J. Antibiot.* **1992,** *45,* 1746.

8.   (a) Kupchan, S. M.; Streelman, D. R.; Sneden, A. T. *J. Nat. Prod.* **1980,** *43,* 296.
     (b) Cassady, J. M. *J. Nat. Prod.* **1990,** *53,* 23.

9.   (a) Kupchan, S. M.; Schubert, R. M. *Science* **1974,** *185,* 791.
     (b) Carter, B. Z.; Mak, D. H.; Schober, W. D.; McQueen, T.; Harris, D.; Estrov, Z.; Evans, R. L.; Andreeff, M. *Blood* **2006,** *108,* 630.

10. Zheng, Y. L.; Lin, J. F.; Lin, C. C.; Xu, Y. *Acta Pharmacol. Sin.* **1994,** *15,* 540.

11. (a) Yang, S. X.; Gao, H. L.; Xie, S. S.; Zhang, W. R.; Long, Z.-Z. *Int. J. Immunopharmacol.* **1992**, *14*, 963.
    (b) Gu, W. Z.; Chen, R.; Brandwein, S.; McAlpine, J.; Burres, N. *Int. J. Immunopharmacol.* **1995**, *17*, 351.
    (c) Qiu, D. M.; Zhao, G. H.; Aoki, Y.; Shi, L. F.; Uyei, A.; Nazarian, S.; Ng, J. C.-H.; Kao, P. N. *J. Biol. Chem.* **1999**, *274*, 13443.
12. (a) Liou, S. S.; Shieh, W. L.; Cheng, T. H.; Won, S. J.; Lin, C. N. *J. Pharm. Pharmacol.* **1993**, *45*, 791.
    (b) Woo, S.; Jung, J.; Lee, C.; Kwon, Y.; Na, Y. *Bioorg. Med. Chem. Lett.* **2007**, *17*, 1163.
13. (a) Xia, Q.-H.; Ge, H.-Q.; Ye, C.-P.; Liu, Z.-M.; Su, K.-X. *Chem. Rev.* **2005**, *105*, 1603.
    (b) Bandini, M.; Cozzi, P. G.; Melchiorre, P.; Umani-Ronchi, A. *J. Org. Chem.* **2002**, *67*, 5386.
    (c) Bandini, M.; Cozzi, P. G.; Melchiorre, P.; Umani-Ronchi, A. *Angew. Chem. Int. Ed.* **2004**, *43*, 84.
    (d) Bertolini, F.; Crotti, P.; Macchia, F.; Pineschi, M. *Tetrahedron. Lett.* **2006**, *47*, 61.
14. (a) Tung, R. D.; Murcko, M. A.; Bhisetti, G. R. *United States Patent 5,558,397* **1996**.
    (b) Bold, G., et al. *J. Med. Chem.* **1998**, *41*, 3387.
15. (a) Yamarylglyoxaluchi, K.; Mori, K.; Mizugaki, T.; Ebitani, K.; Kaneda, K. *J. Org. Chem.* **2000**, *65*, 6897.
    (b) Fraile, J. M.; García, J. I.; Mayoral, J. A.; Sebti, S.; Tahir, R. *Green Chem* **2001**, *3*, 271.
    (c) Lee, S.; MacMillan, D. W. C. *Tetrahedron* **2006**, *62*, 11413 and references cited therein.
    (d) Sheldon, R. A. *Green Chem.* **2007**, *9*, 1273.
16. Weissermel, K.; Arpe, H.-J. completely revised *Industrial Organic Chemistry*, 4th ed.; Wiley-VCH: Weinheim, Germany, 2003.
17. (a) Solladie-Cavallo, A.; Boue' rat, L.; Roje, M. *Tetrahedron Lett.* **2000**, *41*, 7309.
    (b) Bellenie, B. R.; Goodman, J. M. *J. Chem. Soc. Chem. Commun.* **2004**, 1076.
    (c) Arylglyoxalgarwal, V. K.; Ford, J. G.; Thompson, A.; Jones, R. V. H.; Standen, M. C. H. *J. Am. Chem. Soc.* **1996**, *118*, 7004.
18. Fuson, R.; Johnson, R. *J. Am. Chem. Soc.* **1946**, *68*, 1668.
19. (a) Shong, P. D.; Kell, D. A.; Sidler, D. R. *J. Org. Chem* **1985**, *50*, 2309.
    (b) Shong, P. D.; Kell. *Tetrahedron Lett.* **1986**, *27*, 3979.
    (c) Irie, O.; Samizu, K.; Henry, J. R.; Weinreb, S. M. *J. Org. Chem.* **1999**, *64*, 587.
20. (a) Yudin, A. K. *Aziridines and Epoxides in Organic Synthesis; (Ed.);* Wiley-VCH: Weinheim, Germany, 2006.
    (b) Padwa, A. In *Comprehensive Heterocyclic Chemistry III;* Katritzky, A. R., Ramsden, C. A., Scriven, E. F. V., Taylor, R. J. K., Eds.; Vol. 1; Elsevier: Oxford, 2008; pp 1–104.
21. Majumdar, K. C.; Chattopadhyay, S. K. ⓒ, Eds. *Heterocycles in Natural Product Synthesis;* First Edition Wiley-VCH Verlarylglyoxal GmbH & Co. KGaA, 2011Published 2011 by Wiley-VCH Verlarylglyoxal GmbH & Co. KGaA Candice Botuha, Fabrice Chemla, Franck Ferreira and Alejandro P é rez – Luna, Aziridines in Natural Product Synthesis.
22. (a) Katarylglyoxaliri, T.; Takahashi, M.; Fujiwara, Y.; Ihara, H.; Uneyama, K. *J. Org. Chem.* **1999**, *64*, 7323.
    (b) Chuang, T.-H.; Sharpless, K. B. *Org. Lett.* **2000**, *2*, 3555.
    (c) Wu, J.; Hou, X.-L.; Dai, L.-X. *J. Chem. Soc. Perkin. Trans.* **2001**, *1*, 1314.
23. Cardillo, G.; Gentilucci, L.; Tolomelli, A. *Aldrichim. Acta* **2003**, *36*, 39.

24. (a) Watson, I. D. G.; Yu, L.; Yudin, A. K. *Acc. Chem. Res.* **2006**, *39*, 194.
    (b) Kumar, K. S. A.; Chaudhari, V. D.; Puranik, V. G.; Dhavale, D. D. *Eur. J. Org. Chem.* **2007**, 4895.
25. (a) Sweeney, J. B. *Chem. Soc. Rev.* **2002**, *31*, 247.
    (b) Tanner, D. *Angew. Chem. Int. Ed.* **1994**, *33*, 599.
26. (a) Benbow, J. W.; McClure, K. F.; Danishefsky, S. J. *J. Am. Chem. Soc.* **1993**, *115*, 12305.
    (b) Gerhart, F.; Higgins, W.; Tardif, C.; Ducep, J. B. *J. Med. Chem.* **1990**, *33*, 2157.
    (c) Sweeney, J. B. *Chem. Soc. Rev.* **2002**, *31*, 247.
27. (a) Kasai, M.; Kono, M. *Synlett* **1992**, 778.
    (b) Hodgkinson, T. J.; Shipman, M. *Tetrahedron* **2001**, *57*, 4467.
28. (a) Skibo, E. B.; Islam, I.; Heileman, M. J.; Schulz, W. G. *J. Med. Chem.* **1994**, *37*, 78.
    (b) Moonen, K.; Laureyn, I.; Stevens, C. V. *Chem. Rev.* **2004**, *104*, 6177.
    (c) Park, C. S.; Choi, H. G.; Lee, H.; Lee, W. K.; Ha, H.-J. *Tetrahedron: Asymmetry* **2000**, *11*, 3283.
29. Narylglyoxalanawa, H.; Usui, N.; Takita, T.; Hamada, M.; Umezawa, H. (S)-2,3-Dicarboxy-aziridine, a new metabolite from a steptomyces. *J. Antibiot.* **1975**, *28*, 828−829.
30. Molinski, T. F.; Ireland, C. M. Dysidazirine, a cytotoxic azacyclopropene from the marine sponge Dysidea frarylglyoxalilis. *J. Org. Chem.* **1988**, *53*, 2103−2105.
31. Iwami, M.; Kiyoto, S.; Terano, H.; Kohsaka, M.; Aoki, H.; Imanaka, H. A new antitumor antibiotic, FR—900482. *J. Antibiot.* **1987**, *40*, 589−593.
32. Akiyama, T.; Suzuki, T.; Mori, K. *Org. Lett.* **2009**, *11*, 2445.
33. Narylglyoxalayama, S.; Kobayashi, S. *Chem. Lett.* **1998**, 685.
34. Nagaraju, M.; Shahnawaz, K.; Satyanarayana, B.; Manoj, K.; Ajai, P. G.; Qazi, N. A.; Ram, A. V. *Org. Lett.* **2014**, *16*, 1152−1155.
35. Anil, K. P.; Nagaraju, M.; Deepika, S.; Ram, A. V.; Qazi, N. A. *Eur. J. Org. Chem.* **2015**, 3577−3586.
36. (a) Wu, X.; Gao, Q.; Liu, S.; Wu, A. *Org. Lett.* **2014**, *16*, 2888.
    (b) Al-Rashid, Z. F.; Johnson, W. L.; Hsung, R. P.; Wei, Y.; Yao, P.-Y.; Liu, R.; Zhao, K. *J. Org. Chem.* **2008**, *73*, 8780.
37. Nagaraju, M.; Narsaiah, B.; Satyanarayana, B.; Shahnawaz, K.; Ram, A. V.; Qazi, N. A. *Chem. Eur. J.* **2014**, *20*, 1−8.
38. Zhang, C.; Moran, E. J.; Woiwode, T. F.; Short, K. M.; Mjalli, A. M. M. *Tetrahedron Lett.* **1996**, *37*, 751.
39. Moreno, E.; Ancizu, S.; Perez-Silanes, S.; Torres, E.; Aldana, I.; Monge, A. *Eur. J. Med. Chem.* **2010**, *45*, 4418.
40. Sung, K.; Wu, S.-H.; Chen, P.-I. *Tetrahedron* **2002**, *58*, 5599.
41. Khalili, B.; Tondro, T.; Hashemi, M. M. *Tetrahedron* **2009**, *65*, 6882.
42. Amitina, S. A.; Tikhonov, A. Y.; Grigorev, I. A.; Gatilov, Y. V.; Selivanov, B. A. *Chem. Heterocycl. Compd.* **2009**, *45*, 691.
43. Anil, K. P.; Raju, R. K.; Athimoolam, S.; Qazi, N. A. *Org. Lett.* **2016**, *18* (1), 96−99.
44. Sadhanendu, S.; Susmita, M.; Sougata, S.; Golam, K.; Alakananda, H. *J. Org. Chem.* **2016**, *81*, 10088−10093.
45. Kinoshita, E.; Yamakoshi, J.; Kikuchi, M. *Biosci. Biotechnol. Biochem.* **1993**, *57*, 1107.
46. Fushiya, S.; Tanura, T.; Tashiro, T.; Nozoe, S. *Heterocycles* **1984**, *22*, 1039.
47. Sono, I. K.; Asahi, K.; Suzuki, S. *J. Am. Chem. Soc.* **1969**, *91*, 7490.
48. Erickson, B. I.; Carlsson, S.; Halvarsson, M.; Risberg, B.; Mattson, C. *Thromb. Haemost.* **1997**, *78*, 1404.

49. Kumar, A.; Rajput, C. S.; Bhati, S. K. *Bioorg. Med. Chem.* **2007**, *15*, 3089.
50. Gurupadayya, B. M.; Gopal, M.; Padmashali, B.; Manohara, Y. N. *Indian J. Pharm. Sci.* **2008**, *70*, 572.
51. (a) Veinverg, G.; Shestakova, I.; Vorona, M.; Kanepe, I.; Lukevics, E. *Bioorg. Med. Chem. Lett.* **2004**, *14*, 147.
    (b) Banik, B. K.; Becker, F. F.; Banik, I. *Bioorg. Med. Chem.* **2004**, *12*, 2523.
52. (a) Chavan, A. A.; Pai, N. R. *Molecules* **2007**, *12*, 2467.
    (b) Mistry, K.; Desai, K. R. *Indian J. Chem. Sect. B: Org. Chem. Incl. Med. Chem.* **2006**, *45*, 1762.
53. Deiccolo, C. L. M.; Mata, E. G. *Tetrahedron Lett* **2004**, *45*, 4085.
54. Singh, G. S.; Mbukwa, E.; Pheko, T. *Arkivoc* **2007**, *ix*, 80.
55. (a) Georg, G. I., Ed. *The Organic Chemistry of β-Lactams;* VCH: New York, 1993.
    (b) Sammes, P. G. *Chem. Rev.* **1976**, *76*, 113.
56. Pedrosa, R.; Andres, C.; Nieto, J.; del Pozo, S. *J. Org. Chem.* **2005**, *70*, 1408.
57. Lynch, J. E.; Riseman, S. M.; Laswell, W. L.; Tschaen, D. M.; Volante, R. P.; Smith, G. B.; Shinkai, I. *J. Org. Chem.* **1989**, *54*, 3792.
58. Palomo, C.; Aizpurua, J. M.; Iturburu, M.; Urchegui, R. *J. Org. Chem.* **1994**, *59*, 240.
59. Tanaka, H.; Hai, A. K. M. A.; Sadakane, M.; Okumoto, H.; Torii, S. *J. Org. Chem.* **1994**, *59*, 3040.
60. Pearson, N. D.; Smale, T. C.; Southgate, R. *Tetrahedron Lett.* **1995**, *36*, 4493.
61. (a) Wang, Y.; Tennyson, R. L.; Romo, D. *Heterocycles* **2004**, *64*, 605.
    (b) Müller, H.-M.; Seebach, D. *Angew. Chem. Int. Ed.* **1993**, *32*, 477.
    (c) Pommier, A.; Pons, J.-M. *Synthesis* **1995**, 729.
    (d) Rieth, L. R.; Moore, D. R.; Lobkovsky, E. B.; Coates, G. W. *J. Am. Chem. Soc.* **2002**, *124*, 15239.
62. He, L.; Lv, H.; Zhang, Y.-R.; Ye, S. *J. Org. Chem.* **2008**, *73*, 8101.
63. (a) Notz, W.; Tanaka, F.; Barbas, C. F., III *Acc. Chem. Res.* **2004**, *37*, 580.
    (b) Felpin, F. X.; Lebreton, J. *Eur. J. Org. Chem.* **2003**, 3693.
64. (a) Calter, M. A.; Tretyak, O. A.; Flaschenriem, C. *Org. Lett.* **2005**, *7*, 1809.
    (b) Zhu, C.; Shen, X.; Nelson, S. G. *J. Am. Chem. Soc.* **2004**, *126*, 5352.
    (c) Schneider, C. *Angew Chem. Int. Ed.* **2002**, *41*, 744.
65. (a) Bianco, A.; Maggini, M.; Scorrano, G.; Toniolo, C.; Marconi, G.; Villani, C.; Prato, M. *J. Am. Chem. Soc.* **1996**, *118*, 4072.
    (b) Royles, B. J. L. *Chem. Rev.* **1995**, *95*, 1981.
66. (a) Seayad, J.; List, B. *Org. Biomol. Chem.* **2005**, *3*, 719.
    (b) Dalko, P. I.; Moisan, L. *Angew Chem. Int. Ed.* **2004**, *43*, 5138.
67. Edon, V.; David, T. S.; Jon, T. N. *J. Med. Chem.* **2014**, *57*, 10257–10274.
68. Kim, D.; Kowalchick, J. E.; Edmondson, S. D.; Mastracchio, A.; Xu, J. M.; Eiermann, G. J.; Leiting, B.; Wu, J. K.; Pryor, K.; Patel, R. A., et al. *Bioorg. Med. Chem. Lett* **2007**, *17*, 3373–3377.
69. Malherbe, P.; Ballard, T. M.; Ratn, H. Tachykinin neurokinin 3 receptor antagonists: a patent review. *Expert Opin. Ther. Pat.* **2011**, *21*, 637–655.
70. Lilli, A.; David, W.; Banner, J. B.; Katrin, G. Z.; Jacques, H.; Hans, H.; Walter, H.; Bernd, K.; Jean-Luc, M.; Michael, B. O., et al. *Bioorganic Med. Chem. Lett.* **2010**, *20*, 5313–5319.
71. Tsuge, O.; Uneo, K.; Kanemasa, S.; Yorozu, K. *Bull. Chem. Soc. Jpn.* **1986**, *59*, 1809.
72. Grigg, R.; Rankovic, Z.; Thornton-Pett, M.; Somasunderam, A. *Tetrahedron* **1993**, *49*, 8679.
73. Congde, H.; Yong, Y. J. *Org. Chem.* **2015**, *80*, 12704–12710.

74. Felluga, F.; Forzato, C.; Nitti, P.; Pitacco, G.; Valentin, E.; Zangrando, E. *J. Heterocycl. Chem.* **2010,** *47,* 664.

75. (a) Yasumoto, T.; Murata, M. *Chem. Rev.* **1993,** *93,* 1897.
    (b) Alali, F. Q.; Liu, X.; McLaughlin, J. L. *J. Nat. Prod.* **1999,** *62,* 504.
    (c) Collum, D. B.; McDonald, J. H., III.; Still, W. C. *J. Am. Chem. Soc.* **1980,** *102,* 2117 2118, 2120.
    (d) Yamada, O.; Ogasawara, K. *Synlett* **1995,** 427.
    (e) Paolucci, C.; Musiani, L.; Venturelli, F.; Fava, A. *Synthesis* **1997,** 1415.
    (f) Doyle, M. P.; Forbes, D. C.; Protopopova, M. N.; Stanley, S. A.; Vasbinder, M. M.; Xavier, K. R. *J. Org. Chem.* **1997,** *62,* 7210.

76. Garson, M. J. *Chem. Rev.* **1993,** *93,* 1699.

77. (a) Ireland, R. E.; Anderson, R. C.; Badoud, R.; Fitzsimmons, B. J.; McGarvey, G. J.; Thaisrivongs, S.; Wilcox, C. S. *J. Am. Chem. Soc.* **1983,** *105,* 1988.
    (b) Lord, M. D.; Negri, J. T.; Paquette, L. A. *J. Org. Chem.* **1995,** *60,* 191.
    (c) Maurer, B.; Hauser, A.; Ohloff, G. *Helv. Chim. Acta* **1980,** *63,* 2503.
    (d) Ottinger, H.; Soldo, T.; Hofmann, T. *J. Agric. Food. Chem.* **2001,** *49,* 5383.
    (e) Shin, S. S.; Byun, Y.; Lim, K. M.; Choi, J. K.; Lee, K.-W.; Moh, J. H.; Kim, J. K.; Jeong, Y. S.; Kim, J. Y.; Choi, Y. H., et al. *J. Med. Chem.* **2004,** *47,* 792.

78. (a) Semple, J. E.; Wang, P. C.; Lysenko, Z.; Joullie, M. M. *J. Am. Chem. Soc.* **1980,** *102,* 7505.
    (b) Fraga, B. M. *Nat. Prod. Rep.* **1992,** *9,* 217.
    (c) Benassi, R. In *Compr Heterocycl Chem II;* Katritzky, A. R., Rees, C. W., Scrivan, E. F. V., Bird, C. W., Eds.; Vol. 2; Elsevier: Oxford, 1996; p 259.
    (d) Ward, R. S. *Nat. Prod. Rep.* **1999,** *16,* 75.

79. (a) Alvarez, E.; Candenas, M.-L.; Perez, R.; Ravelo, J. L.; Martin, J. D. *Chem. Rev.* **1995,** *95,* 1953.
    (b) Sekido, M.; Aoyagi, K.; Nakamura, H.; Kabuto, C.; Yamamoto, Y. *J. Org. Chem.* **2001,** *66,* 7142.
    (c) Sinha, S. C.; Sinha, A.; Sinha, S. C.; Keinan, E. *J. Am. Chem. Soc.* **1997,** *119,* 12014.
    (d) Sinha, S. C.; Keinan, E.; Sinha, S. C. *J. Am. Chem. Soc.* **1998,** *120,* 9076.
    (e) Lipshutz, B. H.; Barton, J. C. *J. Am. Chem. Soc.* **1992,** *114,* 1084.
    (f) Mihelich, E. D.; Hite, G. A. *J. Am. Chem. Soc.* **1992,** *114,* 7318.
    (g) Chen, J.; Song, Q.; Li, P.; Guan, H.; Jin, X.; Xi, Z. *Org. Lett.* **2002,** *4,* 2269.
    (h) Kang, S.-K.; Baik, T.-G.; Kulak, A. N. *Synlett* **1999,** 324.
    (i) Walkup, R. D.; Guan, L.; Mosher, M. D.; Kim, S. W.; Kim, Y. S. *Synlett* **1993,** 88.
    (j) Yoneda, E.; Kaneko, T.; Zhang, S.-W.; Onitsuka, K.; Takahashi, S. *Org. Lett.* **2000,** *2,* 441.
    (k) Eom, D.; Kang, D.; Lee, P. H. *J. Org. Chem.* **2010,** *75,* 7447.
    (l) Ichikawa, J.; Fujiwara, M.; Wada, Y.; Okauchi, T.; Minami, T. *Chem. Commun.* **2000,** 1887.
    (m) Feldman, K. S.; Wrobleski, M. L. *J. Org. Chem.* **2000,** *65,* 8659.

80. Fuchibe, K.; Aoki, Y.; Akiyama, T. *Chem. Lett.* **2005,** *34,* 538.

81. Mosslemin, M. H.; Anary-Abbasinejad, M.; Fazli Nia, A.; Bakhtiari, S.; Anaraki-Ardakani, H. *J. Chem. Res.* **2009,** 599.

82. Kadhim, S. A.; Bowlin, T. L.; Waud, W. R.; Angers, E. G.; Bibeau, L.; DeMuys, J.-M.; Bednarski, K.; Cimpoia, A.; Attardo, G. *Cancer Res* **1997,** *57,* 4803.

83. Bednarski, K.; Dixit, D. M.; Wang, W.; Evans, C. A.; Jin, H.; Yuen, L.; Mansour, T. S. *Bioorg. Med. Chem. Lett.* **1994,** *4,* 2667.

84. (a) Colobert, F.; Obringer, M.; Solladie, G. *Eur. J. Org. Chem.* **2006,** 1455.
    (b) Williams, D. B. G.; Caddy, J.; Blann, K. *Carbohydr. Res.* **2005,** *340*, 1301.
    (c) Kim, H. O.; Schinazi, R. F.; Shanmuganathan, K.; Jeong, L. S.; Beach, J. W.; Nampalli, S.; Cannon, D. L.; Chu, C. K. *J. Med. Chem.* **1993,** *36*, 519.
85. Clerici, A.; Porta, O. *J. Org. Chem.* **1989,** *54*, 3872.
86. Shao, Q.; Li, C. *Synlett* **2008,** 2412.
87. Agmi, C.; Couty, F.; Prince, B.; Venier, O. *Tetrahedron Lett.* **1993,** *34*, 7061.
88. Shao, Q.; Shi, W.; Li, C. *Synthesis* **2008,** 3925.
89. Pinho e Melo, T. M. V. D.; Soares, M. I. L.; d'A. Rocha Gonsalves, A. M.; Paixao, J. A.; Beja, A. M.; Silva, M. R.; da Veiga, L. A.; Pessoa, J. C. *J. Org. Chem.* **2002,** *67*, 4045.
90. (a) Liu, J.; Steigel, A.; Reininger, E.; Bauer, R. *J. Nat. Prod.* **2000,** *63*, 403–405.
    (b) Ulubelen, A.; Tuzlaci, E.; Atilan, N. *Phytochemistry* **1989,** *28*, 649–650.
91. Talinli, N.; Karliga, B. *J. Heterocycl. Chem.* **2004,** *41*, 205.
92. (a) Kong, Y. C.; Kim, K. *J. Heterocycl. Chem.* **1999,** *36*, 911.
    (b) Williams, R. M.; Glinka, T.; Flanarylglyoxalan, M. E.; Gallegos, R.; Coffman, H.; Pei, D. *J. Am. Chem. Soc.* **1992,** *114*, 733.
    (c) Tomita, F.; Takahashi, K.; Tamaoki, T. *J. Antibiot.* **1984,** *37*, 1268.
93. (a) Williams, R. M.; Glinka, T.; Flanarylglyoxalan, M. E.; Gallegos, R.; Coffman, H.; Pei, D. *J. Am. Chem. Soc.* **1992,** *114*, 733.
    (b) Tomita, F.; Takahashi, K.; Tamaoki, T. *J. Antibiot.* **1984,** *37*, 1268.
94. Russell, O. H.; Pelak, B. A.; Lynn, S. G.; Lynn, L. S.; Frederick, M. K.; Meng-H, C.; Arthur, A. P.; Susan, M. G.; Sheryl, A. H.; Matt, S. A., et al. *Science* **1996,** *274* (5289), 980–982.
95. Arai, T.; Ogino, Y.; Sato, T. *Chem. Commun.* **2013,** *49*, 7776.
96. (a) Pirrung, M. C.; Tumey, L. N. *J. Comb. Chem.* **2000,** *2*, 675.
    (b) Wuts, P. G. M.; Northuis, J. M.; Kwan, T. A. *J. Org. Chem.* **2000,** *65*, 9223.
    (c) Rajaram, S.; Sigman, M. S. *Org. Lett.* **2002,** *4*, 3399.
    (d) Michaelis, D. J.; Ischay, M. A.; Yoon, T. P. *J. Am. Chem. Soc.* **2008,** *130*, 6610.
    (e) Michaelis, D. J.; Williamson, K. S.; Yoon, T. P. *Tetrahedron* **2009,** *65*, 5118.
    (f) Trost, B. M.; Jiang, C. *J. Am. Chem. Soc.* **2001,** *123*, 12907.
97. Agami, C.; Couty, F.; Prince, B.; Venier, O. *Tetrahedron Lett.* **1993,** *34*, 7061.
98. Yang, X.-L.; Xu, C.-M.; Lin, S.-M.; Chen, J.-X.; Ding, J.-C.; Wu, H.-Y.; Su, W.-K. *J. Braz. Chem. Soc.* **2010,** *21*, 37.
99. Hyvl, J.; Srogl, J. *Eur. J. Org. Chem.* **2010,** 2849.
100. Werber, G.; Buccheri, F.; Marino, M. L. *J. Heterocycl. Chem.* **1975,** *12*, 581.
101. Werber, G.; Buccheri, F.; Vivona, N.; Gentile, M. *J. Heterocycl. Chem.* **1977,** *14*, 1433.
102. (a) Aridoss, G.; Amirthagnesan, S.; Jeong, Y. T. *Bioorg. Med. Chem. Lett.* **2010,** *20*, 2242.
     (b) Ishida, J.; Hattori, K.; Yamamoto, H.; Iwashita, A.; Mihara, K.; Matsuoka, N. *Bioorg. Med. Chem. Lett.* **2005,** *15*, 4221.
     (c) Suleman, N. K.; Flores, J.; Tanko, J. M.; Isin, E. M.; Castagnoli, N., Jr. *Bioorg. Med. Chem.* **2008,** *16*, 8557.
     (d) Roberts, E.; Sancon, J. P.; Sweeney, J. B.; Workman, J. A. *Org. Lett.* **2003,** *5*, 4775.
103. (a) Tsukamoto, H.; Kondo, Y. *Angew Chem. Int. Ed.* **2008,** *47*, 4851.
     (b) Spiegel, D. A.; Schroeder, F. C.; Duvall, J. R.; Schreiber, S. L. *J. Am. Chem. Soc.* **2006,** *128*, 14766.
     (c) Han, R.-G.; Wang, Y.; Li, Y.-Y.; Xu, P.-F. *Adv. Synth. Catal.* **2008,** *350*, 1474 1435.
     (d) Aridoss, G.; Amirthagnesan, S.; Jeong, Y. T. *Bioorg. Med. Chem. Lett.* **2010,** *20*, 2242.

(e) Misra, M.; Pandey, S. K.; Pandey, V. P.; Pandey, J.; Tripathi, R.; Tripathi, R. P. *Bioorg. Med. Chem.* **2009**, *17*, 625.

(f) Khan, A. T.; Parvin, T.; Choudhury, L. H. *J. Org. Chem.* **2008**, *73*, 8398.

104. (a) Boger, D. L.; Weinreb, S. M. *Hetero Diels-Alder Methodology in Organic Synthesis;* Academic Press: Orlando, 1987.

(b) Weinreb, S. M. *Acc. Chem. Res.* **1985**, *18*, 16.

(c) Birkinshaw, T. N.; Tabor, A. B.; Holmes, A. B.; Kaye, P.; Mayne, P. M.; Raithby, P. R. *J. Chem. Soc. Chem. Commun.* **1988**, 1599.

105. Kobayashi, S.; Ishitani, H.; Nagyama, S. *Chem. Lett.* **1995**, 423.

Kobayashi, S.; Ishitani, H.; Nagyama, S. *Synthesis* **1995**, 1195.

106. Chou, S.-S. P.; Hung, C.-C. *Synth. Commun.* **2001**, *31*, 1097.

107. (a) Michael, J. P. *Nat. Prod. Rep.* **2008**, *25*, 166.

(b) Bazin, M.; Kuhn, C. *J. Comb. Chem.* **2005**, *7*, 302.

(c) Michael, J. P. *Nat. Prod. Rep.* **2004**, *21*, 650.

(d) Michael, J. P. *Nat. Prod. Rep.* **2007**, *24*, 223.

(e) Chen, Y.-L.; Fang, K.-C.; Sheu, J.-Y.; Hsu, S.-L.; Tzeng, C.-C. *J. Med. Chem.* **2001**, *44*, 2374.

108. (a) Sridharan, V.; Suryavanshi, P. A.; Menéndez, J. C. *Chem. Rev.* **2011**, *111*, 7157−7259.

(b) Matada, B. S.; Pattanashettar, R.; Yernale, N. G. *Bioorganic & Medicinal Chemistry* **2021**, *32*, 115973.

109. (a) Yearick, K.; Ekoue-Kovi, K.; Iwaniuk, D. P.; Natarajan, J. K.; Alumasa, J.; de Dios, A. C.; Roepe, P. D.; Wolf, C. *J. Med. Chem.* **2008**, *51*, 1995.

(b) Ramesh, E.; Manian, R. D. R. S.; Raghunathan, R.; Sainath, S.; Raghunathan, M. *Bioorg. Med. Chem.* **2009**, *17*, 660.

(c) Gholap, A. R.; Toti, K. S.; Shirazi, F.; Kumari, R.; Bhat, M. K.; Deshpande, M. V.; Srinivasan, K. V. *Bioorg. Med. Chem.* **2007**, *15*, 6705.

(d) Smith, H. C.; Cavanaugh, C. K.; Friz, J. L.; Thompson, C. S.; Saggers, J. A.; Michelotti, E. L.; Garcia, J.; Tice, C. M. *Bioorg. Med. Chem. Lett.* **2003**, *13*, 1943.

(e) Samosorn, S.; Bremner, J. B.; Ball, A.; Lewis, K. *Bioorg. Med. Chem.*, 14. ; 2006857.

(f) Foley, M.; Tilley, L. *Pharmacol. Ther.* **1998**, *79*, 55.

(g) Gupta, M. K.; Prabhakar, Y. S. *Eur. J. Med. Chem.* **2008**, *43*, 2751.

(h) Dorey, G.; Lockhart, B.; Lestage, P.; Casara, P. *Bioorg. Med. Chem. Lett.* **2000**, *10*, 935.

110. He, J.; Lion, U.; Sattler, I.; Gollmick, F. A.; Grabley, S.; Cai, J.; Meiners, M.; Schunke, H.; Schaumann, K.; Dechert, U., et al. *J. Nat. Prod.* **2005**, *68*, 1397.

111. Saggiomo, A. J.; Kano, S.; Kikuchi, T.; Okubo, K.; Shinbo, M. *J. Med. Chem.* **1972**, *15*, 989.

112. Nimgirawath, S.; Kaewket, S. *J. Sci. Soc. Thail.* **1982**, *8*, 167.

113. Gilchrist, T. L.; d'A. Rocha Gonsalves, A. M.; Pinho e Melo, T. M. V. D. *Tetrahedron Lett.* **1993**, *34*, 4097.

114. Magesh, C. J.; Makesh, S. V.; Perumal, P. T. *Bioorg. Med. Chem. Lett.* **2004**, *14*, 2035.

115. Kumar, R. S.; Nagrajan, R.; Perumal, P. T. *Synthesis* **2004**, 949.

116. Balalaie, S.; Soleiman-Beigi, M.; Rominger, F. *J. Iran Chem. Soc.* **2005**, *2*, 319.

117. Oi, S.; Terada, E.; Ohuchi, K.; Kato, T.; Tachibana, Y.; Inoue, Y. *J. Org. Chem.* **1999**, *64*, 8660.

118. (a) Luo, H.-K.; Khim, L. B.; Schumann, H.; Lim, C.; Jie, T. X.; Yang, H.-Y. *Adv. Synth. Catal.* **2007**, *349*, 1781.

(b) Luo, H.-K.; Woo, Y.-L.; Schumann, H.; Jacob, C.; van Meurs, M.; Yang, H.-Y.; Tan, Y.-T. *Adv. Synth. Catal.* **2010**, *352*, 1356.

119.   Becker, J. J.; Van Orden, L. J.; White, P. S.; Gagne, M. R. *Org. Lett.* **2002**, *4*, 727.

120.   Tonoi, T.; Mikami, K. *Tetrahedron Lett* **2005**, *46*, 6355.

121.   (a) Shchepin, V. V.; Korzun, A. E.; Nedugov, A. N.; Sazhneva, Y. K.; Shurov, S. N. *Russ. J. Org. Chem.* **2002**, *38*, 248.

   (b) Shchepin, V. V.; Korzun, A. E.; Sazhneva, Y. K.; Nedugov, A. N. *Chem. Heterocycl. Compd.* **2001**, *37*, 374.

122.   (a) Lumma, W. C.; Ma, O. H. *J. Org. Chem.* **1970**, *35*, 2391.

   (b) Adams, D. R.; Bhatnagr, S. P. *Synthesis* **1977**, 661.

   (c) Arundale, E.; Mikeska, L. A. *Chem. Rev.* **1952**, *51*, 505.

   (d) El Gharbi, R.; Delmas, M.; Gaset, A. *Synthesis* **1981**, 361.

   (e) Tateiwa, J.-I.; Hashimoto, K.; Yamauchi, T.; Uemura, S. *Bull. Chem. Soc. Jpn.* **1996**, *69*, 2361.

   (f) Gu, Y.; Karam, A.; Jerô͂me, F.; Barrault, J. *Org. Lett.* **2007**, *9*, 3145.

   (g) Piergentili, A.; Quaglia, W.; Giannella, M.; Del Bello, F.; Bruni, B.; Buccioni, M.; Carrieri, A.; Ciattini, S. *Bioorg. Med. Chem.* **2007**, *15*, 886.

123.   Becerra-Martínez, E.; Velazquez-Ponce, P.; Sa´nchez-Aguilar, M. A.; Rodríguez-Hosteguín, A.; Joseph-Nathan, P.; Tamariza, J.; Zepeda, L. G. *Tetrahedron: Asymmetry* **2007**, *18*, 2727.

124.   Capperucci, A.; Degl'Innocenti, A.; Ricci, A.; Mordini, A.; Reginato, G. *J. Org. Chem.* **1991**, *56*, 7323.

125.   (a) Ingall, A. H. In *In Comprehensive Heterocyclic;* Chemistry, I. I., McKillop, A., Eds.; Vol. 5; Pergamon Press: Oxford, U.K, 1996; pp 501−617.

   (b) Al Nakib, T.; Bezjak, V.; Meegan, M.; Chandy, R. *Eur. J. Med. Chem.* **1990**, *25*, 455.

   (c) Al Nakib, T.; Bezjak, V.; Rashid, S.; Fullam, B.; Meegan, M. *Eur. J. Med. Chem.* **1991**, *26*, 221.

126.   (a) Yamamoto, K.; Yamazaki, S.; Osedo, H.; Murata, I. *Angew Chem. Int. Ed.* **1986**, *25*, 635.

   (b) Yamamoto, K.; Yamazaki, S.; Murata, I.; Fukazawa, Y. *J. Org. Chem.* **1987**, *52*, 5239.

127.   (a) Chen, C. H.; Reynolds, G. A. *J. Org. Chem.* **1980**, *45*, 2449.

   (b) Nakasuji, K.; Nakasuka, M.; Murata, I. *J. Chem. Soc. Chem. Commun.* **1981**, 1143.

   (c) Chen, C. H.; Reynolds, G. A.; Luss, H. R.; Perlstein, J. H. *J. Org. Chem.* **1986**, *51*, 3282.

   (d) Fox, J. L.; Chen, C. H.; Luss, H. R. *J. Org. Chem.* **1986**, *51*, 3551.

128.   Turgut, Z.; Pelit, E.; Koyeu, A. *Molecules* **2007**, *12*, 345.

129.   (a) Kurz, T. *Tetrahedron* **2005**, *61*, 3091.

   (b) Lesher, G. Y.; Surrey, A. R. *J. Am. Chem. Soc.* **1955**, *77*, 636.

130.   Adib, M.; Sheibani, E.; Mostofi, M.; Ghanbary, K.; Bijanzadeh, H. R. *Tetrahedron* **2006**, *62*, 3435.

131.   (a) De Poel, H. V.; Guillaumet, G.; Viaud-Massuard, M.-C. *Tetrahedron Lett.* **2002**, *43*, 1205.

   (b) Remillard, S.; Rebhun, L. I.; Howie, G. A.; Kupchan, S. M. *Science* **1975**, *189*, 1002.

132.   Cocuzza, A. J.; Chidester, D. R.; Cordova, B. C. ;; Jeffrey, S.; Parsons, R. L.; Bacheler, L. T.; Erickson-Viitanen, S.; Trainor, G. L.; Ko, S. S. *Bioorg. Med. Chem. Lett.* **2001**, *11*, 1177.

133.   Kajino, M.; Shibouta, Y.; Nishikawa, K.; Meguro, K. *Chem. Pharm. Bull.* **1991**, *11*, 2896.

134. (a) Sasaki, K.; Kusakabe, Y.; Esumi, S. *J. Antibiot.* **1972,** *25,* 151.
      (b) Kusakabe, Y.; Nagtsu, J.; Shibuya, M.; Kawaguchi, O.; Hirose, C.; Shirato, S. *J. Antibiot.* **1972,** *25,* 44.
      (c) Kupchan, S. M.; Komoda, Y.; Court, W. A.; Thomas, G. J.; Smith, R. M.; Karim, A.; Gilmore, C. J.; Haltiwanger, R. C.; Bryan, R. F. *J. Am. Chem. Soc.* **1972,** *94,* 1354.
135. Buckman, B. O.; Mohan, R.; Koovakkat, S.; Liang, A.; Trinh, L.; Morrissey, M. M. *Bioorg. Med. Chem. Lett.* **1998,** *8,* 2235.
136. Mosher, H. S.; Frankel, M. B.; Gregory, M. *J. Am. Chem. Soc.* **1953,** *75,* 5326.
137. Kerdesky, F. A. J. *Tetrahedron Lett.* **2005,** *46,* 1711.
138. Ko, K.-Y.; Park, J.-Y. *Tetrahedron Lett.* **1997,** *38,* 407.
139. Ko, K.-Y.; Yun, H. *Heterocycles* **2010,** *81,* 2351.
140. Polyak, F.; Dorofeeva, T.; Zelchans, G.; Shustov, G. *Tetrahedron Lett.* **1996,** *37,* 8223.
141. Pedrosa, R.; Andres, C.; Roso´n, C. D.; Vicente, M. *J. Org. Chem.* **2003,** *68,* 1852.
142. (a) Eliel, E. L.; He, X.-C. *Tetrahedron* **1987,** *43,* 4979.
      (b) Eliel, E. L.; He, X.-C. *J. Org. Chem.* **1990,** *55,* 2114.
143. Grivas, J. C. *J. Org. Chem.* **1976,** *41,* 1325.
144. Amitina, S. A.; Grigorev, I. A.; Mamatyuk, V. I.; Tikhonov, A. Y. *Russ. Chem. Bull.* **2007,** *56,* 1190.
145. Tolkunov, A. S.; Bogza, S. L. *Chem. Heterocycl. Compd.* **2010,** *46,* 711.
146. (a) Choi, M.-S.; Yamazaki, T.; Yamazaki, I.; Aida, T. *Angew Chem. Int. Ed.* **2004,** *43,* 150.
      (b) Aratani, N.; Kim, D.; Osuka, A. *Acc. Chem. Res.* **2009,** *42,* 1922.
      (c) Faiz, J. A.; Heitz, V.; Sauvage, J.-P. *Chem. Soc. Rev.* **2009,** *38,* 422.
      (d) Benniston, A. C.; Harriman, A.; Li, P. *J. Am. Chem. Soc.* **2010,** *132,* 26.
      (e) Balaban, T. S. *Acc. Chem. Res.* **2005,** *38,* 612.
147. (a) Jurow, M.; Schuckman, A. E.; Batteas, J. D.; Drain, C. M. *Coord. Chem. Rev.* **2010,** *254,* 2297.
      (b) Otsuki, J. *Coord. Chem. Rev.* **2010,** *254,* 2311.
      (c) Drain, C. M.; Varotto, A.; Radivojevic, I. *Chem. Rev.* **2009,** *109,* 1630.
      (d) Waskitoaji, W.; Hyakutake, T.; Kato, J.; Watanabe, M.; Nishide, H. *Chem. Lett.* **2009,** *38,* 1164.
148. (a) Peng, X.; Nakamura, Y.; Aratani, N.; Kim, D.; Osuka, A. *Tetrahedron Lett.* **2004,** *45,* 4981.
      (b) Wagner, R. W.; Lindsey, J. S.; Seth, J.; Palaniappan, V.; Bocian, D. F. *J. Am. Chem. Soc.* **1996,** *118,* 3996.
      (c) Campbell, W. M.; Burrell, A. K.; Officer, D. L.; Jolley, K. W. *Coord. Chem. Rev.,* 248. ; 20041363.
      (d) Xiang, N.; Liu, Y.; Zhou, W.; Huang, H.; Guo, X.; Tan, Z.; Zhao, B.; Shen, P.; Tan, S. *Eur. Polym. J.* **2010,** *46,* 1084.
149. (a) Montes, V. A.; Perez-Bolívar, C.; Agrwal, N.; Shinar, J.; Anzenbacher, P., Jr. *J. Am. Chem. Soc.* **2006,** *128,* 12436.
      (b) Wang, X.; Wang, H.; Yang, Y.; He, Y.; Zhang, L.; Li, Y.; Li, X. *Macromolecules* **2010,** *43,* 709.
150. Poon, C.-T.; Zhao, S.; Wong, W.-K.; Kwong, D. W. J. *Tetrahedron Lett.* **2010,** *51,* 664.
151. (a) Swamy, N.; James, D. A.; Mhor, S. C.; Hanson, R. N.; Ray, R. *Bioorg. Med. Chem.* **2002,** *10,* 3237.
      (b) Sol, V.; Lamarche, F.; Enache, M.; Garcia, G.; Granet, R.; Guilloton, M.; Blais, J. C.; Krausz, P. *Bioorg. Med. Chem.* **2006,** *14,* 1364.

(c) Ethirajan, M.; Chen, Y.; Joshi, P.; Pandey, R. K. *Chem. Soc. Rev.* **2011**, *40*, 340.

152. (a) Goldberg, I. *Chem. Commun.* **2005**, 1243.

    (b) Goldberg, I. *Cryst. Eng. Comm.* **2008**, *10*, 637.

    (c) Cuesta, L.; Sessler, J. L. *Chem. Soc. Rev.* **2009**, *38*, 2716.

    (d) Matano, Y.; Imahori, H. *Acc. Chem. Res.* **2009**, *42*, 1193.

153. (a) Pasternack, R. F.; Gibbs, E. J.; Villafranca, J. J. *Biochemistry* **1938**, *22*, 2406.

    (b) D'Urso, A.; Mammana, A.; Balaz, M.; Holmes, A. E.; Berova, N.; Lauceri, R.; Purrello, R. *J. Am. Chem. Soc.* **2009**, *131*, 2046.

154. (a) Wagner, R. W.; Lawrence, D. S.; Lindsey, J. S. *Tetrahedron Lett.* **1987**, *28*, 3069.

    (b) Dogutan, D. K.; Ptaszek, M.; Lindsey, J. S. *J. Org. Chem.* **2008**, *73*, 6187.

    (c) Lindsey, J. S. *Acc. Chem. Res.* **2010**, *43*, 300.

    (d) Milgrom, L. R. *J. Chem. Soc. Perkin Trans.* **1984**, *1*, 1483.

    (e) Harmjanz, M.; Bozidarevic, I.; Scott, M. *J. Org. Lett.* **2001**, *3*, 2281.

155. Nishino, N.; Wagner, R. W.; Lindsey, J. S. *J. Org. Chem.* **1996**, *61*, 7534.

156. Ivonin, S. P.; Lapandin, A. V.; Anishchenko, A. A.; Shtamburg, V. G. *Synth. Commun.* **2004**, *34*, 451.

157. Ivonin, S. P.; Mazepa, A. V.; Lapandin, A. V. *Chem. Heterocycl. Compd.* **2006**, *42*, 451.

158. Battini, N.; Padala, A. K.; Mupparapu, N.; Vishwakarma, R. A.; Ahmed, Q. N. *RSC Adv.* **2014**, *4*, 26258−26263.

159. Zhan, Z.; Cheng, X.; Zheng, Y.; Ma, X.; Wang, X.; Hai, L.; Wu, Y. *RSC Adv.* **2015**, *5*, 82800−82803.

160. Mupparapu, N.; Khan, S.; Battula, S.; Kushwaha, M.; Gupta, A. P.; Ahmed, Q. N.; Vishwakarma, R. A. *Org. Lett.* **2014**, *16*, 1152−1155.

161. (a) K Zeller, K. P.; Kowallik, M.; Haiss, P. *Org. Biomol. Chem.* **2005**, *3*, 2310−2318.

    (b) Marziano, N. C.; Ronchin, L.; Tortato, C.; Ronchin, S.; Vavasori, A. *J. Mol. Catal. A: Chem.* **2005**, *235*, 26−34.

    (c) Marziano, N. C.; Ronchin, L.; Tortato, C.; Zingales, A.; Scantamburlo, L. *J. Mol. Catal. A: Chem.* **2005**, *235*, 17−25.

    (d) Jin, S. J.; Arora, P. K.; Sayre, L. M. *J. Org. Chem.* **1990**, *55*, 3011−3018.

162. Li, X.-Y.; Yuan, W.-Q.; Tang, S.; Huang, Y.-W.; Xue, J.-H.; Fu, L.-N.; Guo, Q.-X. *Org. Lett.* **2017**, *19*, 1120−1123.

163. Huo, C.; Yuan, Y. *J. Org. Chem.* **2015**, *80*, 12704−12710.

164. Gong, J.; Peshkov, A. A.; Yu, J.; Amandykova, S.; Gimnkhan, A.; Huang, J.; Kashtanov, S.; Pereshivko, O. P.; Peshko, V. A. *RSC Adv.* **2020**, *10*, 10113−10117.

165. Battula, S.; Battini, N.; Singh, D.; Ahmed, Q. N. *Org. Biomol. Chem.* **2015**, *13*, 8637−8641.

166. Battini, N.; Battula, S.; Ahmed, Q. N. *Eur. J. Org. Chem.* **2016**, 658−662.

167. Palmieri, A.; Gabrielli, S.; Sampaolesi, S.; Ballini, R. *RSC Adv.* **2015**, *5*, 36652−36655.

168. Mupparapu, N.; Khushwaha, M.; Gupta, A. P.; Singh, P. P.; Ahmed, Q. N. *J. Org. Chem.* **2015**, *80* (22), 11588−11592.

169. Lili, T.; Xiang, L.; Pengfeng, G.; Yue, Y.; Hua, C. *Org. Lett.* **2020**, *22*, 3841−3845.

170. Khan, S.; Ahmed, Q. N. *Eur. J. Org. Chem.* **2016**, 5377−5385.

171. Khan, S.; Kumar, A.; Gupta, R.; Ahmed, Q. N. *Chem. Sel.* **2017**, *2*, 11336−11340.

172. Ma, G.-H.; Jiang, B.; Tu, X.-J.; Ning, Y.; Tu, S.-J.; Li, G. *Org. Lett.* **2014**, *16*, 4504−4507.

173. Kumar, A.; Battini, N.; Kumar, R. R.; Athimoolam, S.; Ahmed, Q. N. Air-assisted 2-oxo-driven dehydrogenative α, α-diamination of 2-oxo aldehydes to 2-oxo acetamidines. *Eur. J. Org. Chem.* **2016**, 3344−3348.

174. Padala, A. K.; Saikam, V.; Ali, A.; Ahmed, Q. N. *Tetrahedron* **2015**, *71*, 9388–9395.
175. Kumar, A.; Ahmed, Q. N. *Eur. J. Org. Chem.* **2017**, 2751–2756.
176. Kumar, A.; Khan, S.; Ahmed, Q. N. *Org. Lett.* **2017**, *19*, 4730–4733.
177. Zheng, K.; Shi, J.; Liu, X.; Feng, X. *J. Am. Chem. Soc.* **2008**, *130* (47), 15770–15771.
178. Xianming, H.; Kellogg, R. M. *Recl. Trav. Chim. Pays-Bas* **1996**, *115*, 407.
179. O'Brien, P.; Warren, S. *Tetrahedron Lett.* **1995**, *36*, 2681–2684.
180. Pedrosa, R.; Sayalero, S.; Vicente, M.; Maestro, A. *J. Org. Chem.* **2006**, *71*, 2177–2180.
181. Shekouhy, M. *Catal. Sci. Technol.* **2012**, *2*, 1010–1020.
182. Singh, R. P.; Twamley, B.; Shreeve, J. M. *J. Org. Chem.* **2002**, *67*, 1918–1924.
183. Battula, S.; Kumar, A.; Gupta, A. P.; Ahmed, Q. N. *Org. Lett.* **2015**, *17*, 5562–5565.
184. Attanasi, O. A.; Crescentini, L. D.; Foresti, E.; Gatti, G.; Giorgi, R.; Perrulli, F. R.; Santeusanio, S. *J. Chem. Soc. Perkin. Trans.* **1997**, *1*, 1829.
185. Lalezari, I.; Levy, Y. *J. Heterocycl. Chem.* **1974**, *11*, 327.
186. Ueda, T.; Adachi, T.; Nagai, S.-I.; Sakakibara, J.; Murata, M. *J. Heterocycl. Chem.* **1988**, *25*, 791.
187. Cheng, Y.; Huang, Z. T.; Wang, M. X. *Curr. Org. Chem.* **2004**, *8*, 325.
188. Braekman, J. C.; Daloze, D. In *Studies in Natural Products Chemistry;* Atta-ur-Rahman, Ed.; Vol. 6; Elsevier: Amsterdam, 1990; pp 421–466.
189. (a) Bianco, A.; Maggini, M.; Scorrano, G.; Toniolo, C.; Marconi, G.; Villani, C.; Prato, M. *J. Am. Chem. Soc.* **1996**, *118*, 4072.
     (b) Royles, B. J. L. *Chem. Rev.* **1995**, *95*, 1981.
190. (a) Seayad, J.; List, B. *Org. Biomol. Chem.* **2005**, *3*, 719.
     (b) Dalko, P. I.; Moisan, L. *Angew Chem. Int. Ed.* **2004**, *43*, 5138.
191. (a) Huryn, D. M. In *Comprehensive Organic Synthesis;* Trost, B. M., Fleming, I., Eds.; Vol. 1; Pergamon: Oxford, U.K, 1991; pp 64–71.
     (b) Enders, D.; Klatt, M. *Synthesis* **1996**, 1403.
192. (a) Corey, E. J.; Yuen, P. W.; Hannon, F. J.; Wierda, D. A. *J. Org. Chem* **1990**, *55*, 784.
     (b) Kim, B. H.; Lee, H. B.; Hwang, J. K.; Kim, Y. G. *Tetrahedron: Asymmetry* **2005**, *16*, 1215.
193. Bossio, R.; Marcaccini, S.; Pepino, R. *Synthesis* **1994**, 765.
194. Jones, R. A., Ed. *Pyrroles, Part II;* Wiley: New York, 1992.
195. Gilchrist, T. L. *J. Chem. Soc. Perkin. Trans.* **1999**, *1*, 2849.
196. (a) Higgins, S. A. *Chem. Soc. Rev.* **1997**, *26*, 247.
     (b) Lee, C.-F.; Yang, L.-M.; Hwu, T.-Y.; Feng, A.-S.; Tseng, J.-C.; Luh, T.-Y. *J. Am. Chem. Soc.* **2000**, *122*, 4992.
     (c) Naji, A.; Cretin, M.; Persin, M.; Sarrazin, J. *J. Membr. Sci.* **2003**, *212*, 1.
197. Feliciano, A. S.; Caballero, E.; Pereira, J. A. P.; Puebla, P. *Tetrahedron* **1989**, *45*, 6553.
198. Eftekhari-Sis, B.; Vahdati, S. *Efficient Synthesis of Pyrroles and Pyridazines in Water Under Ultrasound Irradiation.* Second International Conference of Young Scientists— "Chemistry Today," Tbilisi, Georgia, April 21–23, 2012.
199. Eftekhari-Sis, B.; Akbari, M.; Amirabedi, A. *Chem. Heterocycl. Compd.* **2010**, *46*, 1330.
200. Yin, G.; Wang, Z.; Chen, A.; Gao, M.; Wu, A.; Pan, Y. *J. Org. Chem.* **2008**, *73*, 3377.
201. Feng, X.; Wang, Q.; Lin, W.; Dou, G.-L.; Huang, Z.-B.; Shi, D.-Q. *Org. Lett.* **2013**, *15*, 2542–2545.
202. (a) Smith, T. A.; Croker, S. J.; Loeffler, R. S. T. *Phytochemistry* **1986**, *25*, 683.
     (b) Anderson, W. K.; Milowsky, A. S. *J. Med. Chem.* **1987**, *30*, 2144.
     (c) Butler, M. S. *J. Nat. Prod.* **2004**, *67*, 2141.
     (d) Bellina, F.; Rossi, R. *Tetrahedron* **2006**, *62*, 7213.

203. (a) Batey, R. A.; Simoncic, P. D.; Lin, D.; Smyj, R. P.; Lough, A. J. *Chem. Commun.* **1999**, 651.

    (b) Ulbrich, H.; Fiebich, B.; Dannhardt, G. *Eur. J. Med. Chem.* **2002**, *37*, 953.

204. Caballero, E.; Puebla, P.; Domercq, M.; Medarde, M.; Lopez, J.-L.; Feliciano, A. S. *Tetrahedron* **1994**, *50*, 7849.

205. Jiang, B.; Li, Q.-Y.; Zhang, H.; Tu, S.-J.; Pindi, S.; Li, G. *Org. Lett.* **2012**, *14*, 700–703.

206. (a) Lahm, G. P.; Cordova, D.; Barry, J. D. *Bioorg. Med. Chem.* **2009**, *17*, 4127.

    (b) Fustero, S.; Roman, R.; Sanz-Cervera, J. F.; SimonFuentes, A.; Bueno, J.; Villanova, S. *J. Org. Chem.* **2008**, *73*, 8545.

207. Lamberth, C. *Heterocycles* **2007**, *71*, 1467.

208. Li, Y.; Zhang, H.-Q.; Liu, J.; Yang, X.-P.; Liu, Z.-J. *J. Agric. Food Chem.* **2006**, *54*, 3636.

209. El-Emary, T. I.; Abdel-Mohsen, S. A. *Phosphorus, Sulfur Silicon Relat Elem.* **2006**, *181*, 2459.

210. Paulis de, T.; Hemstapat, K.; Chen, Y.; Zhang, Y.; Saleh, S.; Alagille, D.; Baldwin, R. M.; Tamagnan, G. D.; Conn, P. J. *J. Med. Chem.* **2006**, *49*, 3332.

211. Bekhit, A. A.; Abdel, A. T. *Bioorg. Med. Chem.* **2004**, *12*, 1935.

212. Meazza, G.; Zanardi, G.; Piccardi, P. *J. Heterocycl. Chem.* **1993**, *30*, 365.

213. Kowalcyk, R.; Skarzewski, J. *Tetrahedron* **2005**, *61*, 623.

214. Singer, R. A.; Caron, S.; McDermott, R. E.; Arpin, P.; Do, N. M. *Synthesis* **2003**, *1727*.

215. Halcrow, M. A. *Dalton Trans.* **2009**, 2059.

216. Del Buttero, P.; Molteni, G.; Pilati, T. *Tetrahedron* **2005**, *61*, 2413.

217. Begtrup, M.; Nytoft, H. P. *J. Chem. Soc. Perkin. Trans.* **1985**, *1*, 81.

218. Kano, K.; Scarpetti, D.; Anselme, J.-P. *Tetrahedron Lett* **1985**, *26*, 6151.

219. Menezes, E. H. C.; Goes, A. J. S.; Diu, M. B. S.; Galdino, S. L.; Pitta, I. R.; Luu-Duc, C. *Pharmazie* **1992**, *46*, 457.

220. Tan, S.; Evans, R. R.; Dahmer, M. L.; Singh, B. K.; Shaner, D. L. *Pest Manage Sci.* **2005**, *61*, 246.

221. Rodgers, T. R.; LaMontagne, M. P.; Markovac, A. A.; Ash, B. *J. Med. Chem.* **1977**, *20*, 591.

222. Watanabe, K.; Morinaka, Y.; Hayashi, Y.; Shinoda, M.; Nishi, H.; Fukushima, N.; Watanabe, T.; Ishibashi, A.; Yuki, S.; Tanaka, M. *Bioorg. Med. Chem. Lett.* **2008**, *18*, 1478.

223. Silver, P. J.; Hamel, L. T.; Bentley, R. G.; Dillon, K.; Connell, M. J.; Conner, B. O.; Ferrari, R. A.; Pagani, E. D. *Drug Dev. Res.* **1990**, *21*, 93.

224. Gupta, R. J.; Gupta, A. S. *Indian J. Chem. Sect. B* **1979**, *16*, 71.

225. Havera, H.J.; Stryeker, W.G. United States Patent 3,835,151, 1973; Chem. Abstr. 1974, 81, 152224m.

226. Paul, S.; Gupta, M.; Gupta, R.; Loupy, A. *Synthesis* **2002**, 75.

227. Hough, T. L.; Hough, I. R.; Pannell, R. W. *J. Heterocycl. Chem.* **1986**, *23*, 1125.

228. Shtamburg, V. G.; Anishchenko, A. A.; Shtamburg, V. V.; Shishkin, O. V.; Zubatyuk, R. I.; Mazepa, A. V.; Rakipov, I. M.; Kostyanovsky, R. G. *Mendeleev. Commun.* **2008**, *18*, 102.

229. (a) Faulkner, D. *J. Nat. Prod. Rep.* **2000**, *17*, 7.

    (b) Ho, J. Z.; Hohareb, R. M. J.; Ahn, H.; Sim, T. B.; Rapoport, H. *J. Org. Chem.* **2003**, *68*, 109.

    (c) Zhang, L.; Peng, X.-M.; Damu, G. L. V.; Geng, R.-X.; Zhou, C.-H. *Med Res Rev.* **2014**, *34*, 340–347.

230. Grimmett, M. R. In *Comprehensive Heterocyclic Chemistry II;* Katritsky, A. R., Scriven, E. F. V., Eds.; Vol. 3; Pergamon: Oxford, 1996; pp 77–220.

231. Lombardino, J. G.; Wiseman, E. H. *J. Med. Chem.* **1974,** *17,* 1182.
232. Maier, T.; Schmierer, R.; Bauer, K.; Bieringer, H.; Buerstell, H.; Sachser, B. German Patent 317094, 1983.
233. Lo, Y. S.; Nolan, J. C.; Maren, T. H.; Welstead, W. J., Jr.; Gripshover, D. F.; Shamblee, D. A. *J. Med. Chem.* **1992,** *35,* 4790.
234. Lantos, I.; Zhang, W. Y.; Shui, X.; Eggleston, D. S. *J. Org. Chem.* **1993,** *58,* 7092.
235. Bleicher, K. H.; Gerber, F.; Wüthrich, Y.; Alanine, A.; Capretta, A. *Tetrahedron Lett.* **2002,** *43,* 7687.
236. Bratulescu, G. *Synthesis* **2009,** 2319.
237. Zuliani, V.; Cocconcelli, G.; Fantini, M.; Ghiron, C.; Rivara, M. *J. Org. Chem.* **2007,** *72,* 4551.
238. Madhavachary, R.; Zarganes-Tzitzikas, T.; Patil, P.; Kurpiewska, K.; Kalinowska-Tłuscik, J.; Domling, A. *ACS Comb. Sci.* **2018,** *20,* 192–196.
239. Moliner, F. D.; Hulme, C. *Org. Lett.* **2012,** *14,* 1354–1357.
240. Gozalishvili, L. L.; Beryozkina, T. V.; Omelchenko, I. V.; Zubatyuk, R. I.; Shishkin, O. V.; Kolos, N. N. *Tetrahedron* **2008,** *64,* 8759.
241. Waugh, R. C.; Ekeley, J. B.; Ronzio, A. R. *J. Am. Chem. Soc.* **1942,** *64,* 2028.
242. Atwood, J. L.; Barbour, L. J.; Heaven, M. W.; Raston, C. L. *J. Chem. Crystallogr.* **2003,** *33,* 175.
243. Fused Five-and Six-Membered Rings with Ring Junction Heteroatom. In *Comprehensive Heterocyclic Chemistry II;* Jones, G., Ed.; Vol. 8; Pergamon: Oxford, U.K., 1996.
244. (a) Deady, L. W.; Stanborough, M. S. *Aust. J. Chem.* **1981,** *34,* 1295.
     (b) Takahashi, T.; Satake, K. J. *Pharm. Soc. Jpn.* **1955,** *75,* 20.
245. Scholl, H.J.; Klauke, E.; DE Patent 2062347, 1973; Chem. Abstr. **1973,** 77, 152194.
246. Houlihan, W. J.; Munder, P. G.; Handley, D. A.; Cheon, S. H.; Parrino, V. A. *J. Med. Chem.* **1995,** *38,* 234.
247. Kubo, K.; Ito, N.; Isomura, Y.; Sozu, I.; Homma, H.; Murakami, M. *Chem. Pharm. Bull.* **1979,** *27,* 1207.
248. Drach, B. S.; Dolgushina, I. Y.; Sinitsa, A. D. *Chem. Heterocycl. Compd.* **1974,** *10,* 810.
249. Peshkov, V. A.; Peshkov, A. A.; Pereshivko, O. P.; Hecke, K. V.; Zamigaylo, L. L.; Eycken, E. V. V.; Gorobets, D. N. Y. *ACS Comb. Sci.* **2014,** *16,* 535–542.
250. Barlin, G. B.; Brown, D. J.; Kadunc, Z.; Petric, A.; Stanovnik, B. *Tisler M. Aust. J. Chem.* **1983,** *36,* 1215.
251. Devillers, I.; Dive, G.; Tollenaere, C. D.; Falmagne, B.; Wergifosse, B. D.; Rees, J.-F.; Marchand-Brynaert, J. *Bioorg. Med. Chem. Lett.* **2001,** *11,* 2305.
252. Fan, W.-Q.; Katritzky, A. R. In *Comprehensive Heterocyclic Chemistry II;* Katritzky, A. R., Rees, C. W., Scriven, E. F. V., Eds.; Vol. 4; Pergamon Press: Oxford, U.K., 1996; pp 1–126.
253. (a) Costa, M. S.; Boechat, N.; Rangel, E. A.; Silva, Fd. Cd; Souza, A. M. Td; Rodrigues, C. R.; Castro, H. C.; Junior, I. N.; Lourenco, M. C. S.; Wardell, S. M. S. V., et al. *Bioorg. Med. Chem.* **2006,** *14,* 8644.
     (b) Sivakumar, K.; Xie, F.; Cash, B.; Long, M. S.; Barnhill, H. N.; Wang, Q. *Org. Lett.* **2004,** *6,* 4603.
254. Ye, C. F.; Gard, G. L.; Winter, R. W.; Syvret, R. G.; Twamley, B.; Shreeve, J. M. *Org. Lett.* **2007,** *9,* 3841.
255. Whiting, M.; Muldoon, J.; Lin, Y.-C.; Silverman, S. M.; Lindstrom, W.; Olson, A. J.; Kolb, H. C.; Finn, M. G.; Sharpless, K. B.; Elder, J. H., et al. *Angew Chem. Int. Ed.* **2006,** *45,* 1435.

256. (a) Moderhack, D. *J. Prakt. Chem.* **1998**, *340*, 687.
    (b) Ostrovskii, V. A.; Pevzner, M. S.; Kofman, T. P.; Shcherbinin, M. B.; Tselinskii, I. V. *Targets Heterocycl. Syst.* **1999**, *3*, 467.
257. Tang, W.-J.; Hu, Y.-Z. *Synth. Commun.* **2006**, *36*, 2461.
258. Yates, P.; Mayfield, R. J. *Can. J. Chem.* **1977**, *55*, 145.
259. (a) Look, S. A.; Burch, M. T.; Fenical, W.; Qitai, Z.; Clardy, J. *J. Org. Chem.* **1985**, *50*, 5741.
    (b) Fenical, W.; Okuda, R. K.; Bandurraga, M. M.; Culver, P.; Jacobs, R. S. *Science* **1981**, *212*, 1512.
    (c) Williams, D.; Andersen, R. J.; Van Duyne, G. D.; Clardy, J. *J. Org. Chem.* **1987**, *52*, 332.
    (d) Rodríguez, A. D. ;; Shi, J.-G.; Huang, S. D. *J. Org. Chem.* **1998**, *63*, 4425.
260. Onitsuka, S.; Nishino, H. *Tetrahedron* **2003**, *59*, 755.
261. Onitsuka, S.; Nishino, H.; Kurosawa, K. *Tetrahedron Lett* **2000**, *41*, 3149.
262. Yang, Y.; Gao, M.; Wu, L.-M.; Deng, C.; Zhang, D.-X.; Gao, Y.; Zhu, Y.-P.; Wu, A.-X. *Tetrahedron* **2011**, *67*, 5142.
263. Li, M.; Kong, X.-J.; Wen, L.-R. *J. Org. Chem.* **2015**, *80*, 11999−12005.
264. Li, J.; Liu, L.; Ding, D.; Sun, J.; Ji, Y.; Dong, J. *Org. Lett.* **2013**, *15*, 2884−2887.
265. Mupparapu, N.; Khan, S.; Bandhoria, P.; Athimoolam, S.; Ahmed, Q. N. *ACS Omega* **2018**, *3*, 5445−5452.
266. (a) Ireland, R. E.; Anderson, R. C.; Badoud, R.; Fitzsimmons, B. J.; McGarvey, G. J.; Thaisrivongs, S.; C. Wilcox, S. *J. Am. Chem. Soc.* **1983**, *105*, 1988.
    (b) Lord, M. D.; Negri, J. T.; Paquette, L. A. *J. Org. Chem.* **1995**, *60*, 191.
267. (a) Collum, D. B.; McDonald, J. H., III; Still, W. C. *J. Am. Chem. Soc.* **1980**, *102*, 2117 2118, 2120.
    (b) Marshall, J. A.; Yu, B. *J. Org. Chem.* **1994**, *59*, 324.
    (c) Yamada, O.; Ogasawara, K. *Synlett* **1995**, 427.
268. (a) Semple, J. E.; Wang, P. C.; Lysenko, Z.; Joullie, M. M. *J. Am. Chem. Soc.* **1980**, *102*, 7505.
    (b) Ward, R. S. *Nat. Prod. Rep.* **1999**, *16*, 75.
269. Beck, B.; Magnin-Lachaux, M.; Herdtweck, E.; Domling, A. *Org. Lett.* **2001**, *3*, 2875.
270. Bossio, R.; Marcaccini, S.; Pepino, R.; Torroba, T. *Synthesis* **1993**, 783.
271. Anary-Abbasinejad, M.; Shams, N.; Hassanabadi, A. *Phosphorus, Sulfur Silicon Relat Elem* **2010**, *185*, 1823.
272. Peter, M.; Gleiter, R.; Rominger, F.; Oeser, T. *Eur. J. Org. Chem.* **2004**, 3212.
273. Ritchie, E.; Taylor, W. C. *Aust. J. Chem.* **1971**, *24*, 2137.
274. Masubuchi, M.; Ebiike, H.; Kawasaki, K.; Sogabe, S.; Morikami, K.; Shiratori, Y.; Tsujii, S.; Fujii, T.; Sakata, K.; Hayase, M.; Shindoh, H., et al. *Bioorg. Med. Chem.* **2003**, *11*, 4463.
275. Abdel-Aziz, H. A.; Mekawey, A. A. I.; Dawood, K. M. *Eur. J. Med. Chem.* **2009**, *44*, 3637.
276. Kraus, G. A.; Kim, I. *Org. Lett.* **2003**, *5*, 1191.
277. Sun, M.; Zhao, C.; Gfesser, G. A.; Thiffault, C.; Miller, T. R.; Marsh, K.; Wetter, J.; Curtis, M.; Faghih, R.; Esbenshade, T. A., et al. *J. Med. Chem.* **2005**, *48*, 6482.
278. Oter, O.; Ertekin, K.; Kirilmis, C.; Koca, M.; Ahmedzade, M. *Sens. Actuators B* **2007**, *122*, 450.
279. Karatas, F.; Koca, M.; Kara, H.; Servi, S. *Eur. J. Med. Chem.* **2006**, *41*, 664.
280. Habermann, J.; Ley, S. V.; Smits, R. *J. Chem. Soc. Perkin Trans* **1999**, *1*, 2421.
281. Chen, C.-X.; Liu, L.; Yang, D.-P.; Wang, D.; Chen, Y.-J. *Synlett* **2005**, 2047.
282. Fujimaki, T.; Nagase, H.; Yamaguchi, R.; Kawai, K.-I.; Otomasu, H. *Chem. Pharm. Bull.* **1985**, *33*, 2663.

283. Kang, Y. K.; Shin, K. J.; Yoo, K. H.; Seo, K. J.; Hong, C. Y.; Lee, C.-S.; Park, S. Y.; Kim, D. J.; Park, S. W. *Bioorg. Med. Chem. Lett.* **2000**, *10*, 95.

284. Frolund, B.; Jorgensen, A. T.; Tagmose, L.; Stensbol, T. B.; Vestergaard, H. T.; Engblom, C.; Kristiansen, U.; Sanchez, C.; Krogsgaard-Larsen, P.; Liljefors, T. *J. Med. Chem.* **2002**, *45*, 2454.

285. Ko, D.-H.; Maponya, M. F.; Khalil, M. A.; Oriaku, E. T.; You, Z.; Lee. *J. Med. Chem. Res.* **1998**, *8*, 313.

286. (a) Daidone, G.; Raffa, D.; Maggio, B.; Plescia, F.; Cutuli, V. M. C.; Mangano, N. G.; Caruso, A. *Arch. Pharm.* **1999**, *332*, 50.
    (b) Mishra, A.; Jain, S. K.; Asthana, J. G. *Orient. J. Chem.* **1998**, *14*, 151.

287. (a) Kozikowski, A. P.; Chen, Y. Y. *J. Org. Chem.* **1981**, *46*, 5248.
    (b) Müller, I.; Jager, V. *Tetrahedron Lett* **1982**, *23*, 4777.
    (c) Jager, V.; Grund, H. *Angew Chem. Int. Ed.* **1976**, *15*, 50.

288. Juhasz-Toth, E.; Patonay, T. *Eur. J. Org. Chem.* **2002**, 3055.

289. (a) Gomez, M.; Muller, G.; Rocamora, M. *Coord. Chem. Rev.* **1999**, *193*, 769.
    (b) Johnson, J. S.; Evans, D. A. *Acc. Chem. Res.* **2000**, *33*, 325.
    (c) Fache, F.; Schulz, E.; Tommasino, M. L.; Lemaire, M. *Chem. Rev.* **2000**, *100*, 2159.

290. (a) Wipf, P.; Venkatraman, S. *J. Org. Chem.* **1995**, *60*, 7224.
    (b) Ousmer, M.; Braun, N. A.; Ciufolini, M. A. *Org. Lett.* **2001**, *5*, 765.

291. Green, T. W.; Wutz, P. G. M. *Protecting Groups in Organic Synthesis*, 2nd ed.; John Wiley and Sons: New York, 1991.

292. (a) Wipf, P.; Fritch, P. C.; Geib, S. J.; Sefler, A. M. *J. Am. Chem. Soc.* **1998**, *120*, 4105.
    (b) Deng, S.; Taunton, J. *J. Am. Chem. Soc.* **2002**, *124*, 916.

293. Choi, I.-Y.; Lee, H. G.; Chung, K.-H. *J. Org. Chem.* **2001**, *66*, 2484.

294. (a) Agami, C.; Couty, F.; Lequesne, C. *Tetrahedron* **1995**, *51*, 4043.
    (b) Agami, C.; Couty, F.; Lequesne, C. *Tetrahedron Lett.* **1994**, *35*, 3309.

295. Ukaji, Y.; Yamamoto, K.; Fukui, M.; Fujisawa, T. *Tetrahedron Lett.* **1991**, *32*, 2919.

296. Ichiba, T.; Yoshida, W. Y.; Scheuer, P. J.; Higa, T.; Gravalos, D. G. *J. Am. Chem. Soc.* **1991**, *113*, 3173.

297. Brown, P.; Davies, D. T.; O'Hanlon, P. J.; Wilson, J. M. *J. Med. Chem.* **1996**, *39*, 446.

298. D'Ambrosio, M.; Guerriero, A.; Pietra, F.; Debitus, C. *Helv. Chim. Acta* **1996**, *79*, 51.

299. Kean, W. F. *Curr. Med. Res. Opin.* **2004**, *20*, 1275.

300. Delpierre, G. R.; Eastwood, F. W.; Gream, G. E.; Kingston, D. G. I.; Sarin, P. S.; Todd, L.; Williams, D. H. *Tetrahedron Lett.* **1966**, *7*, 369.

301. Belyuga, A. G.; Brovarets, V. S.; Drach, B. S. *Russ. J. Gen. Chem.* **2005**, *75*, 523.

302. Babu, V. N.; Murugan, A.; Katta, N.; Devatha, S.; Sharada, D. S. *J. Org. Chem.* **2019**, *84*, 6631–6641.

303. (a) Roncali, J. *Chem. Rev.* **1992**, *92*, 711.
    (b) Roncali, J. *Chem. Rev.* **1997**, *97*, 173.
    (c) Miller, L. L.; Mann, K. R. *Acc. Chem. Res.* **1996**, *29*, 417.

304. Müller, M.; Mauermann-Dull, H.; Wagner, M.; Enkelmann, V.; Müllen, K. *Angew Chem. Int. Ed.* **1995**, *34*, 1583.

305. Larsen, J.; Bechgaard, K. *Acta Chem. Scand.* **1996**, *50*, 71.

306. Tsuji, K.; Nakamura, K.; Ogino, T.; Konishi, N.; Tojo, T.; Ochi, T.; Seki, N.; Matsuo, M. *Chem. Pharm. Bull.* **1998**, *46*, 279.

307. (a) Kiani, M. S.; Mitchell, G. R. *Synth. Met.* **1992**, *46*, 293.
    (b) Dimitrakopoulos, C. D.; Malenfant, P. R. L. *Adv. Mater.* **2002**, *14*, 99.

308. Schopf, G.; Koßmehl, G. *Adv. Polym. Sci.* **1997**, *129*, 1.
309. Greenham, N. C.; Moratti, S. C.; Bradley, D. D. C.; Friend, R. H.; Holmes, A. B. *Nature* **1993**, *365*, 628.
310. Francesco, R. D.; Migliaccio, G. *Nature* **2005**, *436*, 953.
311. Mac Dowell, D. W. H.; Patrick, T. B. *J. Org. Chem.* **1967**, *32*, 2441.
312. Miyahara, Y.; Inazu, T.; Yoshino, T. *Bull. Chem. Soc. Jpn.* **1980**, *53*, 1187.
313. (a) Roy, R.; Gehring, A. M.; Milne, J. C.; Belshaw, P. J.; Walsh, C. T. *Nat. Prod. Rep.* **1999**, *16*, 249.
    (b) Luesch, H.; Yoshida, W. Y.; Moore, R. E.; Paul, V. J.; Corbett, T. H. *J. Am. Chem. Soc.* **2001**, *123*, 5418.
314. Negwer, M. *Organic-Chemical Drugs and their Synonyms: An International Survey*, 7th ed.; Akademie, VCH: New York, 1994.
315. (a) Cosp, A.; Llacer, E.; Romea, P.; Urpí, F. *Tetrahedron Lett* **2006**, *47*, 5819.
    (b) Baiget, J.; Cosp, A.; Galvez, E.; Gomez-Pinal, L.; Romea, P.; Urpí, F. *Tetrahedron* **2008**, *64*, 5637.
    (c) Yang, J.-H.; Yang, G.-C.; Lu, C.-F.; Chen, Z.-X. *Tetrahedron: Asymmetry* **2008**, *19*, 2164.
316. (a) Ramirez, F.; Patwardhan, A. V.; Smith, C. P. *J. Org. Chem.* **1965**, *30*, 2575.
    (b) Ramirez, F.; Patwardhan, A. V.; Smith, C. P. *J. Org. Chem.* **1966**, *31*, 474.
317. (a) Rohet, F.; Rubat, C.; Coudert, P.; Couquelet, J. *Bioorg. Med. Chem.* **1997**, *5*, 655.
    (b) Tucker, J. A.; Allwine, D. A.; Grega, K. C.; Barbachyn, M. R.; Klock, J. L.; Adamski, J. L.; Brickner, S. J.; Hutchinson, D. K.; Ford, C. W.; Zurenko, G. E., et al. *J. Med. Chem.* **1998**, *41*, 3727.
    (c) Heinisch, G.; Kopelent-Franck, H. In *Progress in Medicinal Chemistry;* Ellis, G. P., Luscombe, D. K., Eds.; Vol. 29; Elsevier: Amsterdam, 1992; pp 141–183.
318. (a) Kolar, P.; Tisler, M. *Adv. Heterocycl. Chem.* **1999**, *75*, 167.
    (b) Matyus, P.; Maes, B. U. W.; Riedl, Z.; Hajos, G.; Lemiere, G. L. F.; Tapolcsanyi, P.; Monsieurs, K.; Elias, O.; Dommisse, R. A.; Krajsovszky, G. *Synlett* **2004**, 1123.
    (c) Carboni, R. A.; Lindesey, R. V., Jr. *J. Am. Chem. Soc.* **1959**, *81*, 4342.
    (d) Sauer, J.; Heldmann, D. K.; Hetzenegger, J.; Krauthan, J.; Sichert, H.; Schuster, J. *Eur. J. Org. Chem.* **1998**, 2885.
319. Rimaz, M.; Khalafy, J. *ARKIVOC* **2010**, *ii*, 110.
320. Ismail, K. A.; El-Tombary, A. A.; AboulWafa, O. M.; Omar, A.-M. M. E.; El-Rewini, S. H. *Arch. Pharm.* **1996**, *329*, 433.
321. Marriner, G. A.; Garner, S. A.; Jang, H.-Y.; Krische, M. J. *J. Org. Chem.* **2004**, *69*, 1380.
322. Contreras, J.-M.; Rival, Y. M.; Chayer, S.; Bourguignon, J.-J.; Wermuth, C. G. *J. Med. Chem.* **1999**, *42*, 730.
323. Bourotte, M.; Pellegrini, N.; Schmitt, M.; Bourguignon, J.-J. *Synlett* **2003**, 1482.
324. (a) Corbett, J. W.; Rauckhorst, M. R.; Qian, F.; Hoffman, R. L.; Knauer, C. S.; Fitzgerald, L. W. *Bioorg. Med. Chem. Lett.* **2007**, *17*, 6250.
    (b) Buron, F.; Turck, A. N.; Bischoff, L.; Marsais, F. *Tetrahedron Lett.* **2007**, *48*, 4327.
325. (a) Brophy, J. J.; Cavill, G. W. K. *Heterocycles* **1980**, *14*, 477.
    (b) Seeman, J. I.; Ennis, D. M.; Sector, H. V.; Clawson, L.; Palen, J. *Chem. Senses* **1989**, *14*, 395.
326. (a) Mikuriya, M.; Yoshioka, D.; Handa, M. *Coord. Chem. Rev.* **2006**, *250*, 2194.
    (b) Crutchley, R. J. *Angew Chem. Int. Ed.* **2005**, *44*, 6452.
327. Hasegawa, M.; Katsumata, T.; Ito, Y.; Saigo, K.; Iitaka, Y. *Macromolecules* **1988**, *21*, 3134.

328. Mahboobi, S.; Sellmer, A.; Burgemeister, T.; Lyssenko, A.; Schollmeyer, D. *Monatsh Chem.* **2004**, *135*, 333.

329. Vogl, O.; Taylor, E. C. *J. Am. Chem. Soc.* **1959**, *81*, 2472.

330. Zhang, W.; Haight, A. R.; Ford, K. L.; Parekh, S. I. *Synth. Commun.* **2001**, *31*, 725.

331. Haight, A. R.; Bailey, A. E.; Baker, W. S.; Cain, M. H.; Copp, R. R.; DeMattei, J. A.; Ford, K. L.; Henry, R. F.; Hsu, M. C.; Keyes, R. F., et al. *Org. Process Res. Dev.* **2004**, *8*, 897.

332. Man, N.-N.; Wang, J.-Q.; Zhang, L.-M.; Wen, L.-R.; Li, M. *J. Org. Chem.* **2017**, *82*, 5566−5573.

333. Zeng, X.-H.; Wang, H.-M.; Ding, M.-W. *Org. Lett.* **2015**, *17*, 2234−2237.

334. Huang, Y.-W.; Li, X.-Y.; Fu, L.-N.; Guo, Q.-X. *Org. Lett.* **2016**, *18*, 6200−6203.

335. Zheng, K.-L.; You, M.-Q.; Shu, W.-M.; Wu, Y.-D.; Wu, A.-X. *Org. Lett.* **2017**, *19*, 2262−2265.

336. Cai, Q.; Li, D.-K.; Zhou, R.-R.; Shu, W.-M.; Wu, Y.-D.; Wu, A.-X. doi: 10.1021/acs.orglett.6b00281.

337. Dada, R.; Sulthan, M.; Yaragorla, S. *Org. Lett.* **2020**, *22*, 279−283.

338. Shen, B.; Liu, W.; Cao, W.; Liu, X.; Feng, X. *Org. Lett.* **2019**, *21*, 4713−4716.

339. Lin, W.; Zheng, Y.-X.; Xun, Z.; Huang, Z.-B.; Shi, D.-Q. *ACS Comb. Sci.* **2017**, *19*, 708−713.

340. Battini, N.; Battula, S.; Kumar, R. R.; Ahmed, Q. N. *Org. Lett.* **2015**, *17*, 2992−2995.

341. Fan, W.; Ye, Q.; Xu, H.-W.; Jiang, B.; Wang, S.-L.; Tu, S.-J. *Org. Lett.* **2013**, *15* (9), 2258−2261.

342. Powell, W. C.; Walczak, M. A. *J. Org. Chem.* **2018**, *83*, 10487−10500.

343. Mishra, R.; Panday, A. K.; Choudhury, L. H.; Pal, J.; Subramanian, R.; Verma, A. *Eur. J. Org. Chem.* **2017**, 2789−2800.

344. A.-Bami, F.; Mehrabi, H.; R.-Karimi, R. *J. Sulfur Chem.* **2019**, *40* (5), 467−478.

345. Nallan, L.; Bauer, K. D.; Bendale, P.; Rivas, K.; Yokoyama, K.; Horney, C. P.; Pendyala, P. R.; Floyd, D.; Lombardo, L. J.; Williams, D. K., et al. *J. Med. Chem.* **2005**, *48*, 3704.

346. Wilkinson, G. P.; Taylor, J. P.; Shnyder, S.; Cooper, P.; Howard, P. W.; Thurston, D. E.; Jenkins, T. C.; Loadman, P. M. *Invest N Drugs* **2004**, *22*, 231.

347. Hunt, J. T.; Ding, C. Z.; Batorsky, R.; Bednarz, M.; Bhide, R.; Cho, Y.; Chong, S.; Chao, S.; Gullo-Brown, J.; Guo, P., et al. *J. Med. Chem.* **2000**, *43*, 3587.

348. Sañudo, M.; García-Valverde, M.; Marcaccini, S.; Delgado, J. J.; Rojo, J.; Torroba, T. *J. Org. Chem.* **2009**, *74*, 2189.

349. Lecinska, P.; Corres, N.; Moreno, D.; García-Valverde, M.; Marcaccini, S.; Torroba, T. *Tetrahedron* **2010**, *66*, 6783.

350. (a) Grinsteiner, T. J.; Kishi, Y. *Tetrahedron Lett.* **1994**, *45*, 8333.
     (b) Grinsteiner, T. J.; Kishi, Y. *Tetrahedron Lett.* **1994**, *45*, 8337.

351. (a) Eliel, E. L.; He, X.-C. *Tetrahedron* **1987**, *43*, 4979.
     (b) Eliel, E. L.; He, X.-C. *J. Org. Chem* **1990**, *55*, 2114.

352. (a) Sano, H.; Noguchi, T.; Miyajima, A.; Hashimoto, Y.; Miyachi, H. *Bioorg. Med. Chem. Lett.* **2006**, *16*, 3068.
     (b) Padwa, A.; Eisenbarth, P. *Tetrahedron* **1985**, *41*, 283.
     (c) Tiecco, M.; Testaferri, L.; Marini, F. *Tetrahedron* **1996**, *52*, 11841.

Chapter 4

Applications of 2-Oxoacids

4.1 Introduction

2-Oxoacids or α-oxocarboxylic acids bear an additional keto group adjacent to the carboxylic acid. They are similar in reactivity as compared to simple carboxylic acids and have broad applications in industry and medicinal chemistry.[1] Due to the presence of an additional 2-oxo group many types of reactions are possible for the generation of new molecules like amides/α-ketoamides, aldehydes, ynones, ynediones, α-sulfanyl-substituted heterocyclic acids, α-benzyl amino acids, propargyl amines, diketones, hydroxyl butyrokactones, chiral carboxylic acids, N-aroylsulfoximines, esters, ketones, and thioesters α,β-unsaturated carbonyls.[2,3] All of these reactions work as surrogates of simple acids, α-oxoaldehydes, acetophenones, acetaldehydes, and benzaldehydes (Fig. 4.1). Furthermore, owing to their availability and better stability, they are frequent used as starting materials for the development of novel methods in organic synthesis for C−C and C−X (X = N, P, S) bond formation reactions.[4] After the synthesis of α-keto acid (pyruvic acid) in 1835 by Berzelius, there has been gradual advancement in the applications of α-keto acids.[5]

Furthermore, 2-oxo acids also play a vital role in the process of providing energy to the cells of animals, plants, and bacteria.[6] Thus, a pyruvate derivative anion is an important metabolite that is used in numerous enzyme-catalyzed intracellular phenomena known as the Krebs cycle to produce adenosine triphosphate (ATP) for plants, animals, and bacteria.[7]

The biochemical importance of the pyruvate derivative anion is the formation of acetyl-CoA by acting as an acylating agent. This transformation involves the decarboxylative acylation reaction of CoA−SH to get acetyl-CoA in the presence of pyruvate dehydrogenase in catalytic amount and NAD + . This acetyl-CoA is used for the conversion of oxaloacetic acid into citrate by undergoing a condensation reaction in the presence of citrate synthase enzyme. This citrate is an important intermediate for the synthesis of ATP during the Krebs cycle (Fig. 4.2).

In the presence of a catalytic amount of pyruvate carboxylase, pyruvate anion undergoes condensation reaction with carbonic acid to form another important intermediate in the Krebs cycle, that is, oxaloacetate dianion. In this process, carbonic acid was driven by ATP which is converted into ADP (Fig. 4.3).

Chemistry of 2-Oxoaldehydes and 2-Oxoacids. DOI: https://doi.org/10.1016/B978-0-12-824285-8.00001-7

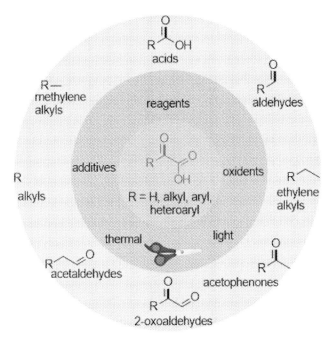

FIGURE 4.1 α-oxoacids as a surrogate of different functional groups.

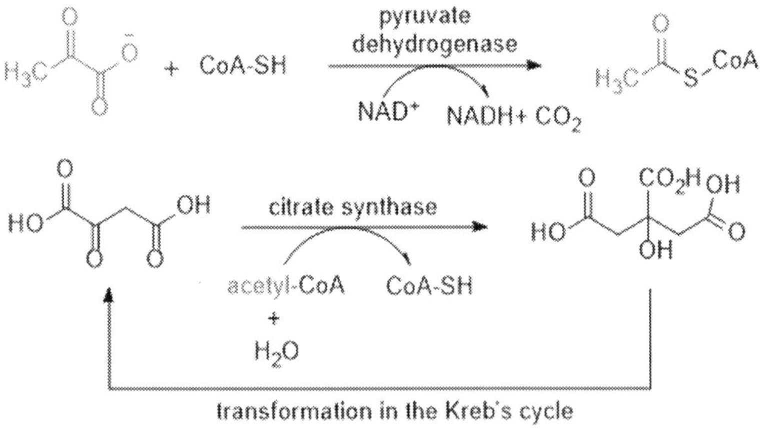

FIGURE 4.2 Formation of citrate by condensation of oxaloacetic acid.

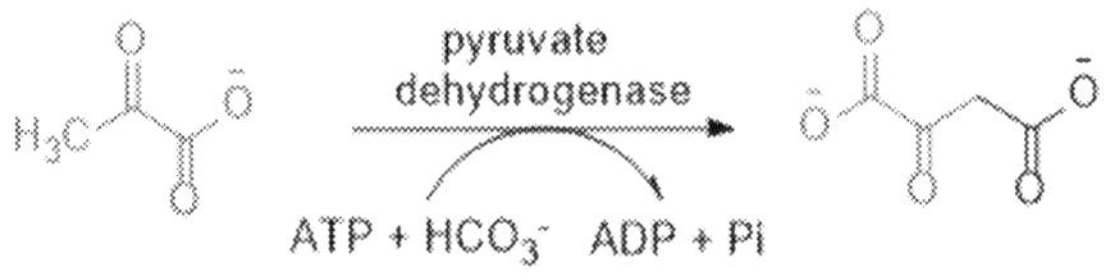

FIGURE 4.3 Formation of oxaloacetate dianion.

In synthetic organic chemistry, 2-oxoacids have widely been used as starting material for the construction of different biologically important scaffolds and in most of the cases, it shows decarboxylative acylation reactions.[8] Different transformative reactions are known in the literature for the synthesis of ynones, ynediones, ester, diketone, amide, thioester, ketone, aldehyde, acid, and heterocyclic compounds by using 2-oxoacids as the starting material. In this chapter, we are mainly focusing on different applications of 2-oxoacids for the construction of valuable scaffolds of synthetic or biological importance.

4.2   Decarboxylative acylation reactions

Decarboxylation is referred as those reactions in which at least one molecule of carbon dioxide is eliminated mostly from carboxylic acid. The release of carbon dioxide from the molecule makes the reaction irreversible in most of the cases. It has been shown that carboxylic acids bearing 2-substituted electron-withdrawing groups undergo loss of carbon dioxide followed by acyl transfer in a single step which is called as decarboxylative acylation reactions. Over the past few decades, chemists have developed serious attention toward the decarboxylative coupling reactions of 2-oxoacids for the formation of C−C, C−N, C−P, and C−S bonds.[9] In most of the decarboxylative reactions of 2-oxoacids, it undergoes decarboxylation and inserts the acyl group in the other molecule. Different transition metals such as Pd or Pd/Ag, Ag and Cu has been used for the decarboxylative reactions of 2-oxoacids with Csp^2-X bonds (C−N, C−P, and C−S) and unactivated Csp^2−H bonds for the synthesis of aryl ketones.[10] Visible-light photoredox-catalysis has also been reported.[11] As far as the mechanism of the decarboxylation of 2-oxoacids is concerned, two different pathways are reported, one is an acyl metal species (RCO−M) while as the other is free acyl radical intermediate (RCO•).

Krishna Nand Singh and his coworkers[12] reported a metal-free protocol for the $C(sp^2)$−H functionalization of isoquinolines **2**. 2-Oxoacids **1** was used as an acylated reagent for the synthesis of C1-benzoylated isoquinolines **3** in the presence of 3 equivalent of $K_2S_2O_8$ in H_2O at 100 °C for 6 h. Different substitution on 2-oxoacids **1** and isoquinolines **2** doesn't affect much on the yields of the C1-benzoylated isoquinolines **3**. This reaction proceeds through the free radical pathway by forming in situ acyl radicals from 2-oxoacids in the presence of persulfate. This acyl radical reacts with the salt of isoquinolines to form intermediate **4**, followed by the reaction with persulfate radical to form the desired C1-benzoylated isoquinolines (Figs. 4.4 and 4.5).

Guan-Wu Wang and his coworker[13] reported decarboxylative ortho-acylation of N-nitrosoanilines reaction in the presence of palladium as a catalyst. By performing a reaction between 2-oxoaldehydes **1** and N-nitrosoanilines **5** in the presence of a catalytic amount of $Pd(OAc)_2$ and 2 equiv of potassium persulfate, which acts as an oxidant to afford acylated N-nitrosoanilines **6**. A mixture of

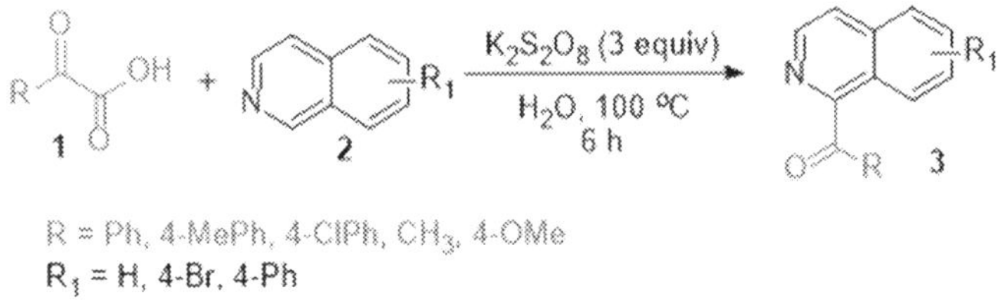

R = Ph, 4-MePh, 4-ClPh, CH$_3$, 4-OMe
R$_1$ = H, 4-Br, 4-Ph

FIGURE 4.4 Formation of isoquinoline derivatives.

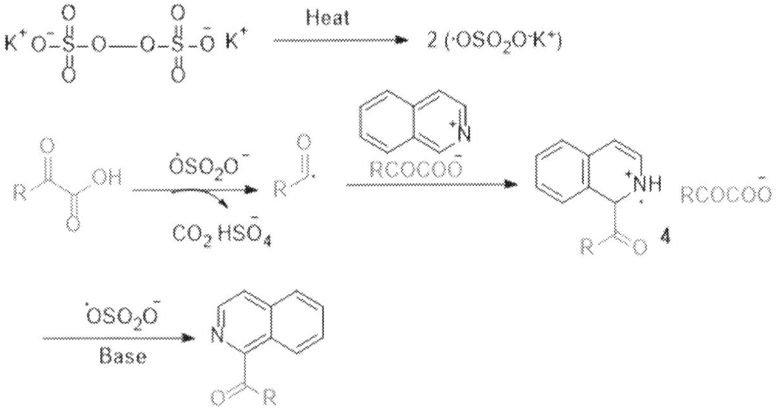

FIGURE 4.5 Mechanism for isoquinoline formation.

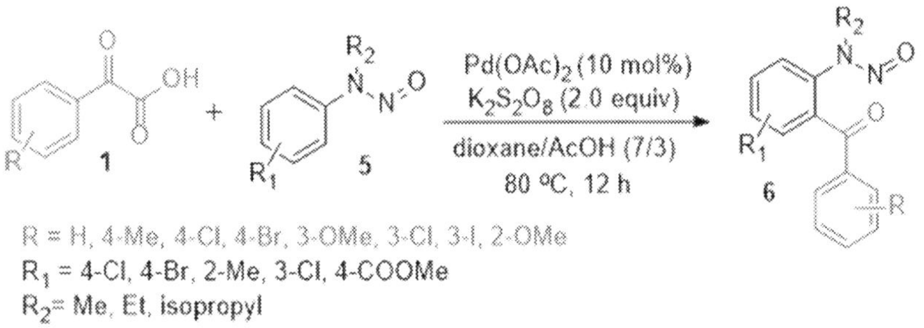

R = H, 4-Me, 4-Cl, 4-Br, 3-OMe, 3-Cl, 3-I, 2-OMe
R$_1$ = 4-Cl, 4-Br, 2-Me, 3-Cl, 4-COOMe
R$_2$ = Me, Et, isopropyl

FIGURE 4.6 Ortho-acylation of N-nitrosoanilines.

dioxane and AcOH in the ratio of 7:3 was used as a solvent and the products were obtained in good to moderate yields with a broad substrate scope (Fig. 4.6).

Later on, the same concept was applied by varying nitrosoanilines **5** with benzamides **7** to obtain ortho-acylated benzamides **8** in good yields in presence of Pd(OAc)$_2$, K$_2$S$_2$O$_8$ and TfOH (20 mol.%) in DCE solvent at 70°C for 24 h (Fig. 4.7).[14]

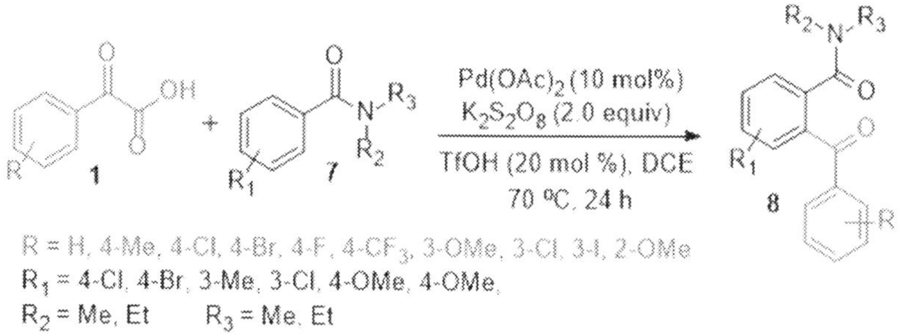

FIGURE 4.7   Ortho-acylation of benzamides.

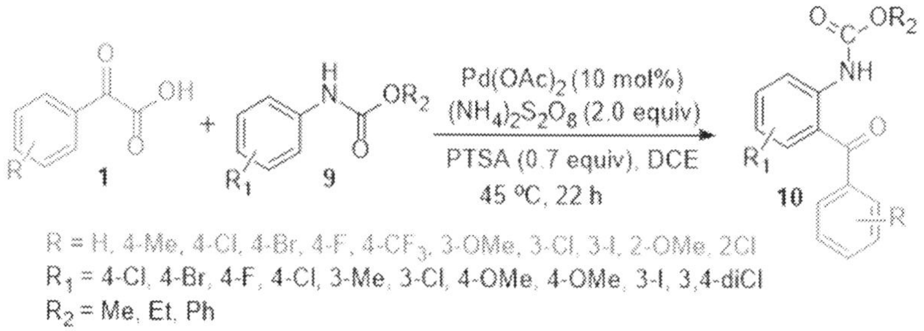

FIGURE 4.8   Ortho-acylation of aniline carbamates.

Furthermore, 2-oxoacids **1** undergo decarboxylative reaction with aniline carbamate **9** to form ortho-acylated aniline carbamate **10** by using Pd(OAc)$_2$ (10 mol.%) as a catalyst (NH$_4$)$_2$S$_2$O$_8$ (2.0 equiv) as an oxidant and p-toluene-sulfonic acid (PTSA) in 1,2-dichloroethane (DCE) at 30°C for 24 h (Fig. 4.8).[15]

Qianqian Wang et al. established[16] an efficient, regioselective synthesis of 2-aminobenzophenones **13** from a three-component reaction between anilines **11**, α-oxocarboxylic acids **1** and tert-butyl nitrite **12** in presence of the catalytic amount of Pd(OAc)$_2$ in DCE at 40°C for 20 h. The main advantage of this reaction is the use of tert-butyl nitrite **12** that plays a dual role as a nitrosation reagent as well as the sustainable oxidant. Moreover, the author has optimized the deprotection conditions of the nitroso group by treating the resulting mixture from the acylation reactions with Fe/NH$_4$Cl at 80°C for 10 h to form 2-aminobenzophenones **14** in one- pot or one-pot two-step reactions (Fig. 4.9).

Palladium (Pd) metal-catalyzed decarboxylative reaction of 2-oxoacid **1a** with aromatic ketones **15** was reported by Wing-Yiu Yu and his coworkers to obtain orthoselective acylated aromatic ketones **16** in good yields. In this

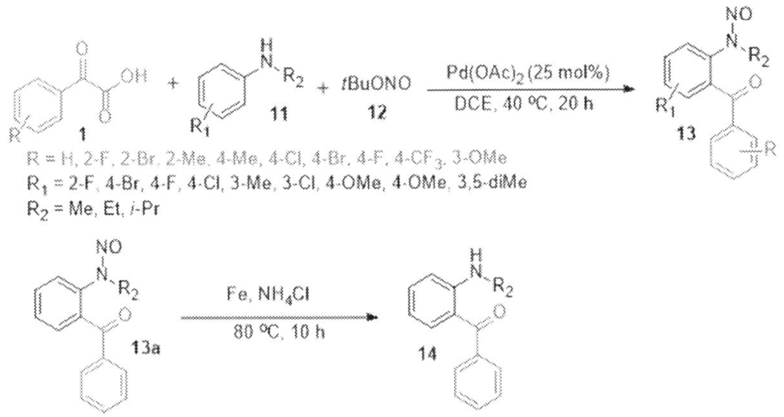

FIGURE 4.9   Ortho-acylation of 2-aminobenzophenones.

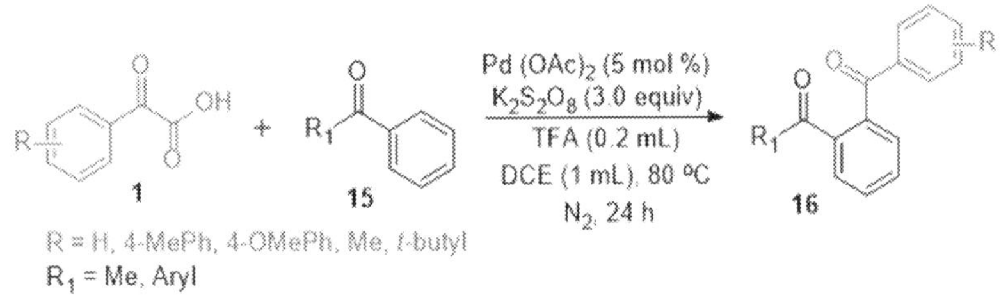

FIGURE 4.10   Ortho-acylation of aryl ketones.

approach $K_2S_2O_8$ (3.0 equiv) was used as an oxidant and trifluoroacetic acid (TFA 0.2 mL) as an additive in dichloroethane (DCE) at 80 °C under nitrogen atmosphere (Fig. 4.10).[17]

Another palladium-catalyzed decarboxylative acylation reaction of 2-oxoacids **1** was reported by Su Kim and his coworkers[18] to form ortho-acylated O-methyl ketoximes **18**. O-methyl ketoximes **17** reacts with 2-keto acids **1** in the presence of Pd(OAc)₂ and $(NH_4)_2S_2O_8$ (2.0 equiv) to undergo decarboxylative cross-coupling reactions in diglyme at 70°C to afford acylated product **18** with high selectivity and broad substrate scope (Fig. 4.11).

Later on, the same group extended the concept of decarboxylative reactions of 2-oxoacids **1** by doing the ortho-acylation of phenylacetamides **19** in presence of Pd(II) catalyst and $(NH_4)_2S_2O_8$ to form 2-acylated phenylacetamides **20** (Fig. 4.12).[19]

Zhiyong Yang et al. reported[20] decarboxylative *ortho*-acylation of O-methyl oximes **21** with 2-oxoacids **1** in presence of palladium (II) catalyst and $(NH_4)_2S_2O_8$ as an oxidant under nitrogen atmosphere at 60°°C for 12 h to access the corresponding acylated product **22** (Fig. 4.13).

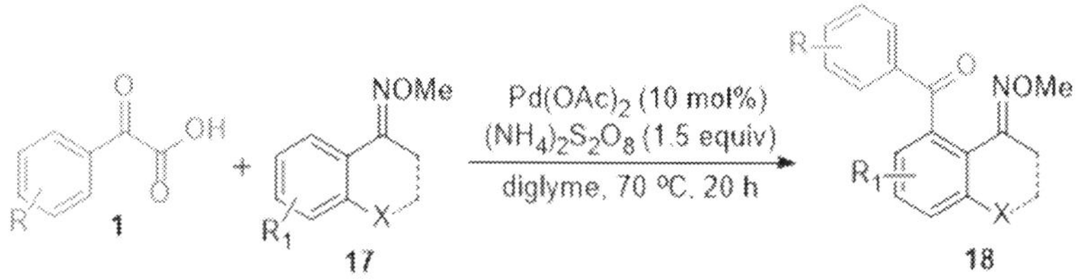

R = H, 4-Me, 4-Cl, 4-Br, 4-F, 4-CF$_3$, 3-OMe, 3-Cl, 3-I, 2-OMe, 2Cl
R$_1$ = 4-Cl, 4-Br, 4-F, 4-Cl, 3-Me, 3-F, 2-F, 4-OMe, 3-I
X = C, O

FIGURE 4.11 Synthesis of o-methyl ketoximes.

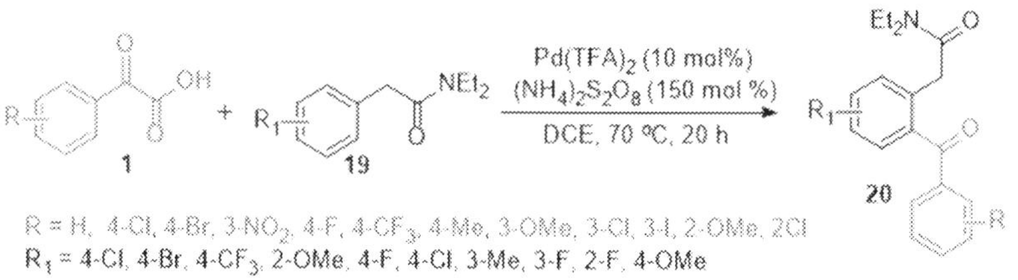

R = H, 4-Cl, 4-Br, 3-NO$_2$, 4-F, 4-CF$_3$, 4-Me, 3-OMe, 3-Cl, 3-I, 2-OMe, 2Cl
R$_1$ = 4-Cl, 4-Br, 4-CF$_3$, 2-OMe, 4-F, 4-Cl, 3-Me, 3-F, 2-F, 4-OMe

FIGURE 4.12 Synthesis of acylated phenyl acetamides.

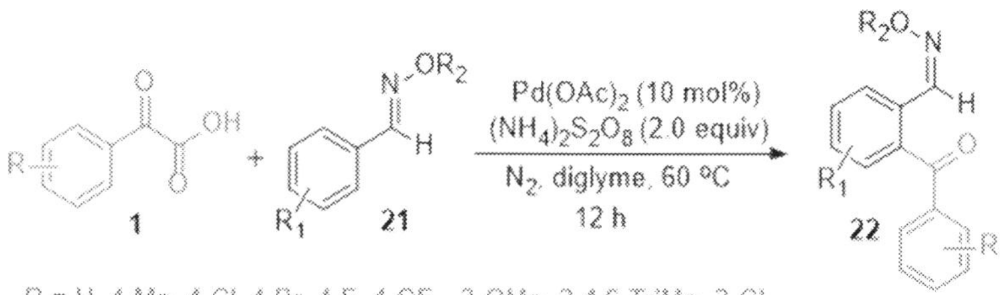

R = H, 4-Me, 4-Cl, 4-Br, 4-F, 4-CF$_3$, 3-OMe, 2,4,6-TriMe, 2-Cl
R$_1$ = 4-Cl, 4-Br, 4-F, 2-Et, 2-OEt, 4-Cl, 3-Me, 3-F, 2-F, 2,3-DiMe
R$_2$ = Me, Bn

FIGURE 4.13 *o*-Acylation of O-methyl oximes.

Palladium-catalyzed C-H activation for the acylation of arenes **23** from 2-oxoacids **1** was established by Mingzong Li et al.[21] Direct aromatic sp^2 C-H bond was activated by Pd/Ag bimetallic system to form aryl ketones **24**. The best-optimized condition for the reaction is reacting 2-oxoacids **1** and 2-phenylpyridine **23** in the presence of Pd(PhCN)$_2$Cl$_2$, K$_2$S$_2$O$_8$, and Ag$_2$O in a mixture of dioxane/AcOH/DMSO (7.5/1.5/1.0) as a solvent at 120°C for 12−16 h (Fig. 4.14).

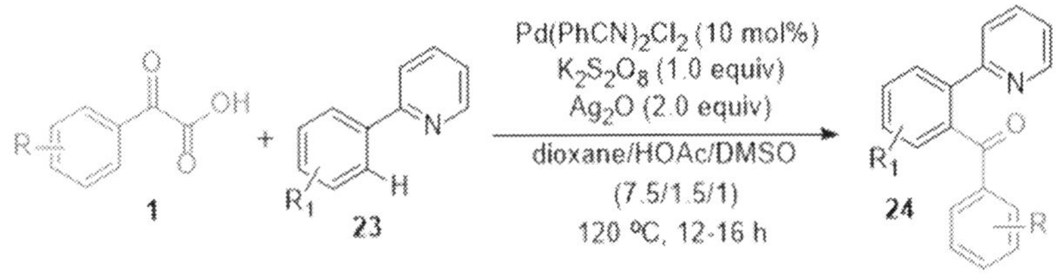

R = H, 4-F, 4-CF$_3$, 3-OMe, 2,4,6-TriMe, 4-Me, 4-Cl, 4-Br
R$_1$ = 4-Me, 4-Br, 3-OMe, 4-Cl, 3-Me, 3-F, 4-OMe

FIGURE 4.14 Synthesis of acyl aryl ketones from 2-phenylpyridine.

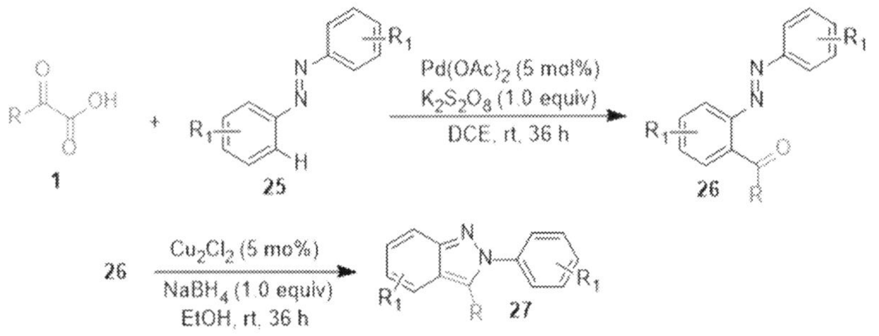

R = Ph, 4-FC$_6$H$_4$, 4-CF$_3$C$_6$H$_4$, 3-OMeC$_6$H$_4$, 4-MeC$_6$H$_4$, 4-ClC$_6$H$_4$, CH$_3$
R$_1$ = 4-Me, 4-Cl, 4-OMe

FIGURE 4.15 Synthesis of indazoles from azobenzene.

Another efficient approach for the ortho-acylation of azobenzenes **25** with 2-oxoacids **1** for the synthesis of acylated azo compounds **26** is palladium-catalyzed decarboxylative reaction reported by Hongji Li et al. The reaction was tested with different substituents on 2-oxoacids **1** without affecting the reaction yields. The reactions were performed in DCE at room temperature for 36 h. The acylated azobenzenes **26** were efficiently converted into indazole derivatives **27** by treating mixture of Cu$_2$Cl$_2$/NaBH$_4$ system in ethanol (Fig. 4.15).[22]

Palladium-catalyzed cross-coupling reaction between 2-oxoacids **1** and benzoic acids **28** for the synthesis of 2-acylbenzoic acid derivative **29** was described by Haibo Ge and coworkers.[23] This work represents the cross-coupling reaction in presence of palladium catalyst and silver carbonate as a base to form 2-acylbenzoic acid **29** (Fig. 4.16).

In 2012, Xin-Hua Duan and coworkers[24] have described an efficient protocol for the acylation of unactivated sp^2 (alkenyl) C-H bonds of cyclic enamides **30** in the presence of palladium catalyst. α-Oxocarboxylic acids **1** undergoes decarboxylation and acts as an acylating reagent to afford β-acyl

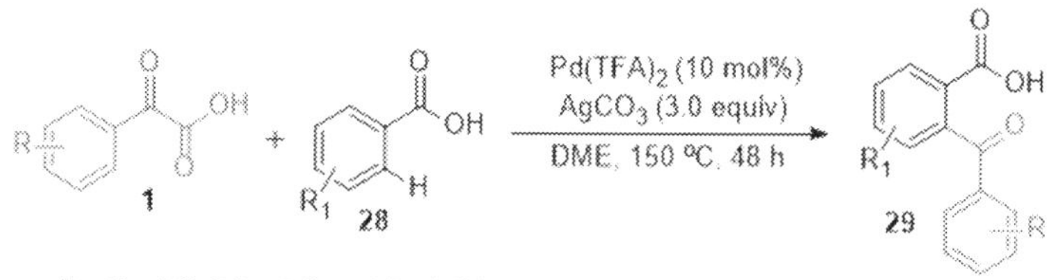

R = H, 4-F, 2-Br, 2-F, 4-CF₃, 3-OMe, 4-Me, 3-Cl, 4-Br
R₁ = 2-Me, 2-Cl, 3-OMe, 4-Cl, 3-F, 3-F, 4-OMe, 3-Cl, 4-CF₃

FIGURE 4.16   Synthesis of 2-acylbenzoic acid.

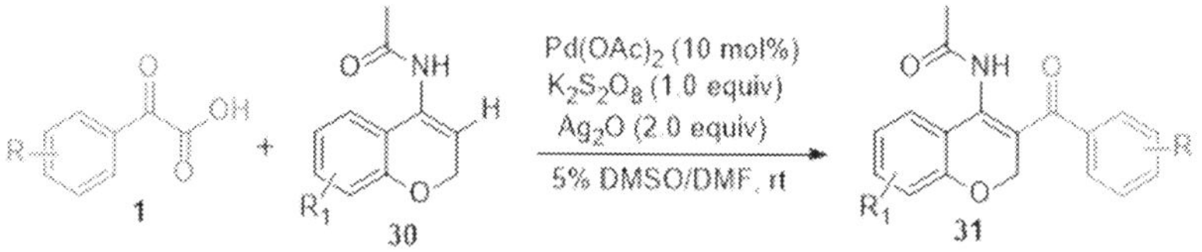

R = H, 2-Cl, 4-F, 2-Br, 2-F, 4-CF₃, 3-OMe
R₁ = 6-Me, 8-Me, 6-Cl

FIGURE 4.17   Synthesis of β-acyl enamide.

enamide **31** in good yields. Diverse acylated enamides **31** were synthesized by using this protocol and substitution on both the reacting partners doesn't hamper much on the yields (Fig. 4.17).

The Pd-catalyzed selective ortho-acylation of 2-phenyl-benzoxazinone and its derivatives **32** was successfully established by Ranu and coworkers[25] in 2016. This protocol involves the reaction between α-keto acids **1** and 2-phenyl-benzoxazinone **32** in the presence of Pd(OAc)₂ (10 mol%) as a catalyst, AgNO₃ (1 equiv), and (NH₄)₂S₂O₈ (2 equiv) as an oxidant in DCE as the solvent at 50°C for 24 h to access the corresponding monoacylated 2-phenyl-benzoxazinone **33** in good yields. Different aryl, heteroaryl, and alkyl-substituted glyoxylic acids were tested against 2-phenyl-benzoxazinone and monoacylated derivatives were successfully prepared in 32% − 86% yields under the optimized reaction conditions. In case of 2-phenyl-benzoxazinone, substituted benzoxazinones gave lower yields as compared to the unsubstituted ones (Fig. 4.18).

A highly stereoselective Pd-catalyzed decarboxylative acylation of unsymmetrically substituted 1,2,3-triazoles **34** with 2-oxoacids **1** was reported by Xie and coworkers[26] in 2018 for the synthesis of N-3-ortho-acylated 1,4-disubstituted 1,2,3-triazoles **35**. A reaction mixture of 1,2,3-triazole **34** with α-keto acid **1** was stirred at 150°C by using catalytic amount of Pd(OAc)₂, K₂S₂O₈ (1 equiv) as an oxidizing agent, and Ag₂O (2 equiv) as a base in a mixture of 1,4-dioxane and AcOH (1:1) as the solvent for 12 h to access the ortho-acylated product **35** in good to moderate yields. Different substituted N-3-ortho-acylated 1,2,3-triazoles

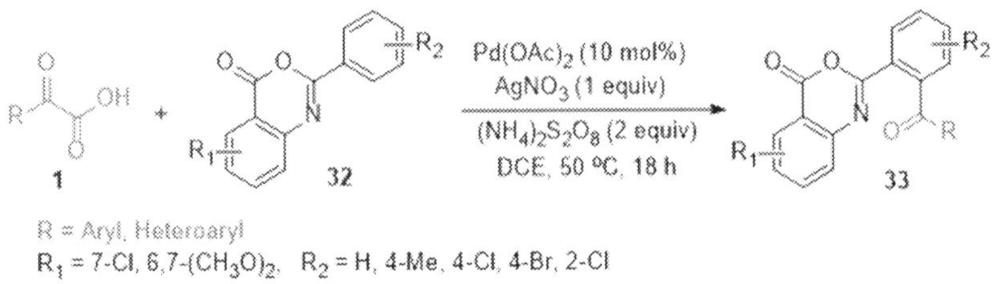

R = Aryl, Heteroaryl
R_1 = 7-Cl, 6,7-(CH$_3$O)$_2$, R_2 = H, 4-Me, 4-Cl, 4-Br, 2-Cl

FIGURE 4.18 Synthesis of 2-phenyl-benzoxazinone.

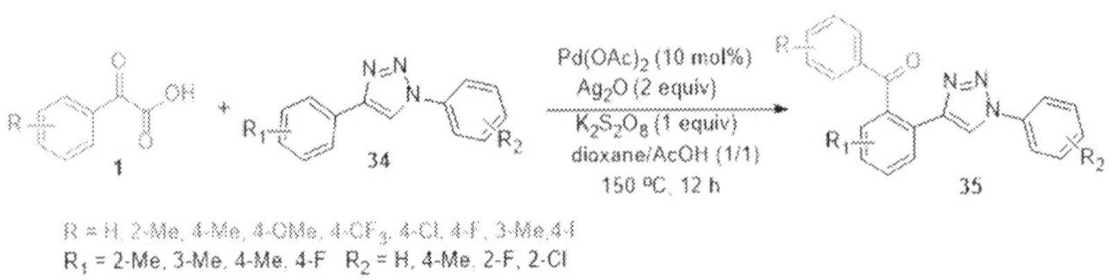

R = H, 2-Me, 4-Me, 4-OMe, 4-CF$_3$, 4-Cl, 4-F, 3-Me,4-f
R_1 = 2-Me, 3-Me, 4-Me, 4-F R_2 = H, 4-Me, 2-F, 2-Cl

FIGURE 4.19 Synthesis of 1,2,3-triazoles.

35 were synthesized with high regioselectivity by using different substituted 2-oxoacids and 1,2,3-triazoles in 47% − 89% yields (Fig. 4.19).

The authors have done radical scavengers (TEMPO or BHT) control experiments in order to confirm the free radical mechanism and was observed that the reaction was not affected by the addition of radical scavengers. Based on these control experiments and literature precedent, a mechanism involving different steps was proposed. The first step is the interaction of Pd(II) and N-3 of triazole **34** to form the five-membered palladacycle **36**. Once the intermediate **36** is formed, it reacts with acyl silver species **37** to access the intermediate **38**. The acyl silver species **37** was obtained from the reaction of phenyl glyoxylic acid **1** with Ag$_2$O with the release of CO$_2$ in the reaction medium. The intermediate **38** undergoes a reductive elimination to form the o-acylated product **35** along with Pd(0), and then oxidation by S$_2$O$_8^{2-}$ to form Pd(II) species to start a new reaction cycle (Fig. 4.20).

3-Acyl-benzofurans **40** and 3-acyl-benzothiophenes **42** are two important classes of heterocyclic compounds that are important in medicinal chemistry because of their biological activities. Wei-Jie Gong et al.[27] in 2015 described the Pd-promoted decarboxylative C3 acylation of 2-pyridyl-benzofurans **39** and 2-pyridyl-benzothiophenes **41** with 2-oxoacids **1**. The synthesis of 3-acyl-benzofurans **40** involves a reaction between 2-pyridyl-benzofurans **39** and 2-oxoacids **1** with Pd(Ph$_3$)$_4$ (10 mol%) as the catalyst, Ag$_2$CO$_3$ (2 equiv), and K$_2$S$_2$O$_8$ (2 equiv) as an oxidant in a mixture of 1,4-dioxane/AcOH/DMF (7.5/1.5/1.0) as a solvent for 21 h at 120°C. For the synthesis of 3-acyl-

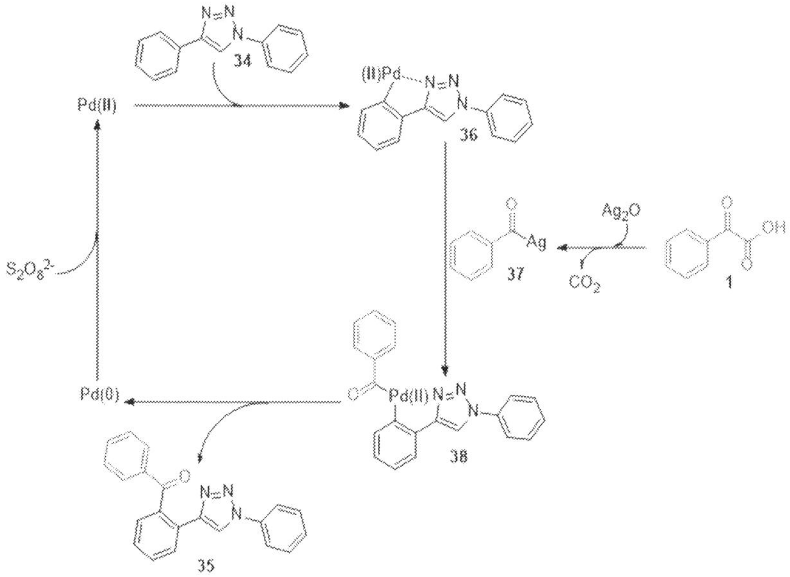

FIGURE 4.20   Proposed mechanism for the synthesis of N-3-ortho-acylated 1,2,3-triazoles.

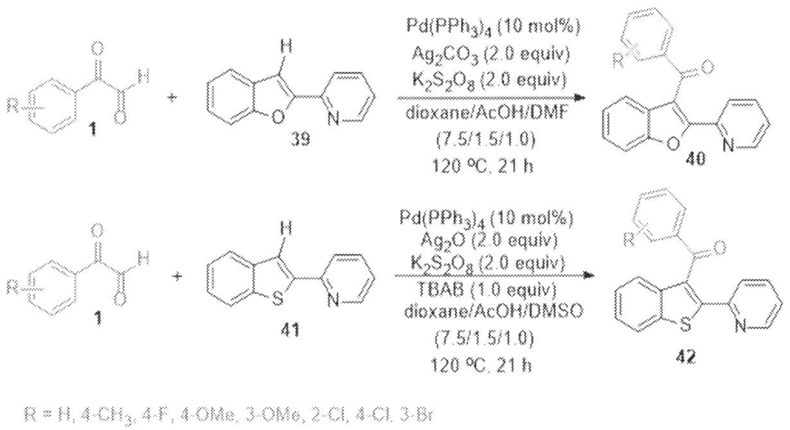

R = H, 4-CH₃, 4-F, 4-OMe, 3-OMe, 2-Cl, 4-Cl, 3-Br

FIGURE 4.21   Synthesis of 3-acyl-benzofurans and 3-acyl-benzothiophenes.

benzothiophenes **42**, tetrabutylammonium bromide (TBAB) was used as an additive and Ag₂O as an oxidant in DMSO (Fig. 4.21).

Chengjian Zhu and his group[28] described C2-acylation of indoles **43** with 2-oxoacids **1**. This protocol was also reported in the presence of a palladium catalyst that can assist the decarboxylation in 2-oxoacids **1** to form

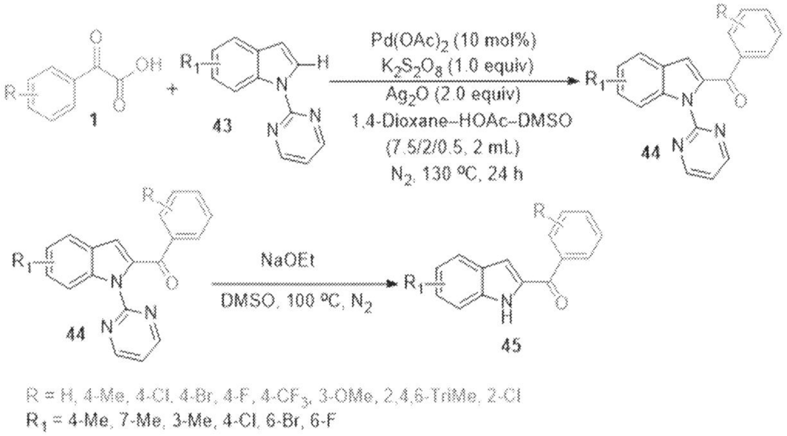

R = H, 4-Me, 4-Cl, 4-Br, 4-F, 4-CF₃, 3-OMe, 2,4,6-TriMe, 2-Cl
R₁ = 4-Me, 7-Me, 3-Me, 4-Cl, 6-Br, 6-F

FIGURE 4.22 Synthesis of 2-aroylindoles.

2-aroylindoles **44** in good yields. The significance of this transformation is the installation of a suitable 2-pyrimidyl group that directs the reaction at the C2 position and is readily removable on the indole nitrogen atom in the presence of NaOEt under nitrogen atmosphere (Fig. 4.22).

The plausible pathway for this approach is the formation of intermediate **46** by coordination of the N atom of **43** to Pd(II) that gives rise to a five-membered palladacycle. The next step involves the addition of **1** that reacts with palladacycle **46** to form another intermediate **47**. The intermediate **47** undergoes decarboxylation to afford intermediate **48** which further undergoes reductive elimination to give product **44** and Pd (0) is regenerated which is oxidized to Pd(II) to start a new reaction cycle (Fig. 4.23).

A novel $R_2(O)P$-directed Pd(II)-catalyzed decarboxylative coupling of 2-phosphorylbiphenyl **49** with 2-oxoacids **1** was described by Shang-Dong Yang and coworkers[29] for the synthesis of phosphorylbiphenyl-2-acyl compounds **50** in good to moderate yields. The reaction proceeds in the presence of 2.5 equiv. $K_2S_2O_8$ as an oxidizing agent and 10 mol% Pd(OAc)₂ as a catalyst at 130°C for 3 h. The protocol shows a broad substrate scope by using different substituted 2-oxoaldehydes **1** and 2-phosphorylbiphenyl **49** to access the corresponding phosphorylbiphenyl-2-acyl compounds **50** in moderate to good yields (Fig. 4.24).

In addition, Lin Yu et al. reported[30] the C3-acylation reaction of N-substituted indoles **51** in the presence of a copper catalyst. On treatment of 2-oxoacids **1** with N-substituted indoles **51** in this reaction Cu(OAc)₂. H_2O acts as an oxidant in MeCN for 10 h that promotes the reaction efficiently without any other transition metal to form C3-acylated indoles **52** in good yields. (Fig. 4.25).

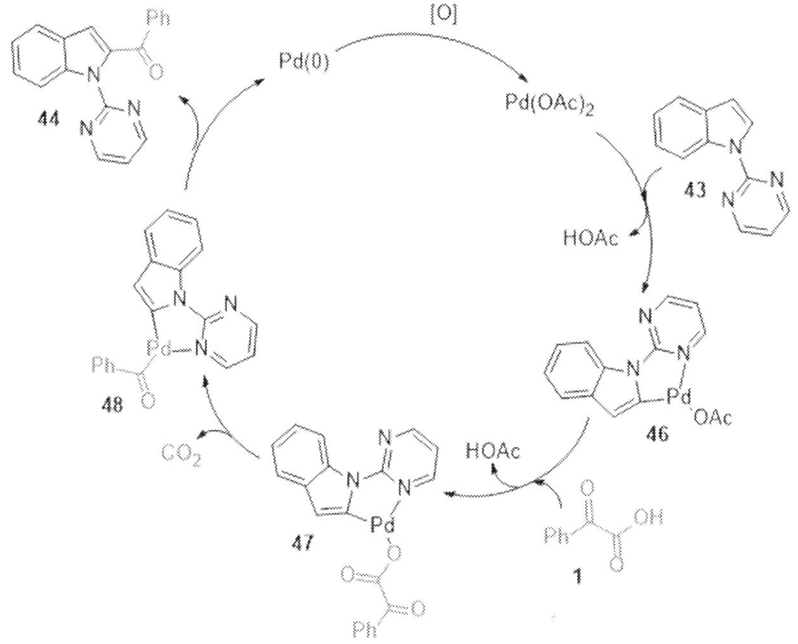
[O]
Pd(0)
$Pd(OAc)_2$
HOAc
HOAc
CO_2
Ph
44
43
46
47
48
1
Pd
OAc
OH

FIGURE 4.23 Plausible Mechanism for the synthesis of C2-acylated indoles.

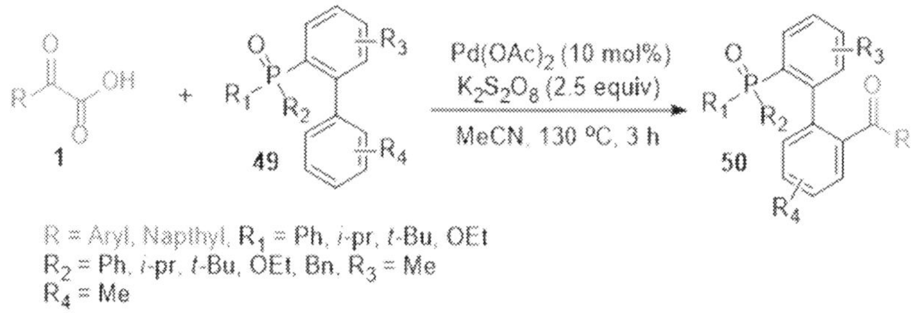
$Pd(OAc)_2$ (10 mol%)
$K_2S_2O_8$ (2.5 equiv)
MeCN, 130 °C, 3 h
1
49
50
R = Aryl, Napthyl, R_1 = Ph, *i*-pr, *t*-Bu, OEt
R_2 = Ph, *i*-pr, *t*-Bu, OEt, Bn, R_3 = Me
R_4 = Me

FIGURE 4.24 Synthesis of phosphorylbiphenyl-2-acyl derrivatives.

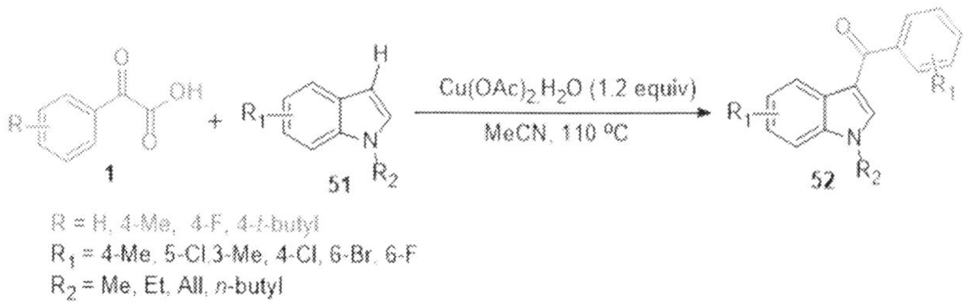
$Cu(OAc)_2.H_2O$ (1.2 equiv)
MeCN, 110 °C
1
51
52
R = H, 4-Me, 4-F, 4-*t*-butyl
R_1 = 4-Me, 5-Cl,3-Me, 4-Cl, 6-Br, 6-F
R_2 = Me, Et, All, *n*-butyl

FIGURE 4.25 Synthesis of C3-acylated indoles.

The plausible mechanism of Cu-catalyzed decarboxylative reaction is proposed in which 2-oxoacid **1** primarily reacts with $Cu(OAc)_2$. H_2O to afford a Cu (II) carboxylate **53** by the loss of AcOH, followed by decarboxylation to form an acyl Cu(II) species **54**. The Cu acetate intermediate **54** is converted into intermediate **55** by reacting at the C3-position of indole followed by rearomatization via C−H bond cleavage of **55** to generate aryl Copper species **56**. The last step is the reductive elimination to obtain the C3-acylation product of indole **52** and Cu(0) is regenerated to complete the next reaction cycle (Fig. 4.26).

In 2018, Tanakorn Kittikool et al.[31] also reported copper promoted decarboxylative C-H acylation of pyrazolones **57** with 2-oxoacids **1** in presence of persulfate as an oxidant to form 4-acylpyrazolones **58**. The reaction proceeds smoothly in the presence of $Cu(OAc)_2$ and $K_2S_2O_8$ in a mixture of MeOH: H_2O (1:1) at 60°C. The copper metal helps in the decarboxylation of 2-oxoacids to form acyl-copper species followed by reductive elimination to give4-acylpyrazolones **58**. Different library of 4-acylpyrazolone compounds was synthesized by using different substituted 2-oxoacids and pyrazolones under mild conditions (Fig. 4.27).

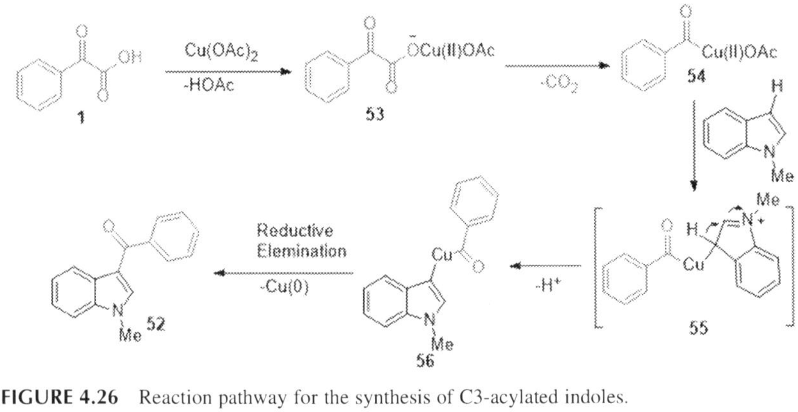
Cu(OAc)2
-HOAc
OCu(II)OAc
-CO2
Cu(II)OAc
Reductive
Elemination
-Cu(0)
-H+

FIGURE 4.26 Reaction pathway for the synthesis of C3-acylated indoles.

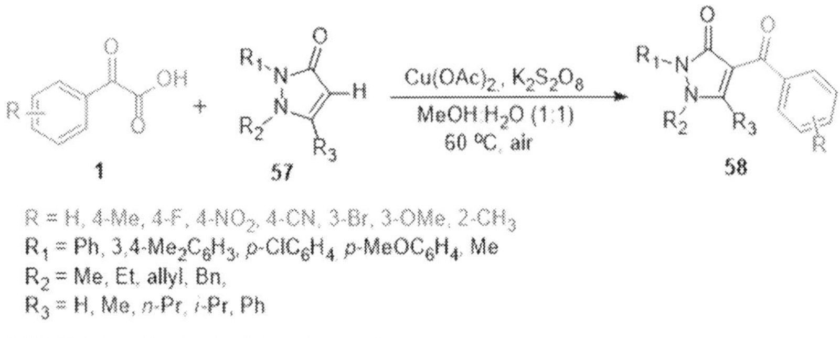
Cu(OAc)2, K2S2O8
MeOH H2O (1:1)
60 °C, air
R = H, 4-Me, 4-F, 4-NO2, 4-CN, 3-Br, 3-OMe, 2-CH3
R1 = Ph, 3,4-Me2C6H3, p-ClC6H4, p-MeOC6H4, Me
R2 = Me, Et, allyl, Bn,
R3 = H, Me, n-Pr, i-Pr, Ph

FIGURE 4.27 Synthesis of 4-acyl pyrazolones.

Cobalt catalyst has been efficiently used for the direct decarboxylative cross-coupling of oxazole and thiazole derivatives **59** with 2-oxoacids **1** to form 2-benzoylated oxazole and thiazole **60** in the presence of Ag_2CO_3 at 170°C for 24 h. Different substitutions on 2-oxoacids **1** and oxazoles/thiazoles **59** don't affect much on the yields of the 2-benzoylated oxazoles and thiazoles **60**. The plausible mechanism for the synthesis of 2-benzoylated oxazoles and thiazoles is explained below. In the presence of Ag_2CO_3 cobalt (II) is converted into Cobalt (III) followed by the insertion of oxazoles/thiazoles to form complex **61**. The Cobalt (III) complex is converted into Cobalt (IV) complex **62** by the insertion of an acyl radical of 2-oxoacids, followed by the reductive elimination to obtain 2-benzoylated oxazoles/thiazoles **60** and cobalt (II) complex (Figs. 4.28 and 4.29).[32]

Functionalization reactions of aromatic compounds has been developed since past a few years.[33] In 2018, Kun Jing et al.[34] established decarboxylative ruthenium-catalyzed *meta*-selective acylation of arenes **63**. 2-Oxoacids **1** transfers acyl group to 2-phenyl pyridine **63** in presence of $Ru_3(CO)_{12}$ (5 mol.%),

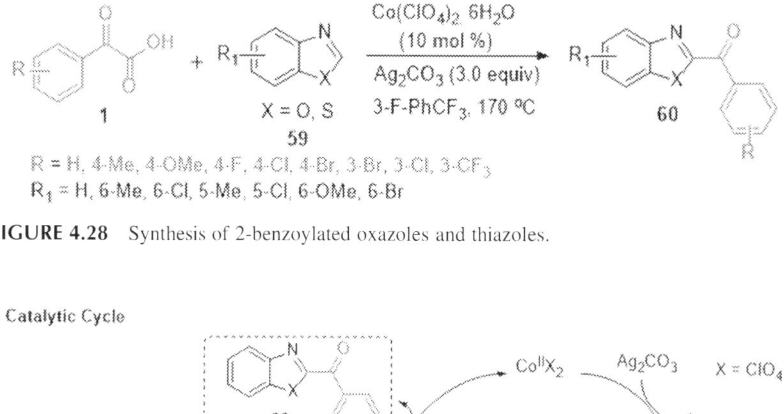

R = H, 4-Me, 4-OMe, 4-F, 4-Cl, 4-Br, 3-Br, 3-Cl, 3-CF$_3$
R$_1$ = H, 6-Me, 6-Cl, 5-Me, 5-Cl, 6-OMe, 6-Br

FIGURE 4.28 Synthesis of 2-benzoylated oxazoles and thiazoles.

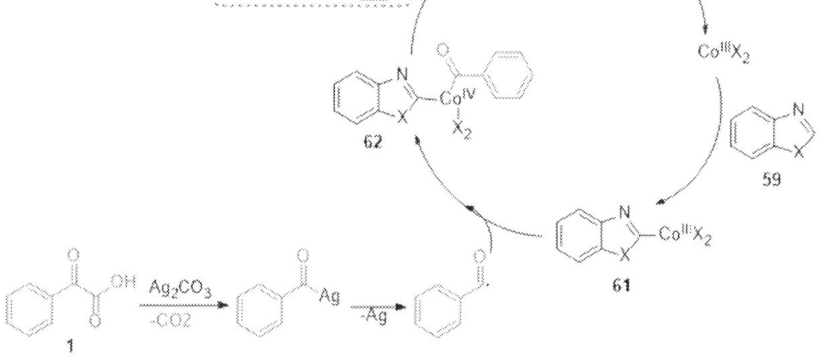

FIGURE 4.29 Mechanism for the formation for the synthesis of 2-benzoylated oxazoles and thiazoles.

$Na_2S_2O_8$ (2.0 equiv), Ag_2CO_3 (2.5 equiv), D-camphorsulfonic acid (D-CSA) (0.5 equiv), and tert-butyl methyl ether (TBME) in DCM at 100°C for 48 h to form **64**. Mechanistically, the reaction proceeds through a radical pathway with the formation of an 18e-octahedral ruthenium species followed by reductive elimination to obtain the desired *meta*-acylated product.

The catalyst $Ru_3(CO)_{12}$ reacts with 2-phenyl pyridine **63** to form Ru complex **65** followed by the electrophilic attack of an acyl radical at the para-position of the Ru − C bond to form complex **66**. The acyl radical was generated from 2-oxoacid **1** with the help of $Na_2S_2O_8$ and Ag_2CO_3. The complex **67** was formed by subsequent oxidative deprotonation of complex **66** in the presence of Ag_2CO_3 and/or $Na_2S_2O_8$. Finally, the reductive elimination and protonation furnishes the desired *meta*-acylated product **64** and complex **65** is regenerated (Figs. 4.30 and 4.31).

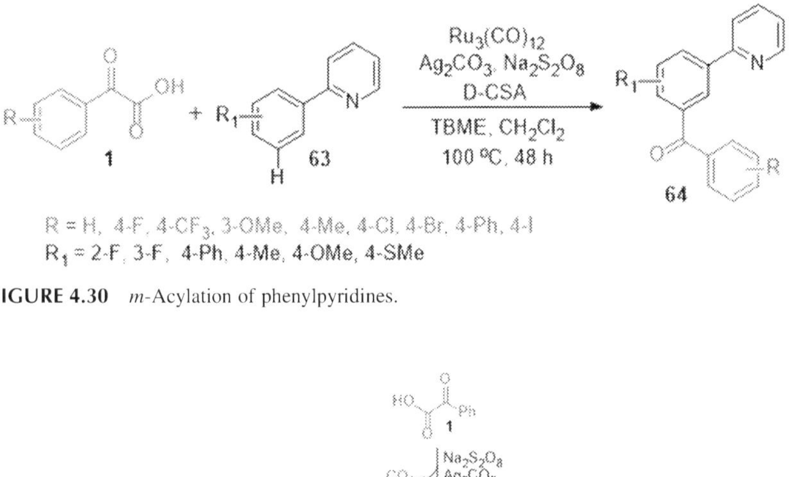

R = H, 4-F, 4-CF$_3$, 3-OMe, 4-Me, 4-Cl, 4-Br, 4-Ph, 4-I
R$_1$ = 2-F, 3-F, 4-Ph, 4-Me, 4-OMe, 4-SMe

FIGURE 4.30   *m*-Acylation of phenylpyridines.

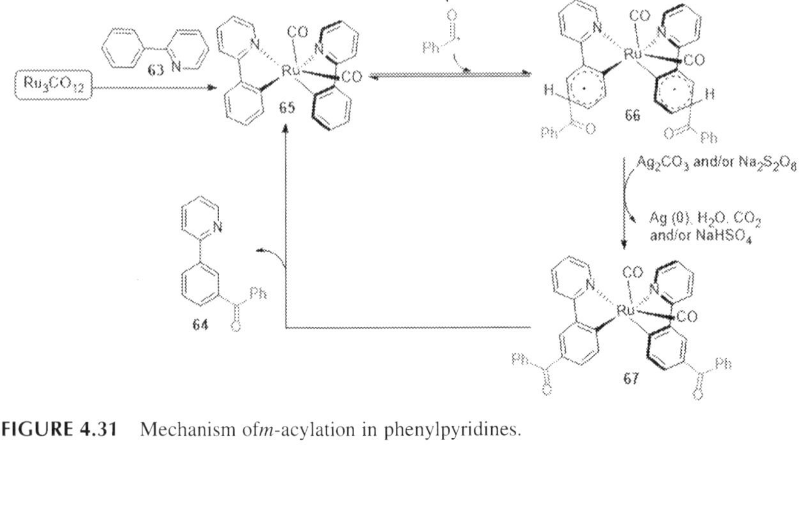

FIGURE 4.31   Mechanism of *m*-acylation in phenylpyridines.

Fontana and his group[35] described the Ag-catalyzed decarboxylative C-H functionalization of pyridines **68** and pyrazines **69** with 2-oxoacids **1**. The author has given the concept of using 2-oxoacids **1** as acylating reagent for the first time which could be used for the selective acylation of different heteroaromatic species. In this reaction, AgNO$_3$ was used as an oxidant along with (NH4)$_2$S$_2$O$_8$ in a mixture of (H$_2$O/CH$_2$Cl$_2$) or water to give a mixture of mono- and diacyl derivatives of pyridines (**71a** and **71b**) and pyrazines (**72a** and **72b**). The author has applied the same reaction on quinoline **70** in order to compare the selectivity between mono- and diacylations of (**73a** and **73b**). Different acids were also screened for better selectivity and it was observed that H$_2$SO$_4$ or CF$_3$CO$_2$H gives better selectivity than others (Fig. 4.32).

In the above approach, undesired formation of polyacylated products in most of the cases occurs and to overcome this issue Zeng and his group[36] in 2017 reported an efficient electrocatalytic scheme for the synthesis of selective monoacylated N-heteroarenes. Different heteroarenes like pyrazines **69**, quinolines **70**, and pyridines **68** were used as the starting material along with 2-oxoacids **1** to form monoacylated N-heteroarenes (**71a**, **72a**, and **73a**) in the presence of NH$_4$I (15 mol%) as a redox catalyst, 0.1 M solution of LiClO$_4$/CH$_3$CN, and hexafluoroisopropanol (HFIP) for 6 h at 70°C. The reaction was carried out in an undivided electrochemical cell with constant current electrolysis of 3 mA cm^{-2} (Fig. 4.33).

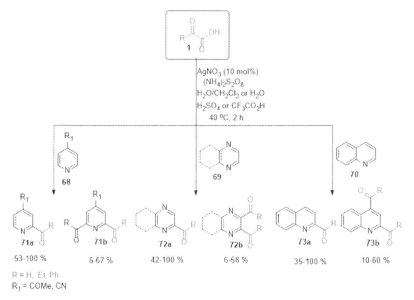

FIGURE 4.32 Functionalization of pyridines and pyrazines using 2-oxoacids.

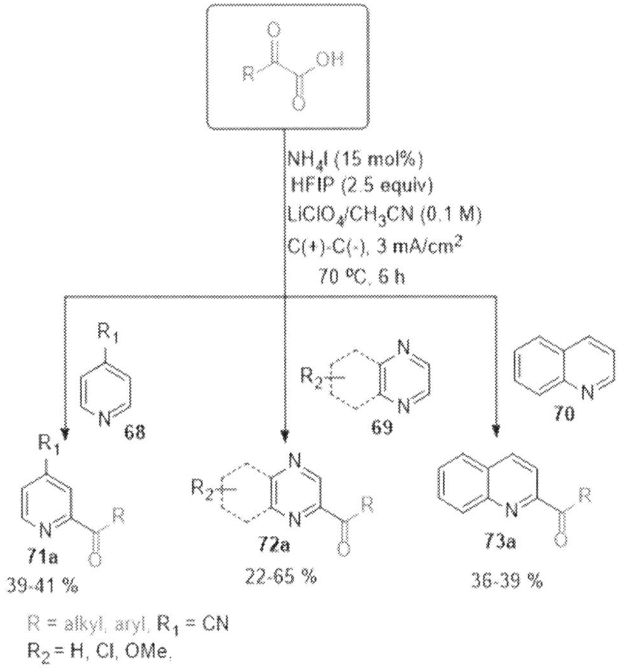

FIGURE 4.33   Acylation of pyridines, pyrazines and quinolines.

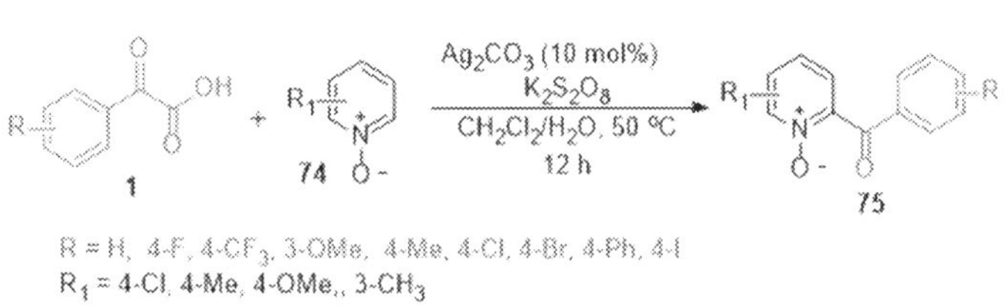

FIGURE 4.34   C2-acylation of pyridine-N-oxides.

In 2014, Muthusubramanian and coworkers[37] reported a similar work that was proposed by Fontana in 1991. The authors have done selective C2-monoacylation of pyridine-N-oxides **74** with 2-oxoacids **1** in the presence of Ag_2CO_3 (10 mol%) as a catalyst and $K_2S_2O_8$ as an oxidant in a mixture of DCM/H_2O solvent at 50°C for 12 h. Different derivatives of acylated pyridine-N-oxides **75** were prepared in 43% − 81% yields (Fig. 4.34).

Wu Zhao, and coworkers[38] have developed a Ag(I)-catalyzed approach for the coupling of pyrazines **76** with 2-oxoacids (benzoyl source) **1**. This protocol is almost similar to that of the Fontana and Minisci approach that was proposed in 1991. This method involves a reaction between pyrazines

76 and 2-oxoacids **1** in the presence of a catalytic amount of Ag_3PO_4 (10 mol%) and $K_2S_2O_8$ (2 equiv) as an oxidant. The reaction was done in a mixture of DCM/H_2O (1.4:0.6) as the solvent and stirred at 40°C for 24 h to afford acylated pyrazines **77** in good yields. Both electron-donating and electron-withdrawing substitution on 2-oxoacids and pyrazines don't affect much on the yields of the desired acylated products (Fig. 4.35).

In 2018, Ganganna Bogonda et al.[39] described Ag(I)-catalyzed synthesis of acylated 2H-indazoles **79** from 2H-indazoles **78** with α-keto acid derivatives **1** as an acyl source through a cross-coupling reaction. Different reaction conditions were screened for the synthesis of C3-acylated 2H-indazole **79**. 2H-indazole **78** was reacted with α-keto acid **1** (3 equiv) in presence of catalytic amount of $AgNO_3$ (20 mol%), and $Na_2S_2O_8$ (3 equiv) as oxidant corresponding acylated 2H-indazole 79 in an equimolar ratio of acetone:H_2O as the solvent, at room temperature for 24 h (Fig. 4.36).

Another Ag(I)-catalyzed decarboxylative acylation approach was established by Wang Hu, and coworkers[40] in 2017 to access 3-acyl quinoxaline-2 (1H)-ones **81** by performing a reaction between quinoxaline-2 ones **80** and α-keto acids **1** in an optimized condition. 3-Acyl quinoxaline-2(1H)-one **81**was obtained after stirring a reaction mixture with quinoxaline-2-one **80** and α-keto acid (2 equiv) **1** in presence of a catalytic amount of $AgNO_3$ (10 mol.%) and $K_2S_2O_8$ (2 equiv) as the oxidant in a (1:1) mixture of MeCN:H_2O as the solvent at 100°C for 3 h. In order to confirm the effect of substitution on the yield of the desired products different substituted 2-oxoacids **1** and quinoxaline-2-ones **80** were used but it was found that

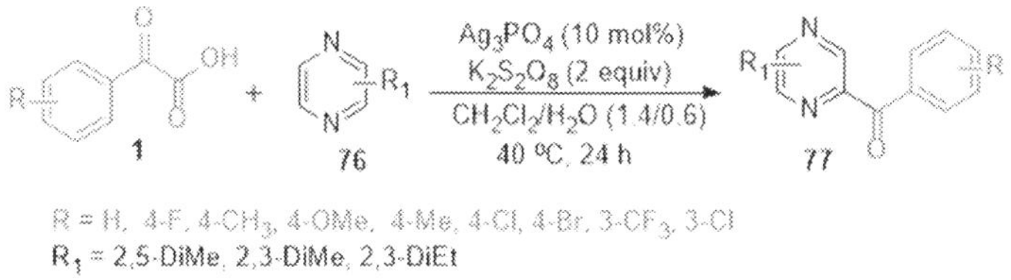

Ag3PO4 (10 mol%)
K2S2O8 (2 equiv)
CH2Cl2/H2O (1.4/0.6)
40 °C, 24 h
1
76
77
R = H, 4-F, 4-CH3, 4-OMe, 4-Me, 4-Cl, 4-Br, 3-CF3, 3-Cl
R1 = 2,5-DiMe, 2,3-DiMe, 2,3-DiEt

FIGURE 4.35 Scylation of pyrazines.

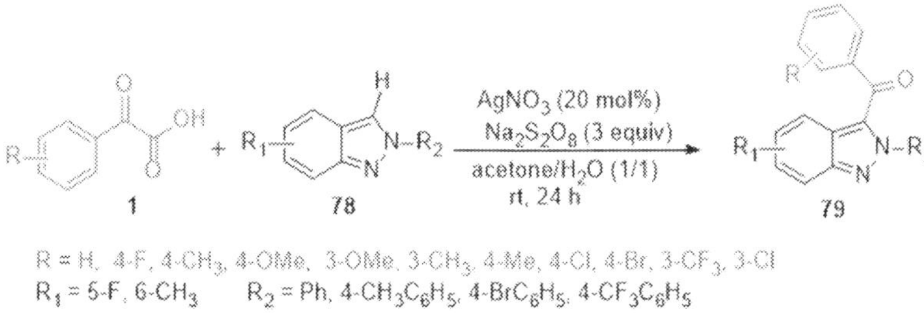

AgNO3 (20 mol%)
Na2S2O8 (3 equiv)
acetone/H2O (1/1)
rt, 24 h
1
78
79
R = H, 4-F, 4-CH3, 4-OMe, 3-OMe, 3-CH3, 4-Me, 4-Cl, 4-Br, 3-CF3, 3-Cl
R1 = 5-F, 6-CH3
R2 = Ph, 4-CH3C6H5, 4-BrC6H5, 4-CF3C6H5

FIGURE 4.36 Acylation of indazoles.

substitution has little effect on the yields of the final products. The yields of the electron donor substitution on the aryl glyoxylic acid and quinoxaline-2-ones have better yields as compared to the electron-withdrawing ones (Fig. 4.37).

The application of this approach was used for the synthesis of anticancer agents (3-benzoyl-2-piperazinyl-quinoxaline derivative) **83** in three steps. The first step involves the formation of 3-acyl quinoxalin-2(1H)-one **81** followed by reaction with thionyl chloride (SOCl$_2$) at 90°C to form 3-chloroquinoxaline **82**. The 3-chloroquinoxaline **82** then treated with 1-phenylpiperazine at 100°C for 2.5 h to yield desired compound **83** in 45% after three steps (Fig. 4.38).

A mild and direct diacylation of coumarins **84** with 2-oxoacids **1** was derived by Xin-Hua Duan and coworkers[41] in the presence of a silver catalyst in DMSO/H$_2$O (1/1) at room temperature. This protocol gives direct access to a variety of 3,4-diacylcoumarins **85a** or 3-acylcoumarins **85b** of different functional groups in moderate to excellent yields with good selectivities (Fig. 4.39).

4.1.2 Synthesis of ketone derivatives

2-Oxoacids are also used for the direct acylation on aromatic compounds via a decarboxylative cross-coupling reaction. Different arene sources are used against 2-oxoacids for the construction of C−C bonds in presence of different

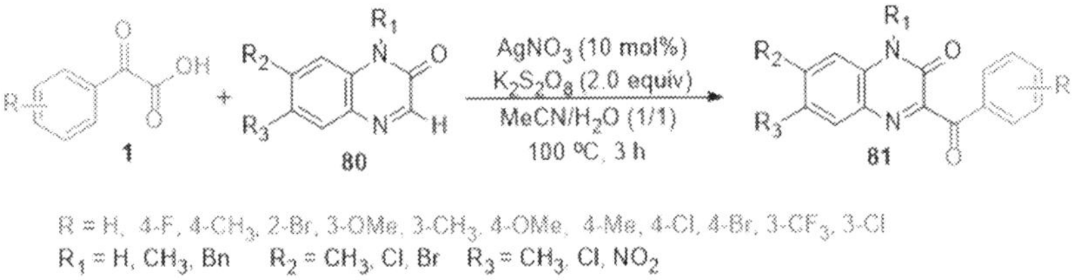

R = H, 4-F, 4-CH$_3$, 2-Br, 3-OMe, 3-CH$_3$, 4-OMe, 4-Me, 4-Cl, 4-Br, 3-CF$_3$, 3-Cl
R$_1$ = H, CH$_3$, Bn   R$_2$ = CH$_3$, Cl, Br   R$_3$ = CH$_3$, Cl, NO$_2$

FIGURE 4.37 Synthesis of quinoxaline-2(1H)-one derrivatives.

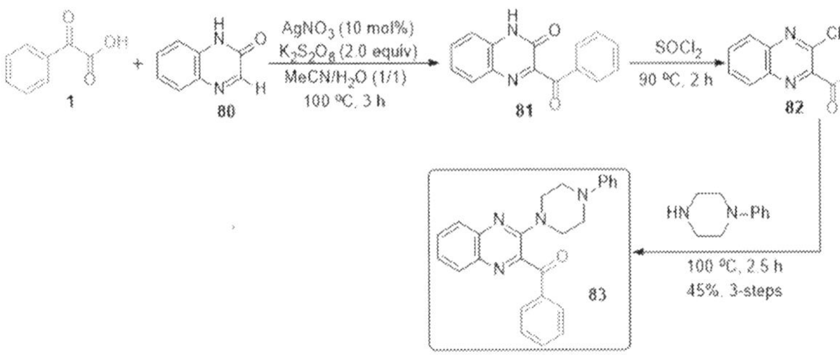

FIGURE 4.38 Synthesis of 3-benzoyl-2-piperazinyl-quinoxalines.

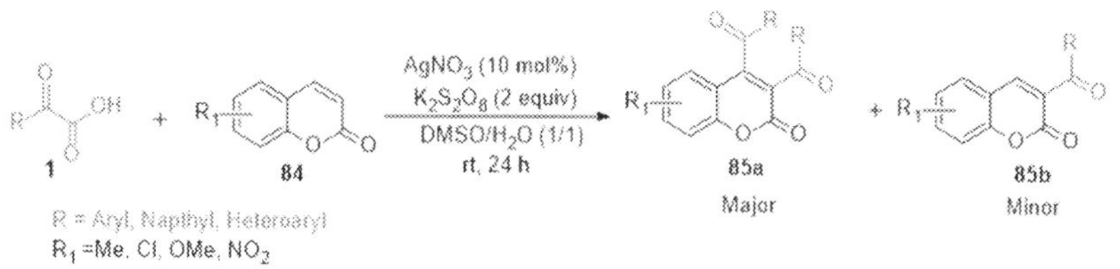

FIGURE 4.39   Synthesis of 3,4-diacylcoumarins.

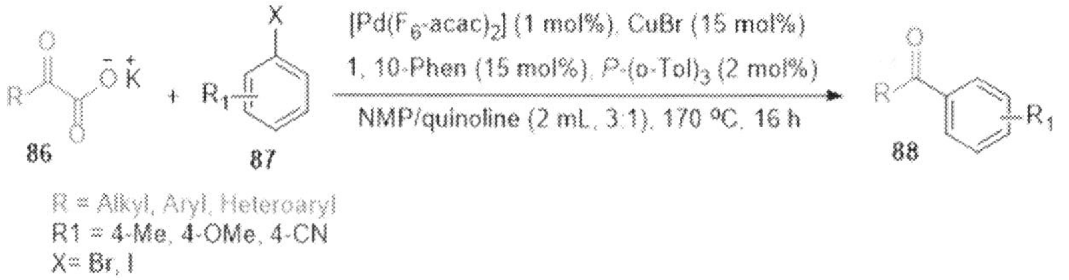

FIGURE 4.40   Acylation of aryl bromides.

transition metals as catalyst. In 2008, Lukas J. Gooßen et al. demonstrated[42] a single step decarboxylative reaction for the synthesis of symmetrical as well as unsymmetrical aryl ketones **88** from α-oxocarboxylates **86** and aryl bromides **87** via a cross-coupling approach. Different substituted aryl ketones **88** were synthesized in 5% − 99% yields in presence of Cu/Pd as a catalyst, P-(o-Tol)$_3$ and 1,10-phenanthroline as ligands after heating at 170°C for 16−36 h in NMP/quinoline (3:1) as a solvent (Fig. 4.40).

A bicyclic plausible mechanism was proposed by the authors in which copper salt plays an essential role in the decarboxylation of the α-keto acid potassium salt **86** to form intermediate **89** followed by CO_2 elimination to generate the stable copper complex **80**. Where as in another catalytic cycle, the oxidative addition of palladium to aryl halides **87** takes place leading to the formation of intermediate **91**. Intermediates **90** and **91** reacts to form intermediate **92** via transmetalation reaction followed by the reductive elimination to afford desired product **88** and regeneration of Pd(0) (Fig. 4.41).

The same group screened different coupling partners against 2-oxoacids for the synthesis of unsymmetrical ketones. In 2009, the authors have successfully reported a decarboxylative cross-coupling reaction between aryl triflates **93** and α-keto acids **86** in the presence of a Pd/Cu bimetallic system to generate biaryl ketones **94**. The reaction was successfully carried out at heating and in microwave conditions in the polar aprotic solvent (NMP). The reaction shows a broad substrate scope with carboxylate salts and aryl triflates for the construction of ketones (Fig. 4.42).[43]

An extension of Gooßen and coworkers work on the direct acylation of aryl halides **95** was carried out in 2014 by Ji and coworkers.[44] This approach

89
-CO2
decarb-
oxylation
90
[Cu]
ArBr
87
L2Pd0
oxidative
addition
91
Ar
86
-KBr
salt
exchange
trans-
metalation
reductive
elemination
88
CuBr
92
PdL2

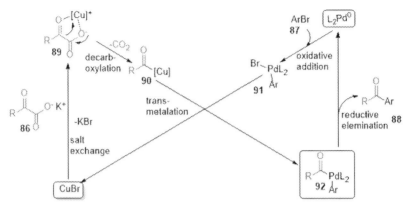

FIGURE 4.41 Proposed mechanism for the synthesis of unsymmetrical aryl ketones.

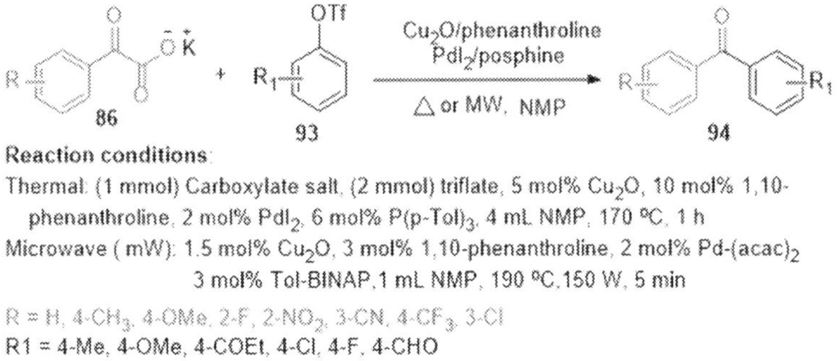

86
93
OTf
Cu2O/phenanthroline
PdI2/posphine
△ or MW, NMP
94
Reaction conditions:
Thermal: (1 mmol) Carboxylate salt, (2 mmol) triflate, 5 mol% Cu2O, 10 mol% 1,10-phenanthroline, 2 mol% PdI2, 6 mol% P(p-Tol)3, 4 mL NMP, 170 °C, 1 h
Microwave (mW): 1.5 mol% Cu2O, 3 mol% 1,10-phenanthroline, 2 mol% Pd-(acac)2
3 mol% Tol-BINAP,1 mL NMP, 190 °C,150 W, 5 min
R = H, 4-CH3, 4-OMe, 2-F, 2-NO2, 3-CN, 4-CF3, 3-Cl
R1 = 4-Me, 4-OMe, 4-COEt, 4-Cl, 4-F, 4-CHO

FIGURE 4.42 Pd/Cu bimetallic system to generate biaryl ketones.

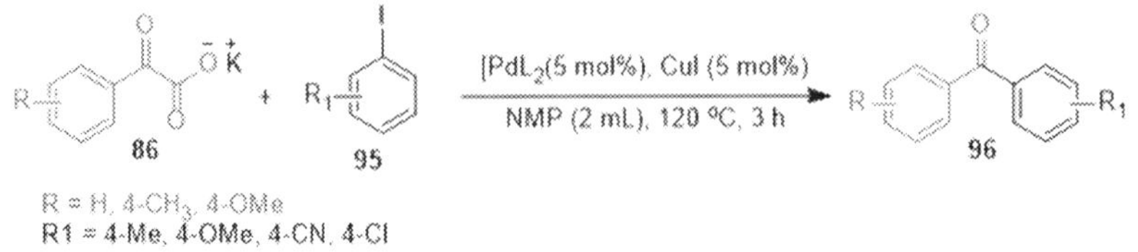

86
95
[PdL2(5 mol%), CuI (5 mol%)
NMP (2 mL), 120 °C, 3 h
96
R = H, 4-CH3, 4-OMe
R1 = 4-Me, 4-OMe, 4-CN, 4-Cl

FIGURE 4.43 Acylation of aryl halides using 2-oxoacid potassium salts.

was also done in the presence of bicatalytic system Pd/Cu and no external P-based and N-based ligands were used. Instead of aryl bromides, aryl iodides **95** furnished the best results in presence of PdI_2/CuI (5 mol.%) in NMP as the solvent at 120°C for 3 h to obtain symmetrical as well as unsymmetrical ketones **96** in 72% − 93% yields (Fig. 4.43).

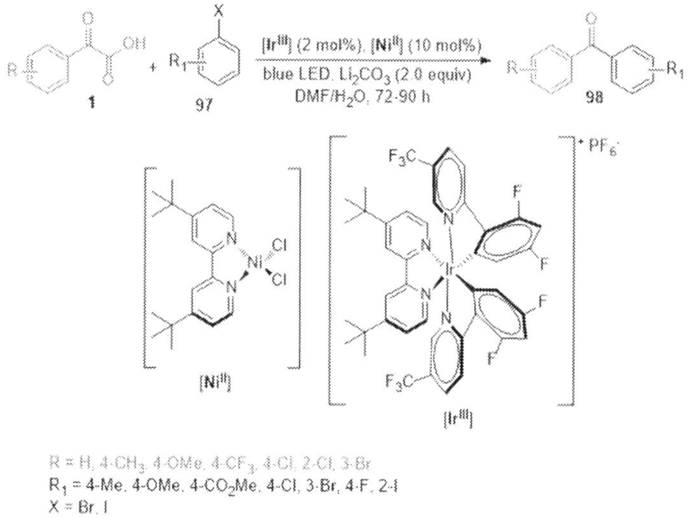
[Ir^III] (2 mol%), [Ni^II] (10 mol%)
blue LED, Li2CO3 (2.0 equiv)
DMF/H2O, 72-90 h
1
97
98
[Ni^II]
[Ir^III]
R = H, 4-CH3, 4-OMe, 4-CF3, 4-Cl, 2-Cl, 3-Br
R1 = 4-Me, 4-OMe, 4-CO2Me, 4-Cl, 3-Br, 4-F, 2-I
X = Br, I

FIGURE 4.44 Ir and Ni catalyzed acylation of ketones.

A new area of photoredox decarboxylation has been developed in the year 2014 and 2015 by using visible-light for the decarboxylation of 2-oxoacids and serves to be an efficient synthetic tool for the acylation reactions of Csp^3 and Csp^2 carbon via the formation of an acyl radical.

MacMillan and his coworkers in 2015 reported[45] the direct acylation of aryl halides **97** by using both aromatic and aliphatic derivatives of α-keto acids **1** in the presence of a bimetallic system under photoredox conditions. In this protocol, an iridium polypyridyl complex $[Ir^{III}]$ and nickel catalysts ($[Ni^{II}]$) were used that shows the synergistic effect of visible-light-mediated photoredox reaction in presence of Li_2CO_3 and DMF/H_2O as a solvent to afforded symmetrical and unsymmetrical ketones **98** in good to excellent yields (Fig. 4.44).

Wan-Min Cheng et al. have established[46] a similar approach that of the MacMillan's work and published in the same year by using photoredox catalyst $[Ir^{III}]$. The authors have also used the Pd-catalyst instead of $[Ni^{II}]$ and this method was used for the direct acylation and amidation of aryl halides **97** with α-keto acids **1** and potassium oxalate monoamides **99** that leads to the formation of unsymmetrical ketones **98** and amides **100** in good yields. The authors took advantage of the synergistic effect of the iridium visible-light photoredox- and palladium-catalyzed activation of the aryl halides. Different substituted ketones **98** and amides **100** were synthesized in 10% − 97% yield by employing Ni-xantphos as a ligand, NaOAc, and 36 W blue LED light in DMF at 25°C for 20 h (Fig. 4.45).

The C-C bond formation reactions using transition metals have proved to be an efficient and most important area in organic synthesis. Most of the C-C bond formation reactions such as Sonogashira, Heck, Suzuki − Miyaura,

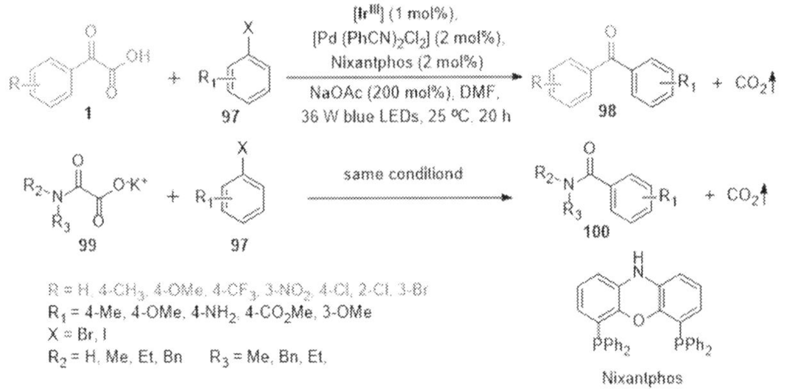

FIGURE 4.45   Ir(III) catalyzed synthesis of substituted ketones.

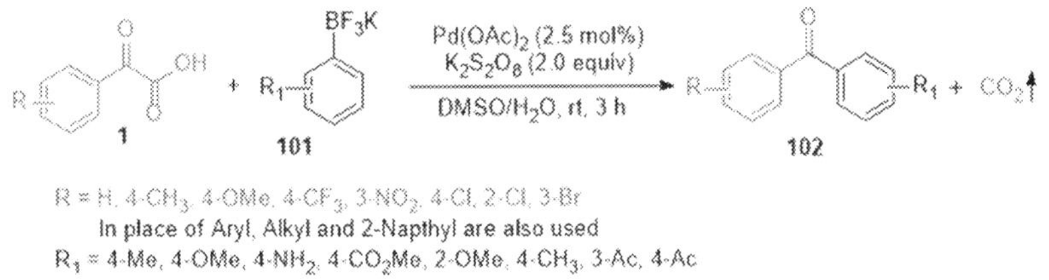

FIGURE 4.46   Formation of ketones from 2-oxoacids.

Stille, and Buchwald − Hartwig are used for the synthesis of biologically active molecules. In 2011, Ge and coworkers have reported[47] the Pd-catalyzed decarboxylative cross-coupling reactions of potassium aryltrifluoroborates **101** with 2-oxoacids **1** in the presence of 2.0 equiv of $K_2S_2O_8$ as an oxidant in a mixture of DMSO/H_2O as the solvent at room temperature for 3 h. Library of substituted symmetrical and unsymmetrical ketones **102** were synthesized upto 41% − 98% yields (Fig. 4.46).

Inspired by Ge and coworkers work of C-C bond formation reaction between potassium aryltrifluoroborates **101** and α-oxocarboxylates **86** in the presence of Pd (II) to form different substituted ketones, Chang and coworkers[48] demonstrated the same concept in presence of Ag(I) species as a catalyst. α-oxocarboxylates **86** is treated with potassium aryltrifluoroborates **101** which undergoes cross-coupling reaction in presence of a catalytic AgNO₃ (5 mol.%) and $K_2S_2O_8$ as an oxidizing agent in water as a solvent at room temperature for 1 h to obtain the corresponding ketones **102**. Both electron-donating and electron-withdrawing groups were tested at different positions against the optimized condition and found that substitution/position doesn't affect much on the yields of the desired product (Fig. 4.47).

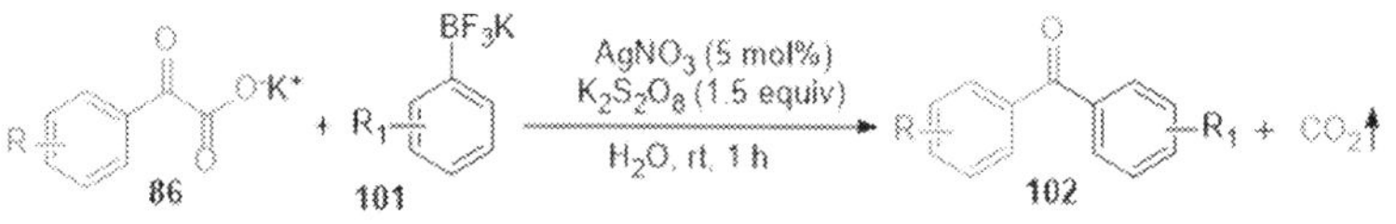

R = H, 4-CN, 4-CH_3, 4-OMe, 4-CF_3, 4-NO_2, 4-Cl, 2-Cl, 3-Br, 2-NO_2
R_1 = 4-Me, 4-OMe, 4-Cl, 4-Br, 4-BnO, 4-CH_3, 4-I, 4-NO_2, 2-CH_3, 3-CH_3

FIGURE 4.47 Formation of ketones from potassium aryltrifluoroborate salt.

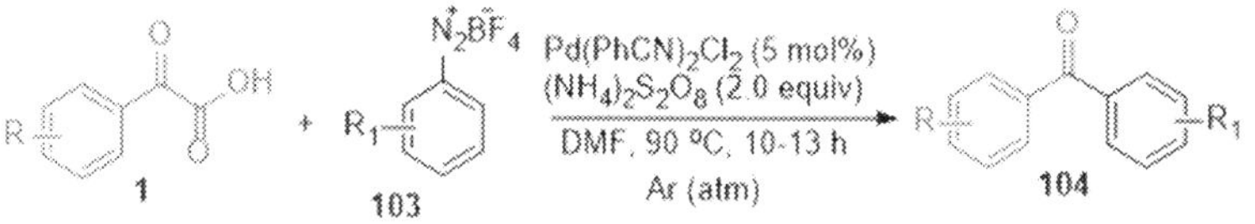

R = H, 4-CH_3, 4-OMe, 4-CF_3, 4-NO_2, 4-Cl, 2-Cl, 3-Br, 2-NO_2, 4-C_2H_5
R_1 = 4-Me, 4-OMe, 4-C_2H_5, 4-Cl, 4-Br, 4-CO_2Me, 3-CN, 4-CH_3, 4-I, 4-NO_2
2-CH_3, 3-CH_3, 2-NO_2

FIGURE 4.48 Acylation of diazonium tetrafluoroborate.

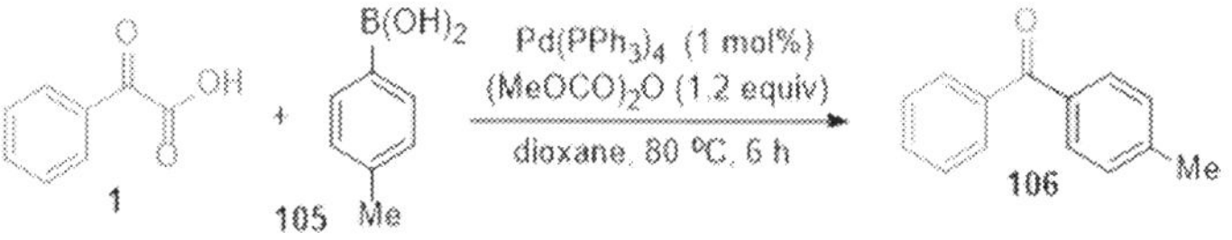

FIGURE 4.49 Acylation of arylboronic acids.

Recently, Subir Panja et al. reported[49] an efficient, ligand-free Pd(II)-catalyzed decarboxylative cross-coupling reaction between α-keto acids **1** and aryl diazonium tetrafluoroborate **103** to generate unsymmetrical diaryl ketones **104**. Various unsymmetrical diaryl ketones **104**were obtained ontreatment of aryl-substituted glyoxylic acid **1** with aryl diazonium tetrafluoroborate **103** in the presence of 5 mol% of Pd(PhCN)$_2$Cl$_2$ and 2.0 equiv of $(NH_4)_2S_2O_8$ in DMF under argon atmosphere after heating at 90°C for 10–13 h (Fig. 4.48).

Ryuki Kakino et al. reported[50] a unique protocol for the decarboxylative C-C bond formation of nucleophilic organoboron compound **105** with phenyl glyoxylic acid **1** to form 4- methylbenzophenone **106** in 75% yields. The reaction was carried between *p*-tolylboronic acid and phenyl glyoxylic acid in the presence of Pd (PPh$_3$)$_4$ (1 mol%) as a catalyst, and anhydride as an activator in dioxane solvent at 80 °C for 6 h to produce 4- methylbenzophenone **106** (Fig. 4.49).

Inspired by the work of Ryuki Kakino et al. Qi and coworkers[51] in 2014, described Ag-catalyzed decarboxylative cross-coupling reactions between aryl boronic acids **107** and aryl-substituted glyoxylic acids **1** to gave unsymmetrical ketones **108**. The reactions were stirred at 60°C for 1 h in the presence of Ag_2CO_3 (10 mol%) as a catalyst. This protocol was efficient to construct a library of substituted unsymmetrical ketones **108** by using a variety of substituted electron-donating, electron-withdrawing, heterocyclic, and sterically hindered glyoxylic acids **1** and aryl boronic acid **107** in 69% − 95% yields (Fig. 4.50).

The authors have shown the application of this protocol by conducting three different experiments as shown below. In one experiment, 4-(carboxycarbonyl) benzoic acid **1b** was directly coupled with phenylboronic acid **107a** in presence of an Ag$_2$CO$_3$ catalyst to access *p*-benzoyl benzoic acid **108a** in 75% yield (**Aa**). This experiment shows the preference of the decarboxylative acylation over arylation under the optimized conditions. In the second experiment, the author has done one-pot, two-step reaction for the synthesis of fluorenone **109** (**Ab**). 2-(carboxycarbonyl) benzoic acid 1c reacts with phenylboronic acid **108a** to form decarboxylative acylation product **108b** followed by the cyclization reaction in presence of K$_2$S$_2$O$_8$ to form fluorenone **109** in 62% yield (Fig. 4.51A).

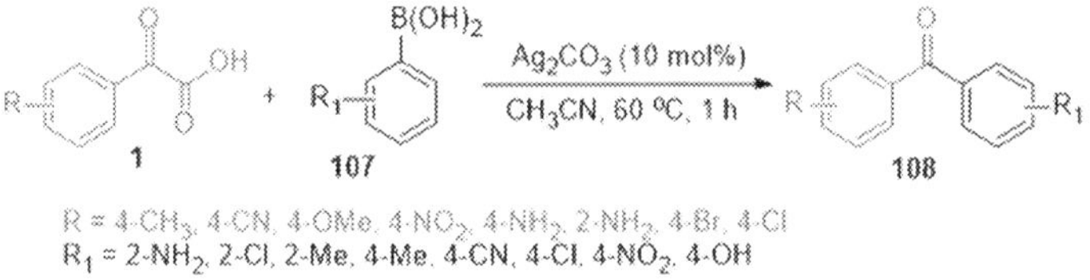

R = 4-CH$_3$, 4-CN, 4-OMe, 4-NO$_2$, 4-NH$_2$, 2-NH$_2$, 4-Br, 4-Cl
R$_1$ = 2-NH$_2$, 2-Cl, 2-Me, 4-Me, 4-CN, 4-Cl, 4-NO$_2$, 4-OH

FIGURE 4.50   Acylation of arylboronic acids.

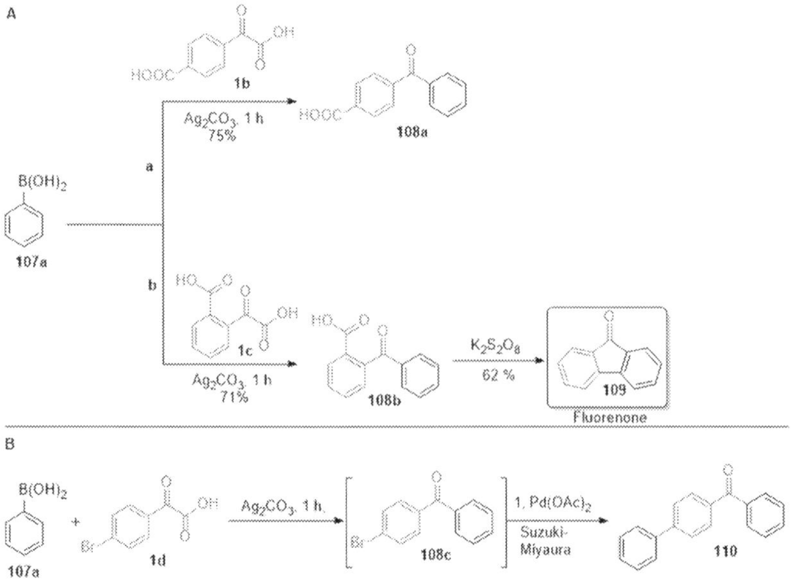

FIGURE 4.51   Acylation of arylboronic acids to form carbonyl compounds.

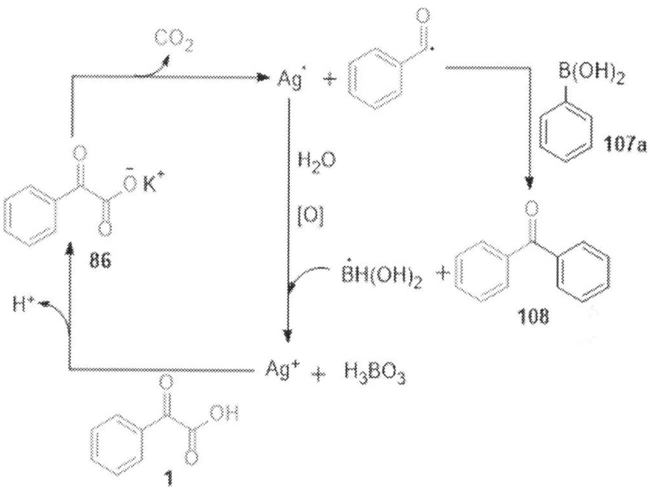

FIGURE 4.52 Proposed mechanism for the synthesis of aryl ketones.

In the third experiment (**B**), the authors have shown the decarboxylative acylation reaction in combination with Pd-catalyzed Suzuki—Miyarau cross-coupling reaction. A reaction between 4-bromo-phenyl glyoxylic acid **1d** and phenylboronic acid **107a** in presence of Ag_2CO_3 as a catalyst to yield acylated product **108c** which after Suzuki cross-coupling in presence of Pd (OAc)$_2$ affords 4-biphenylyl-benzophenone **110** in 87% yield (Fig. 4.51B).

The authors have proposed a free radical pathway and this was confirmed by the radical trapping experiments with hydroquinone and TEMPO (2,2,6,6-tetramethylpiperidine-1-oxyl). The proposed mechanism involves a silver catalyst to form acyl radical from 2-oxoacids **1** followed by reaction with boronic acid **107a** to form the desired product **108** (Fig. 4.52).

4.1.3 Synthesis of thioesters

In recent times, chemists have shown key interest in C—S bond formation reactions because sulfur-containing scaffolds like cysteine, allicin, salacinol, leinamycin, (-)-chaetocin are extensively found in various natural products while as many others like Penicillin, Epivir, and Casodex are used as drugs. Therefore, various methods have been reported for the development of C—S bond by using different sulfur-containing starting materials.[52] Thioesters are the carboxylic acid derivatives that are present in inactivated form and considered important due to their significant applications as versatile building blocks in synthetic chemistry to produce biologically active scaffolds.[53] 2-oxoacids are used as acylating reagents for the synthesis of different substituted thioesters using two different thiol sources.

Guangwei Rong et al. established[54] an efficient and mild copper-catalyzed decarboxylative cross-coupling approach for the synthesis of thioesters (**113** or **114**) through C—S bond formation from α-keto acids **1** and diphenyl disulfides **112** or thiophenols **111**. The reaction between 2-oxoacids **1** and diphenyl disulfides **112** or thiophenols **111** gave the corresponding thioesters (**113** and **114**) in good to moderate yields in presence of 20 mol% of CuO, $(NH_4)_2S_2O_8$ as an oxidizing reagent in a mixture of DMSO/H_2O as a solvent at 80° for 12 h. Different substituted 2-oxoacids and thiols were tested against the optimized condition and a total of 25 thiol esters (**113** or **114**) were synthesized (Fig. 4.53).

The authors have shown a free radical pathway initiated by the copper catalyst for the synthesis of thioesters (**113** or **114**). The first step involves the formation of benzoyl radical from benzoylformic acid **1** by copper(II) catalyst with the release of CO_2 and Cu(I). The benzoyl radical then reacts with thiophenol/disulfide (**111/112**) to form thioester (**113** or **114**). The copper(I) ion is oxidized by ammonium persulfate to give copper(II) which repeats the reaction cycle. (Fig. 4.54).

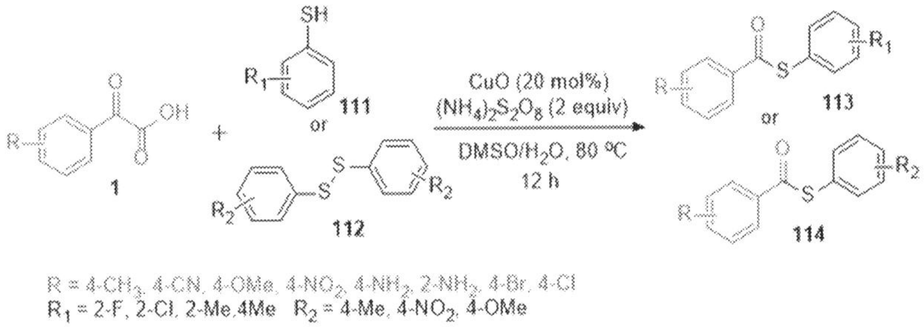

R = 4-CH₃, 4-CN, 4-OMe, 4-NO₂, 4-NH₂, 2-NH₂, 4-Br, 4-Cl
R₁ = 2-F, 2-Cl, 2-Me,4Me   R₂ = 4-Me, 4-NO₂, 4-OMe

FIGURE 4.53 Synthesis of thioesters from 2-oxoacids.

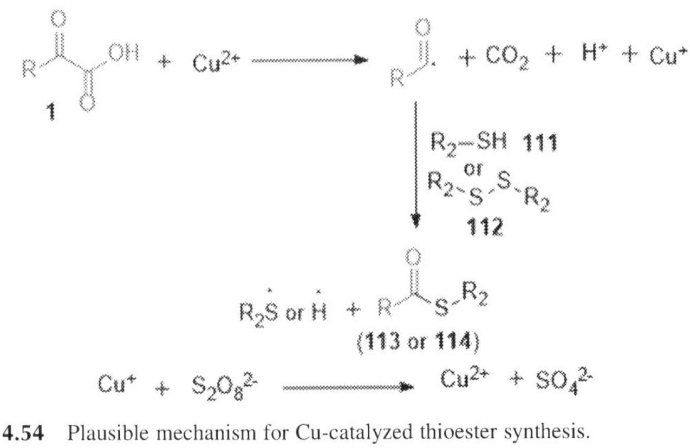

FIGURE 4.54 Plausible mechanism for Cu-catalyzed thioester synthesis.

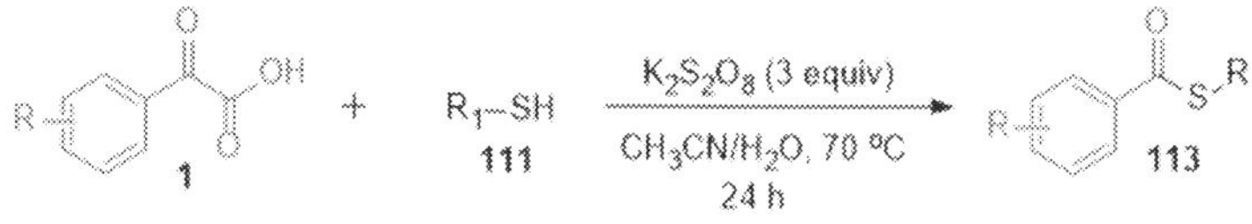

R = H, 4-CH₃, 4-Br, 4-OMe, 4-Br, 4-Cl
R₁ = Ph, 2-MePh, 3-MePh, 4-ClPh, 4-BrPh, n-butyl

FIGURE 4.55 Synthesis of thioesters.

Another C-S bond formation reaction between 2-oxoacids and thiophenols was reported by Hua Wang and his coworkers[55] in a metal-free condition. The authors have described an efficient, novel, and catalyst-free approach for the synthesis of thioesters **113** from α-keto acids **1** and thiols **111** in presence of 3 equiv of $K_2S_2O_8$ as an oxidant. The reaction was carried out in a mixture of CH_3CN/H_2O as a solvent at 70 °C for 24 h. A library of thioesters **113** was synthesized through this method by using different substituted α-keto acids **1** and thiols **111** (Fig. 4.55).

4.1.4 Synthesis of α,β-unsaturated carbonyl compounds

α,β-Unsaturated carbonyls are an important class of compounds that are naturally occurring or synthetically prepared and are showing different pharmaceutical and biological activities such as antimicrobial, anticancer, anti-inflammatory, antioxidant, antimutagenic, antianginal, antimalarial, anti-allergic, antihepatotoxic, and antimitotic activities.[56] These compounds also serve as an attractive intermediate for the synthesis of different biologically active heterocycles and in functional materials.[57] Due to these broader applications in medicine, biology, and materials science, various efforts have been made to establish different synthetic schemes for the synthesis of α,β-unsaturated carbonyl compounds.[58] Traditionally, aldol condensation reaction between ketones and aldehydes is widely used for the synthesis of α,β-unsaturated ketones. In recent couple of years, 2-oxoacids have been used for the synthesis of different substituted α,β-unsaturated carbonyl compounds under different optimized conditions.

One of the simple approaches for the construction of α,β-unsaturated carbonyl compounds was established by Shang Wu and coworkers[59] in 2017. The authors have described a decarboxylative cross-coupling reaction between 2-oxoacids **1** and alkenes **115** in presence of 10 mol.% of Ag_2CO_3 as a catalyst, and 3 equiv of $K_2S_2O_8$ as an oxidant in CH_3CN at 100°C for 24 h to form α,β-unsaturated carbonyl compounds **116** in moderate to high yields. This approach has a wide range of substrate tolerance on both 2-oxoacids **1** and alkenes **115** (Fig. 4.56).

The authors have proposed a free radical mechanism that was initiated by a low valent Ag(I) species in the presence of $K_2S_2O_8$ to oxidize it into a

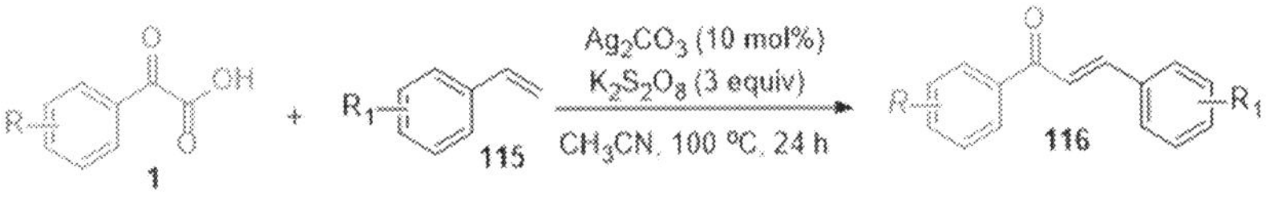

R = H, 2-Me, 4-Me, 2-Cl, 2-Br, 4-Cl, 2-OMe
R₁ = H, 4-Me, 2-Br, 2-Cl, 4-Br

FIGURE 4.56   Synthesis of chalcones.

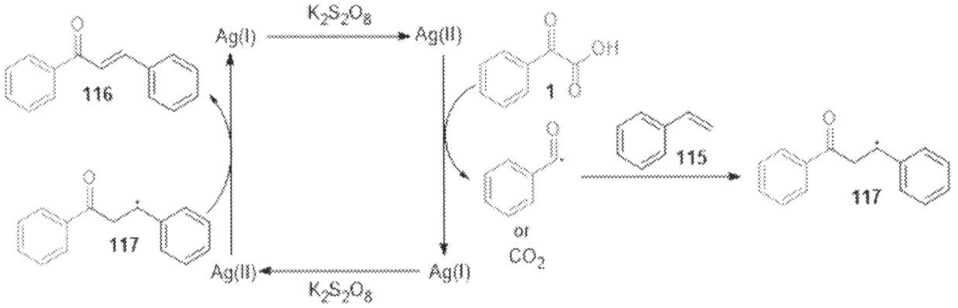

FIGURE 4.57   A plausible mechanism for α,β-unsaturated carbonyl compound synthesis.

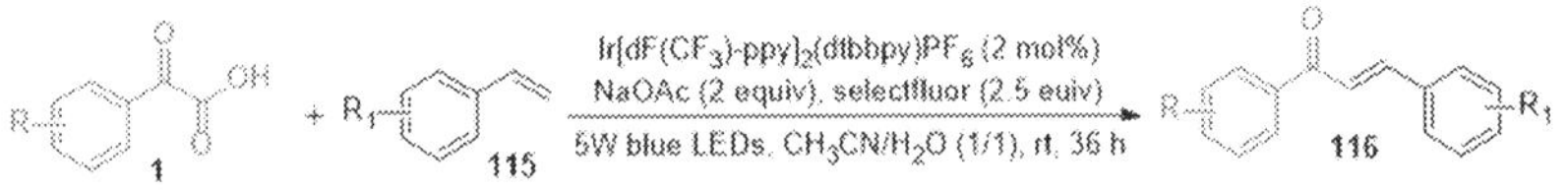

R = H, 4-OMe, 4-CF3, 2-Me, 4-Me, 2-Cl, 2-Br, 3-Br, 4-Cl, 2-OMe
R₁ = H, 4-Ph, 4-Me, 2-Br, 2-Cl, 4-Br, 3-Cl, 4-Br, 3-Me

FIGURE 4.58   Photocatalyzed synthesis of chalcones.

higher valent Ag(II) species. 2-oxoacids **1** undergoes decarboxylation from Ag (II) species to form acyl radical along with Ag(I) species. The acyl radical reacts with arenes **115** to form a radical intermediate **117** followed by the oxidation with Ag(II) species to form the final product **116** along with the low valent silver species (Fig. 4.57).

A decarboxylative cross-coupling reaction between α-oxocarboxylic acids **1** and styrene **115** to form α,β-unsaturated ketones (chalcones) **116** has been reported.[60] This domino-fluorination−protodefluorination decarboxylative coupling reaction occurs in presence of Ir[dF(CF₃)-ppy]₂(dtbbpy)PF₆ as a photocatalyst at room temperature in 36 h to form the corresponding α,β-unsaturated ketones **116**. The reaction is carried in presence of 5 W blue LED light. The important step in this reaction is the carbon − fluorine (C − F) bond without the formation of side products. In this strategy a variety of functional groups are tested on both α-keto acids **1** and styrene **115** to generate the corresponding α,β-unsaturated ketones **116** (Fig. 4.58).

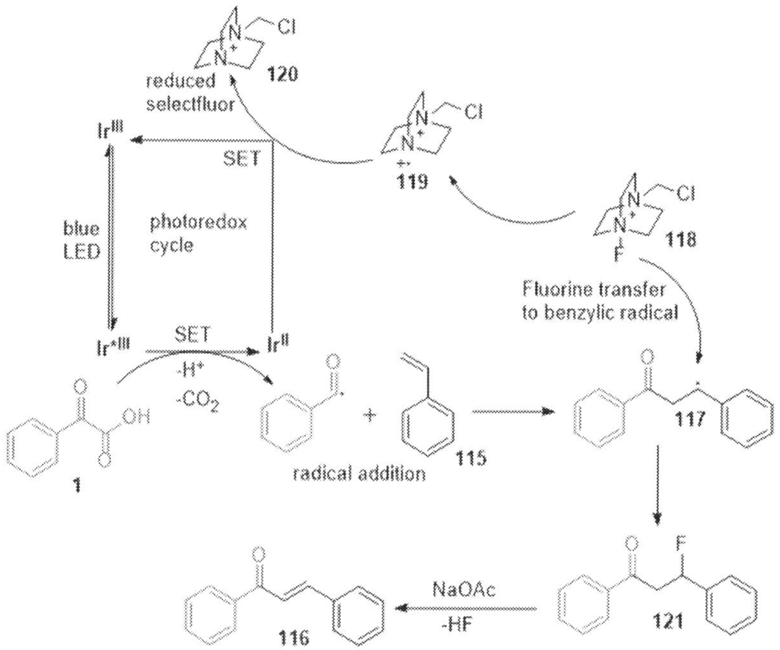

FIGURE 4.59 Single electron transfer mechanism for chalcone synthesis.

The authors have proposed the Single Electron Transfer (SET) mechanism under the visible-light irradiation. The photocatalyst Ir[dF(CF$_3$)-ppy]$_2$(dtbbpy)PF$_6$ in presence of blue LED leads to the formation of an excited state Ir*^{III} by the process of metal-to-ligand charge transfer. α-keto acid **1** oxidizes in the presence of Ir*^{III} and Ir^{II} species. The carboxyl radical loses CO$_2$ to form the acyl radical species followed by reaction with styrene **115** to generate benzylic radical **117**. At this step, C−F bond formation takes place by the direct fluorine transfer from selectfluor **118** to the benzylic radical and radical cation **119**. This radical cation **119** behaves as an oxidant to convert reduced-state Ir^{II} into the ground-state Ir^{III} (1) species and completes the catalytic cycle along with **120**. The acylfluorinated product **121** is unstable in presence of NaOAc and eliminates HF to form α,β-unsaturated ketone **116** (Fig. 4.59).

Another decarboxylative cross-coupling reaction of 2-oxoacids was reported by Ning Zhang et al.[61] In this protocol, the authors described the silver-catalyzed double-decarboxylative cross-coupling reaction of α-keto acids **1** with cinnamic acids **122** for the synthesis of substituted α,β-unsaturated carbonyls **123** in good yields. The reaction was carried out in presence of 10 mol.% of AgNO$_3$, 2 equiv of Na$_2$S$_2$O$_8$, and 1 equiv of K$_2$CO$_3$ in an aqueous medium at 100°C for 24 h. A series of α,β-unsaturated compounds **123** was synthesized in moderate to good yields under the mild optimized conditions (Fig. 4.60).

The authors have proposed the free radical catalytic mechanism for the above reaction. Initially, peroxodisulfate oxidizes Ag(I) into Ag(II) cation followed by the reaction of α-keto acid anion **124** with Ag(II) cation to form the acyl radical along with CO_2 and Ag(I) cation. Later, the acyl radical reacts with the double bond of cinnamate anion **125** to form the intermediate **126**. Finally, intermediate **126** is converted into the desired product **123** along with the loss of carbon dioxide and Ag(I) cation (Fig. 4.61).

One more decarboxylative cross-coupling reaction between 2-oxoacids **1** and acrylic acids **127** in the presence of Fe(II) as a catalyst was established by Can-Cheng Guo and coworkers[62] for the synthesis of α,β-unsaturated carbonyls **128**. On stirring a reaction between 2-oxoacids **1** and acrylic acids **127** in the presence of 10 mol.% of $FeCl_2$ along with 2.5 equiv of $K_2S_2O_8$ as an oxidant in a mixture of $DMSO/H_2O$ at 120°C for 15 h to form the corresponding α,β-unsaturated carbonyls **128** in good yields. This method shows wide substrate scope and good functional group tolerance to access an important class of the corresponding α,β-unsaturated carbonyls (Fig. 4.62).

Furthermore, Lukas J. Gooßen and his coworkers[63] successfully established an efficient Pd-catalyzed decarboxylative allylation of α-oxocarboxylates **129**

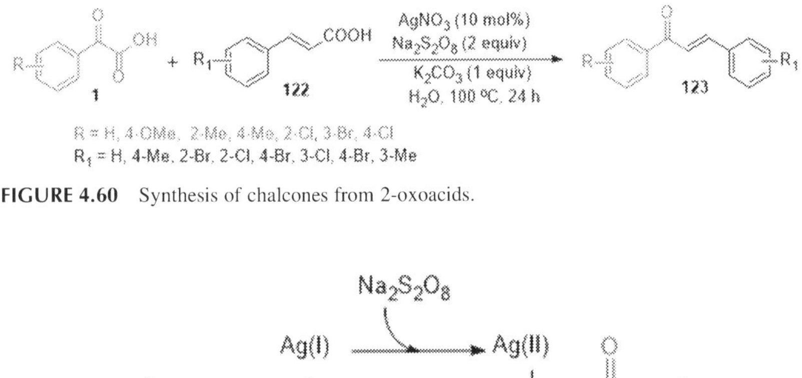

R = H, 4-OMe, 2-Me, 4-Me, 2-Cl, 3-Br, 4-Cl
R_1 = H, 4-Me, 2-Br, 2-Cl, 4-Br, 3-Cl, 4-Br, 3-Me

FIGURE 4.60 Synthesis of chalcones from 2-oxoacids.

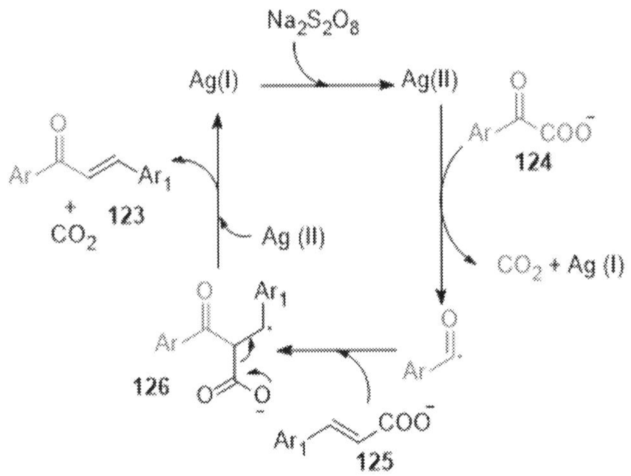

FIGURE 4.61 Proposed mechanism for silver-catalyzed double-decarboxylative reaction.

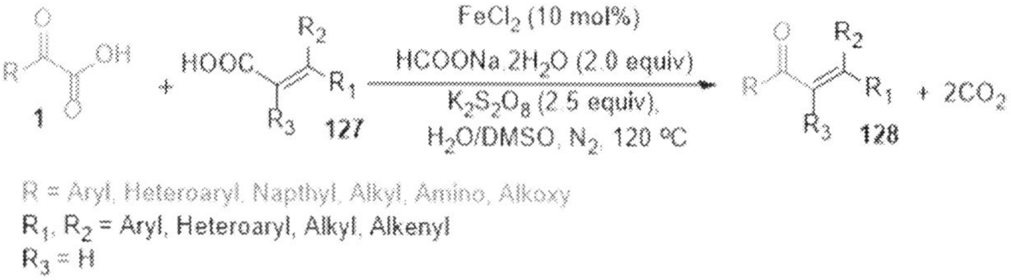

R = Aryl, Heteroaryl, Napthyl, Alkyl, Amino, Alkoxy
R_1, R_2 = Aryl, Heteroaryl, Alkyl, Alkenyl
R_3 = H

FIGURE 4.62   Synthesis of carbonyl compounds from acrylic acid.

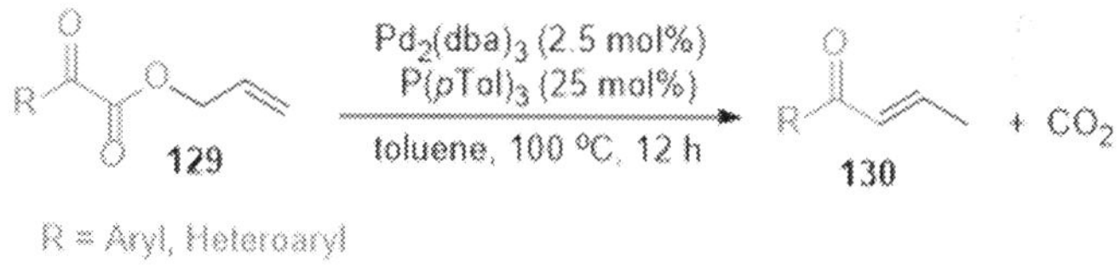

R = Aryl, Heteroaryl

FIGURE 4.63   Substituted aliphatic α,β-unsaturated ketones.

R = Aryl, Heteroaryl, Napthyl

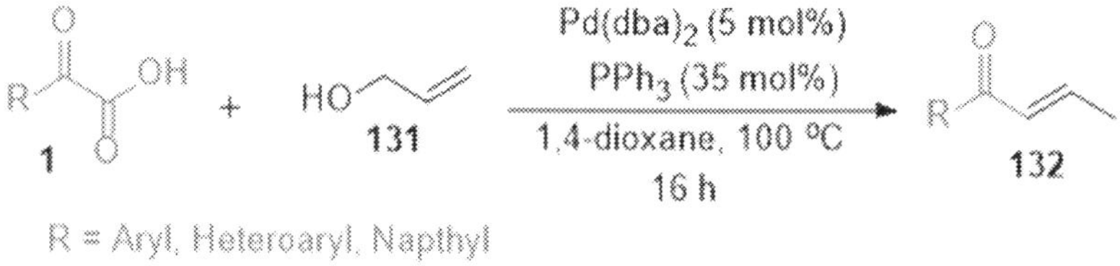

FIGURE 4.64   Synthesis of aliphatic unsaturated ketones via Pd catalysis.

in presence of $Pd_2(dba)_3$ (2.5 mol.%)and P(pTol)$_3$ (25 mol.%) in toluene at 100°C for 12 h. Different substituted α,β-unsaturated ketones 130 were synthesized by using this protocol (Fig. 4.63).

Later on, the same group has described the decarboxylative allylation of 2-oxoacids 1 with allyl alcohol 131 in presence of 5 mol.% of Pd(dba)$_2$ as a catalyst and 35 mol.% of PPh$_3$ as a ligand in 1,4-dioxane at 100°C for 16 h. Different substituted 2-oxoacids 1 were tested against the optimized condition to generate the corresponding α, β-unsaturated ketones 132 (Fig. 4.64).[64]

Furthermore, the same concept was applied with diallyl carbonate 133 for the synthesis of α,β-unsaturated ketones 134. On stirring a reaction mixture of 2-oxoacids 1 and diallyl carbonate 133 in presence of a catalytic amount of $Pd_2(dba)_3$ and P(pTol)$_3$ in 1,4-dioxane at 100°C for 12 h afforded α,β-unsaturated ketones 134 in good yields along with the loss of CO_2 (Fig. 4.65).[65]

Enaminones consist of an amino group bonded to carbonyl group via C = C bond. They serve as synthetic intermediates that link ambident nucleophilicity of enamines and enones. They are stable due to carbonyl group,

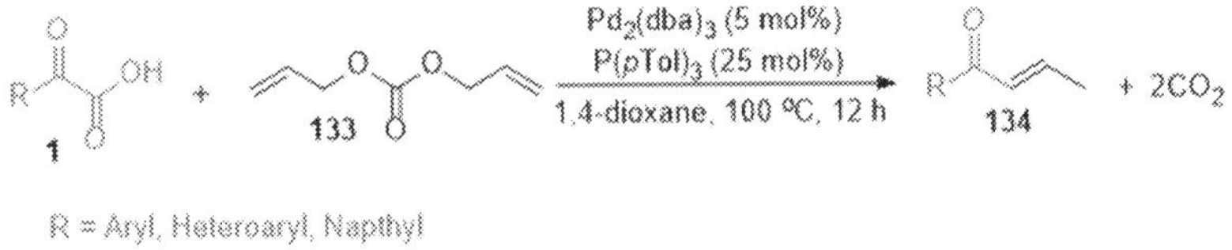

R = Aryl, Heteroaryl, Napthyl

FIGURE 4.65 Synthesis of α,β-unsaturated ketones from diallyl carbonates.

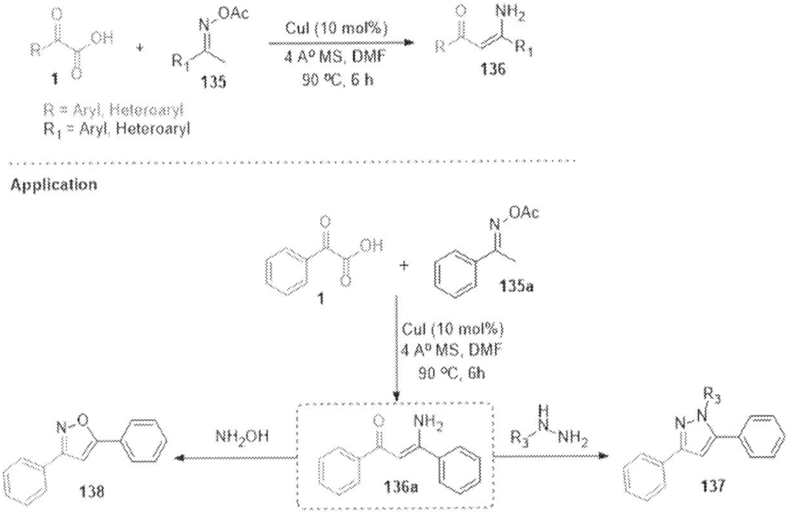

R = Aryl, Heteroaryl
R₁ = Aryl, Heteroaryl

- -

Application

FIGURE 4.66 Synthesis of enaminones.

conjugated to the enamine moiety, and hence can be easily prepared, isolated, and stored. A new Cu(I)-catalyzed decarboxylative cross-coupling approach was reported by Deng, Jiang, and coworkers[66] to access the corresponding enaminones **136** from α-keto acids **1** and oxime acetate derivatives **135**. The reaction is carried out in a sealed tube by mixing α-keto acid **1** with oxime acetate **135** in presence of a catalytic amount of CuI and (4 Å Ms) molecular sieves in DMF at 90°C for 6 h. By using this optimized condition, different derivatives of enaminones **136** were prepared in good o moderate yields. This methodology was successfully applied for the gram-scale synthesis of enaminones.

The obtained enaminone is used for the synthesis of different heterocyclic compounds. These enaminones undergo cyclization reaction in presence of hydrazine derivatives and hydroxylamine hydrochloride to generate the corresponding pyrrole **137** and 3,5-diphenylisoxazole **138** derivatives (Fig. 4.66).

4.1.5   Amide synthesis

Amides are an important class of organic functional groups and are present in various natural products and biologically active molecules like Caffeine, Nicotinamide, Ergovaline, Plitidepsin and many drugs, for example, Paracetamol, Afatinib, Methotrexate, and so on.[67] Due to the importance of the amide group, various strategies have been developed from time to time by using different starting materials.[68] The most frequently used strategy for the construction of different substituted amides is the coupling of amines with carboxylic acids or carboxylic acid derivatives in presence of different coupling reagents.[69] In recent years, α-keto acids have been repeatedly used for the construction of C−C, C−S, C−O, and C−N bonds through the decarboxylative approach. From time to time, various research groups have been working for the synthesis of different amides by conducting a coupling between α-keto acids and amines in different optimized conditions.

In 2014, Aiwen Lei and his coworkers reported[70] the decarboxylative and oxidative amidation of 2-oxoacids **1** with amines **139** in presence of visible-light under the oxygen atmosphere. A visible-light-mediated reaction was conducted between α-oxoacids **1** and amines **139** in presence of 1 mol.% of [Ru(Phen)$_3$]Cl$_2$ as a photoredox catalyst in DMSO under the oxygen atmosphere (balloon) for 36 h to furnish corresponding amides **140**. This protocol has a wide range of functional group tolerance on both 2-oxoacids and amines (Fig. 4.67).

Two pathways (Path **A** and **B**) have been proposed by the authors for the synthesis of amides. In the presence of visible light, the ruthenium complex **I** gets excited and forms intermediate **III**. In Path **A**, the intermediate **III** is quenched by amine **139** to form intermediate **II** by a reductive quenching mechanism along with a radical cation of amine **141** followed by a SET from O$_2$ to regenerate intermediate **I** along with superoxide radical-anion of oxygen. This radical-anion then abstracts an electron from **142** to form two intermediates **144** and **143** followed by the decarboxylation of **143** to form the acyl radical **145**. This acyl radical then reacts with amine **139** in presence of **144** to generate amide radical-anion **146** along with the loss of H$_2$O$_2$ and finally undergoes SET to afford the desired amide product **140**. Whereas in Path **B**, compound **III** forms compound **IV** by an oxidative quenching

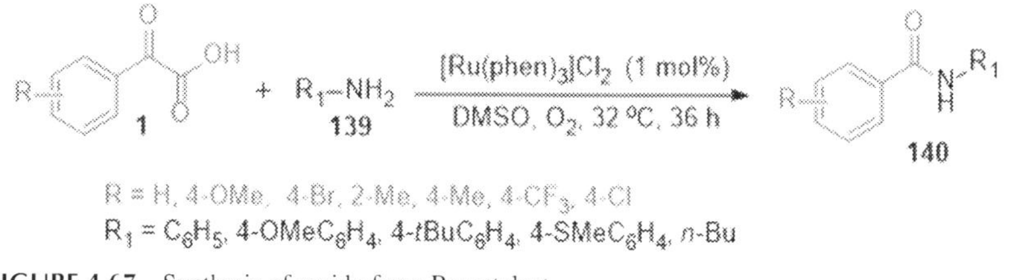

R = H, 4-OMe, 4-Br, 2-Me, 4-Me, 4-CF$_3$, 4-Cl
R$_1$ = C$_6$H$_5$, 4-OMeC$_6$H$_4$, 4-tBuC$_6$H$_4$, 4-SMeC$_6$H$_4$, n-Bu

FIGURE 4.67   Synthesis of amide from Ru catalyst.

process in presence of O2 and abstracts electron from **142** to form **143** along with the regeneration of compound **I** (Fig. 4.68).

Inspired by Lei's work for the direct synthesis of amides from 2-oxoacids and amines, Xiao-Lan Xu et al.[71] have reported silver-initiated decarboxylative amidation reaction of α-keto acids **1** with amines **139** to form different substituted amides **140**. This reaction between α-keto acids **1** and amines **139** in presence of 2 equiv of $Ag(OTf)_2$ proceeded smoothly in a mixture of $MeCN/H_2O$ to afford the corresponding amides **140** in good yield under air at 60°C for 24 h and also shows good functional group tolerance (Fig. 4.69).

Another direct decarboxylative amidation approach for the synthesis of imides was successfully established by Ning Xu et al.[72] in 2016. The authors have conducted a reaction between N-substituted N-heteroarene-2 carboxamides **147** with 2-oxoacids **1** in presence of $Pd(OAc)_2$ as a catalyst and $K_2S_2O_8$ as an oxidant under reflux at 90°C for 24 h to form the corresponding imides **148**. This approach shows tolerance against both electron-

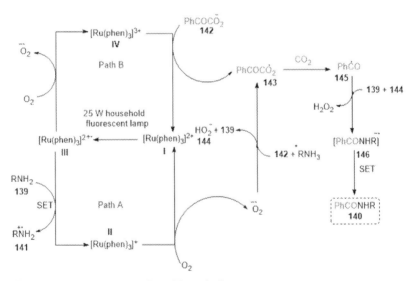

FIGURE 4.68 SET mechanism for amide synthesis.

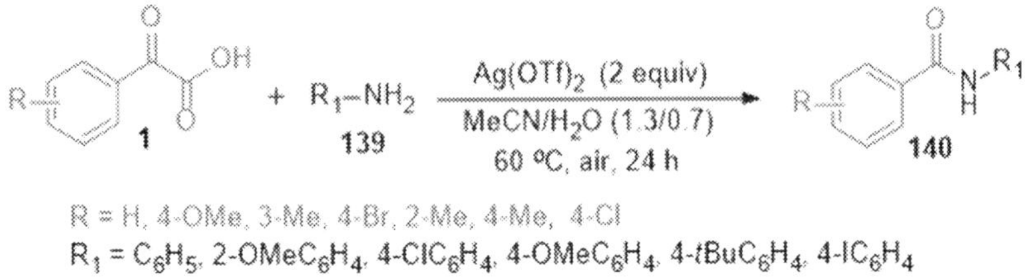

R = H, 4-OMe, 3-Me, 4-Br, 2-Me, 4-Me, 4-Cl
$R_1 = C_6H_5$, 2-OMeC$_6$H$_4$, 4-ClC$_6$H$_4$, 4-OMeC$_6$H$_4$, 4-tBuC$_6$H$_4$, 4-IC$_6$H$_4$

FIGURE 4.69 Synthesis of amides from 2-oxoacids.

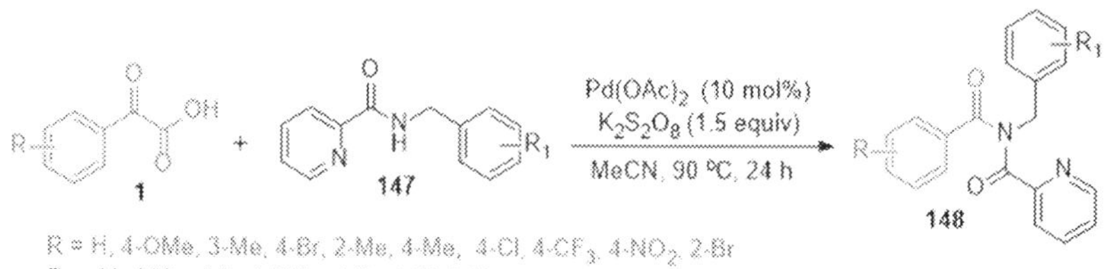

FIGURE 4.70 Synthesis of amides from 2-oxoacids and carboxamides.

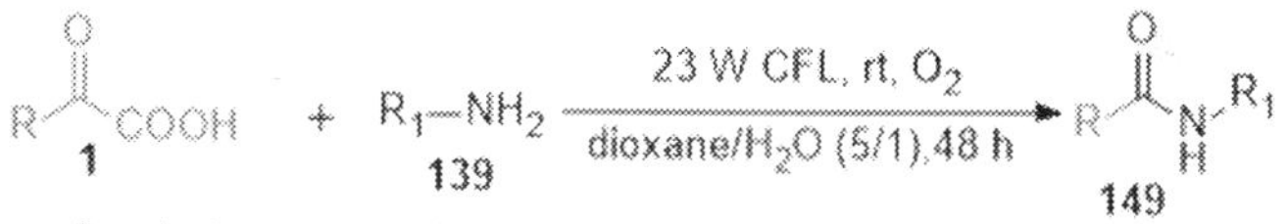

FIGURE 4.71 Decarboxylative amidation reaction of 2-oxoacids with amines.

releasing and electron-withdrawing substitution on both 2-oxoacids **1** and N-substituted N-heteroarene-2-carboxamides **147** to form the corresponding imides in good yields (Fig. 4.70).

A catalyst-free decarboxylative amidation reaction of 2-oxoacids with amines in the presence of CFL light was reported by Hua-Jian Xu and his coworkers.[73] The scope of the reaction was thoroughly investigated by using different substitution at different positions of 2-oxoacids **1** and amines **139** which showed that the substitution doesn't affect much on the yield of final product **149**. Different controlled experiments were performed to know the importance of visible light and O_2. Deuterium labeling experiment was also conducted in order to confirm that the oxygen in final product is derived from from a water molecule (Fig. 4.71).

Reaction of α-ketoacid **1** with primary amine **139** leads to the condensation reaction to form α-iminoacids **150**. In presence of CFL, singlet oxygen is generatedfrom oxygen which captures an electron from α-imino acids **150** to produce **152** along with **151** followed by the decarboxylation of **152** to generate N-arylimidoyl radicals **153**. The radical **153** reacts with water in presence of **151** to form the enol product **154**, followed by tautomerization to afford corresponding amides **149** (Fig. 4.72).

4.1.6 Ester synthesis

The unactivated C(sp3)−H bond functionalization reaction of formamides **155** and ethers **157** with 2-oxoacids **1** was achieved by Guo Duan and coworkers[74] in metal-free catalytic systems. This reaction was catalyzed by 20 mol.% of tetrabutylammonium iodide (TBAI) in the presence of TBHP (2.0 equiv) as an

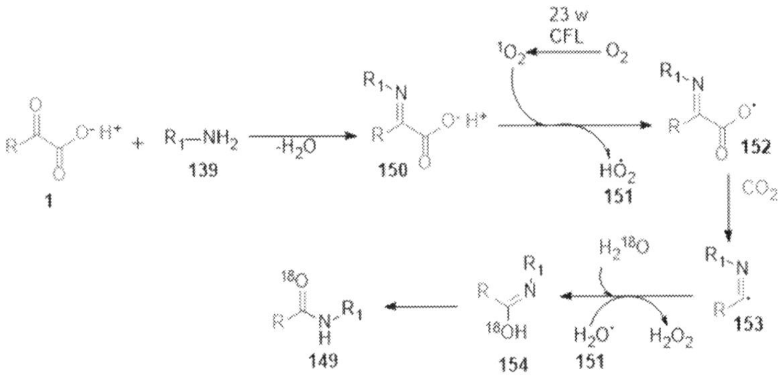

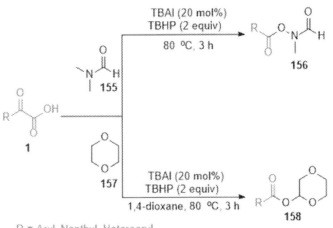

FIGURE 4.72 Proposed mechanism for catalyst-free decarboxylative amidation reaction.

FIGURE 4.73 Formation of acyloxymethyl amides and acyloxy esters.

oxidizing reagent to generate N-acyloxymethylamides **156** and α-acyloxy ethers **158** in moderate to good yields. A variety of aryl, heteroaryl, and naphthyl based derivatives of N-acyloxymethylamides **156** and α-acyloxy ethers **158** were synthesized under the optimized conditions. In these approaches, the new C–O bond was easily formed via the decarboxylative cross-coupling of 2-oxoacids **1** with formamides **145** and ethers **157** (Fig. 4.73).

In 2016, Gao-Qing Yuan and his coworkers described[75] an efficient, TBAI-catalyzed, decarboxylative cross-coupling reaction of 2-oxoacids **1** with carbonyl compounds **159** to generate α-acyloxycarbonyl compounds **160** with broad substrate scope. A reaction mixture of 2-oxoacids **1** and carbonyl compounds **159** was stirred at 80° in the presence of *tert*-butyl hydroperoxide (TBHP) as the oxidant, TBAI as a catalyst in EtOAc for 24 h to access the corresponding α-acyloxycarbonyl compounds **160**. The scope of 2-oxoacids

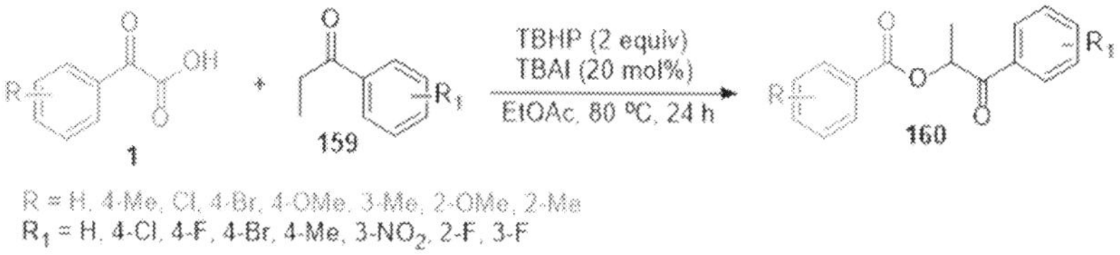

FIGURE 4.74 Formation of α-acyloxycarbonyl compound.

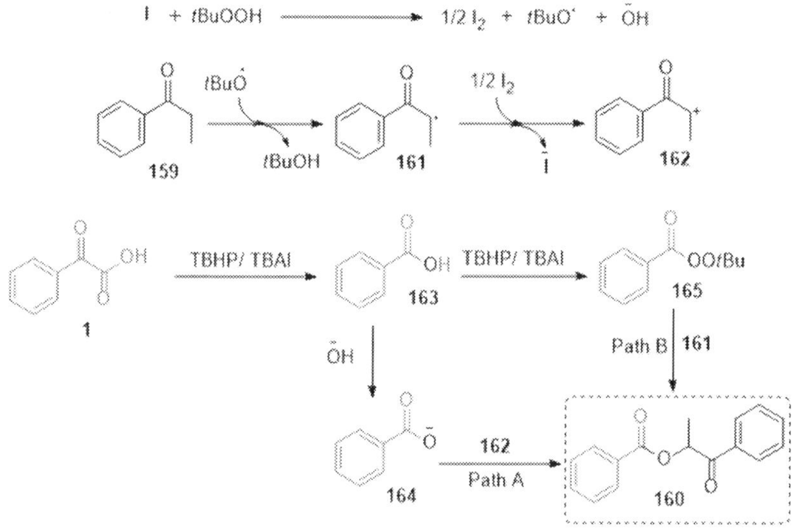

FIGURE 4.75 Tetrabutylammonium iodide (TBAI)-catalyzed ester synthesis.

and carbonyl compounds was thoroughly examined and observed that substitutions don't impact much on the yields of the final product (Fig. 4.74).

The authors have given the radical and ionic pathway for the reaction. TBHP reacts with TBAI to produce tert-butoxyl radicals along with iodine. The tert-butoxyl radicals abstract the α-hydrogen of the ketones **159** to generate the α-carbonyl radical **161** followed by iodine participation to produce the cation intermediate **162**. Additionally, the 2-oxoacid **1** undergoes oxidative decarboxylation in the presence of TBHP/TBAI to form benzoic acid **163**. This benzoic acid undergoes deprotonation to form a benzoate anion **164** followed by the reaction of benzoate anion with the intermediate cation **162** to generate the final product **160** (Path **A**). On the other hand, benzoic acid **163** in the presence of TBHP/TBAI system forms the tert-butyl perester **165**. The obtained tert-butyl perester reacts with **165** with α-carbonyl radical **161** to form the final product **160** (Path **B**) (Fig. 4.75).

Hypervalent iodine (III)-promoted oxidative decarboxylation of α-ketoacids **1** with alcohols **166** to form the corresponding esters **167** was described by Yoshiji Takemoto and his coworkers.[76] α-ketoacid **1** react with hypervalent iodine(III) species to form a highly reactive acylating agent **168** followed by the alcohol participation to form the final product **167**. This esterification protocol is carried under the mild condition and shows high chemoselectivity as compared to the general acylation reactions. Different types of 2-oxoacids **1** and alcohols **166** were tested against the optimized conditions to access the corresponding esters **167** in moderate to good yields (Fig. 4.76).

Mechanistically, alcohol **166** reacts with the keto group of the α-ketoacid **1** to form hemiacetal **169**. The tertiary hydroxy or carboxy groups of the hemiacetal **169** reacts with the hypervalent iodine to generate the reactive intermediate **170a** or **170b**. This reactive intermediate undergoes oxidative decarboxylation to give the desired ester **167** (Fig. 4.77).

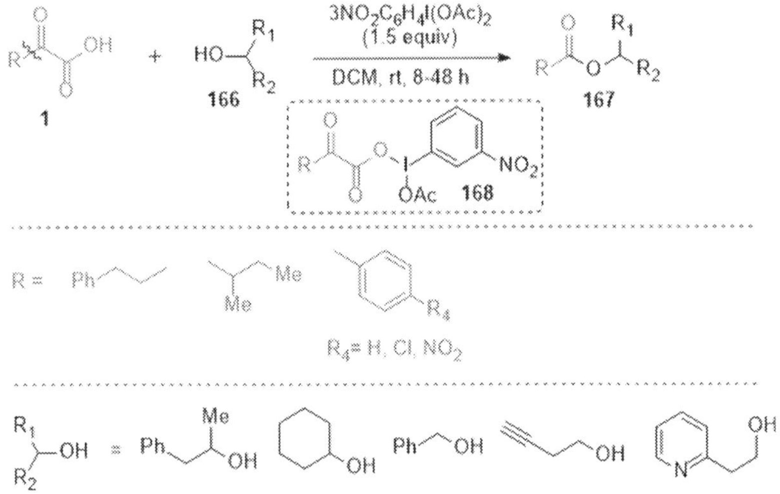

FIGURE 4.76 Formation of esters.

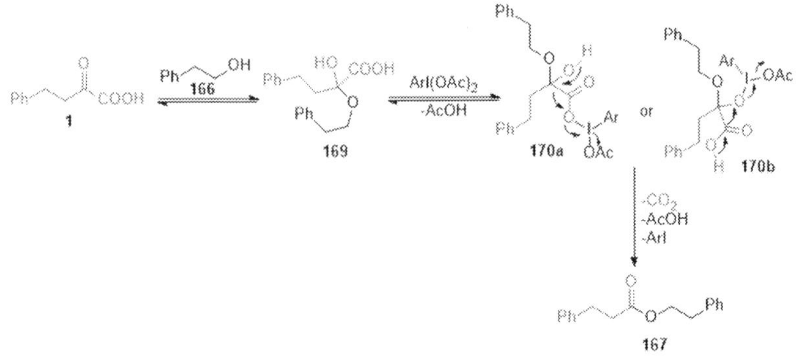

FIGURE 4.77 Hypervalent iodine (III)-promoted oxidative decarboxylative reaction.

Ahmed and his coworkers[77] described an efficient, mild, and cost-effective protocol for the generation of esters **171** from 2-oxoacids **1** by using oxone as a catalyst in MeOH under refluxing conditions. The protocol shows good tolerance against different substitutions (both electron-releasing and electron-donating) on the phenyl ring of 2-oxoacids (Fig. 4.78).

4.1.7   Synthesis of N-aryl sulfoximines

Sirilata Yotphan and her coworkers[78] have reported copper-catalyzed decarboxylative cross-coupling of α-keto acids **1** with NH-sulfoximines **172** for the construction of C−N bond to give N-aroylsulfoximine products **173** quickly under mild conditions. The optimized conditions for the reaction is; 1 equiv. of α-oxo carboxylic acid **1**, 2 equiv. of NH-sulfoximine substrate **172**, 10 mol.% of CuBr as a catalyst, 1.1 equiv. of $K_2S_2O_8$ as an oxidant, CH_3CN as a solvent at 75°C for 1 h to access the corresponding N-aroylsulfoximine products **173**. The substrate scope and limitation of this reaction under the optimized conditions were thoroughly investigated. This reaction proceeds well with aryl, naphthyl, heteroaryl substituted α-oxo carboxylic acids and fails with the alkyl-substituted ones. Different alky and aryl-substituted NH-sulfoximines were tested against phenyl glyoxylic acid and the reaction works smoothly under the given conditions (Fig. 4.79).

The authors have proposed two plausible pathways (Path A and Path B) for the synthesis of N-aroylsulfoximine products **173**. In pathway **A**, the Cu (I) catalyst is oxidized to form Cu(II) species, followed by the interaction with sulfoximines **172** leading to the generation of Cu-sulfoximines complex.

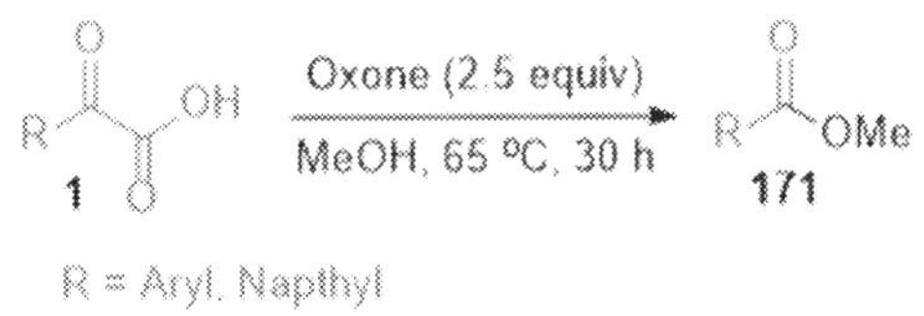

R = Aryl, Napthyl

FIGURE 4.78   Oxone promoted formation of esters.

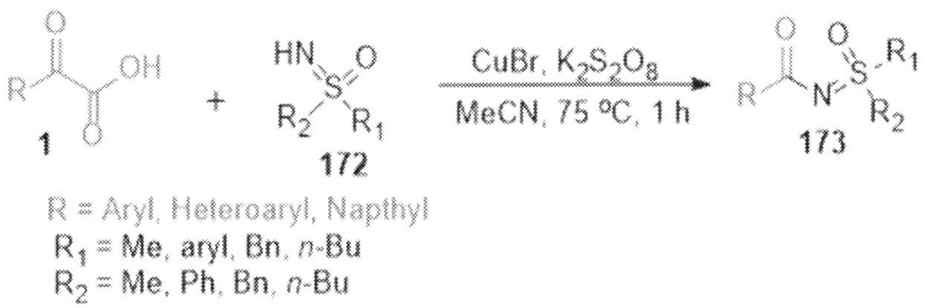

R = Aryl, Heteroaryl, Napthyl
R_1 = Me, aryl, Bn, n-Bu
R_2 = Me, Ph, Bn, n-Bu

FIGURE 4.79   Cu catalyzed synthesis of N-sulfoximine.

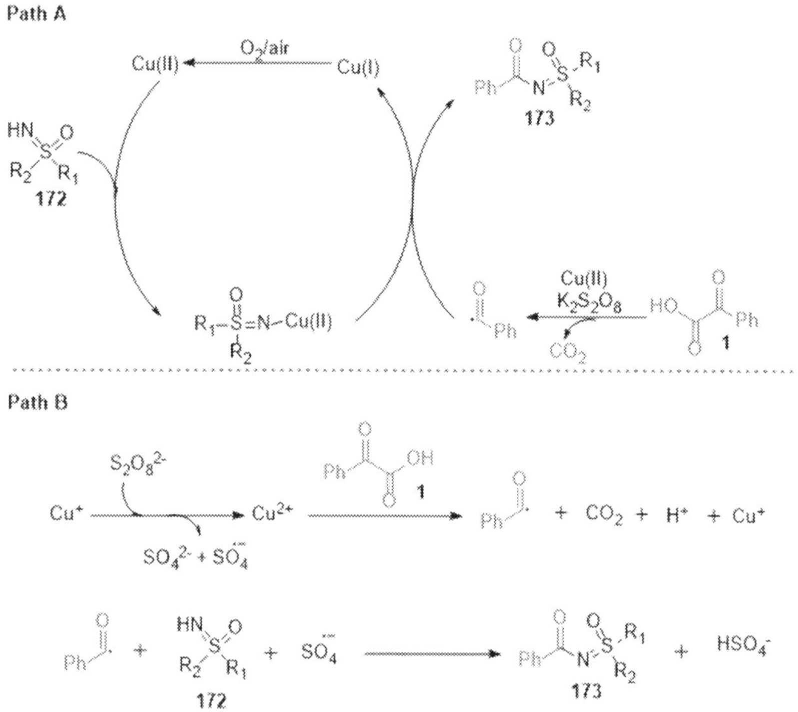

Path A

Path B

FIGURE 4.80   Proposed mechanism for copper-catalyzed N-aroylsulfoximine synthesis.

α-keto acid **1** in the presence of potassium persulfate and Cu(II) undergoes decarboxylation to form an aroyl radical. This aroyl radical reacts with Cu-sulfoximines complex to form aroylsulfoximine product **173** along with the regeneration of Cu(I) species to resume the catalytic cycle. In pathway **B**, the Cu(I) catalyst is oxidized to Cu(II) species in the presence of potassium persulfate. α-keto acid **1** in presence of Cu(II) undergoes the decarboxylation process resulting in the formation of a highly reactive aroyl radical along with Cu(I) ion. The last step is the interaction of reactive aroyl radical with sulfoximines **172** in presence of sulfate radical-anion to afford the final aroylsulfoximine product **173** (Fig. 4.80).

4.1.8   Synthesis of aldehydes and carboxylic acids from 2-oxoacids

Aldehydes and acids are important organic functional groups that are used for numerous synthetic transformations because of the active nature of the formyl and the carboxyl group.[79] In addition, both groups are also used for the construction of various biologically active compounds and play one of the major roles in pharmaceutical and agricultural chemistry.[80] Various transformations are reported for the synthesis of aldehydes and acids from 2-oxoacids in different optimized conditions.

Theodore Cohen has described[81] decarboxylative approach of 2-oxoacids **1** to access the corresponding benzaldehydes **174** in presence of acid anhydrides and pyridine. For phenylglyoxalic acid, three substrates like phenylacetic acid, acetic acid and benzoic anhydrides were used. Out of these three, benzoic anhydride gives a better transformation under refluxing conditions. The reaction is carried out by mixing phenylglyoxylic acid, benzoic anhydride and pyridine in benzene under reflux to obtain benzaldehyde in 75% yield (Fig. 4.81).

The reaction pathway involves the nucleophilic attack of pyridine on the keto group of phenylglyoxalic acid **1** and simultaneously the acyl protection on oxygen atom to form intermediate **175** followed by the loss of proton and CO_2 to access the intermediate **176**. The intermediate **176** in the presence of H^+ forms the intermediate **177** which undergoes deacylpyridination to form benzaldehyde **174** along with anhydride and pyridine (Fig. 4.82).

Lukas J. Gooßen and coworkers[63] have reported the decarboxylative approach of phenylglyoxalic acid **1** in the presence of tri-p-tolylphosphine $(P(p\text{Tol})_3)$ to access benzaldehyde **180** in moderate yields. In this reaction, phosphine acts as an organocatalyst for the decarboxylation step of phenylglyoxalic acid. The reaction is carried out in toluene solvent at 100°C to obtain benzaldehyde in 56% yield (Fig. 4.83).

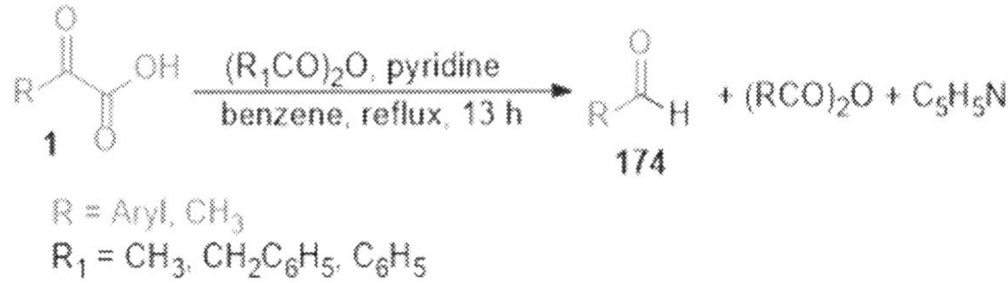

FIGURE 4.81   Formation of aldehydes from 2-oxoacids.

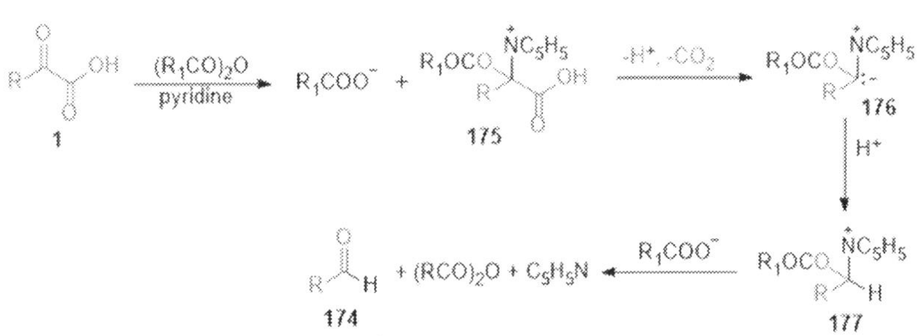

FIGURE 4.82   Proposed mechanism for benzaldehyde synthesis.

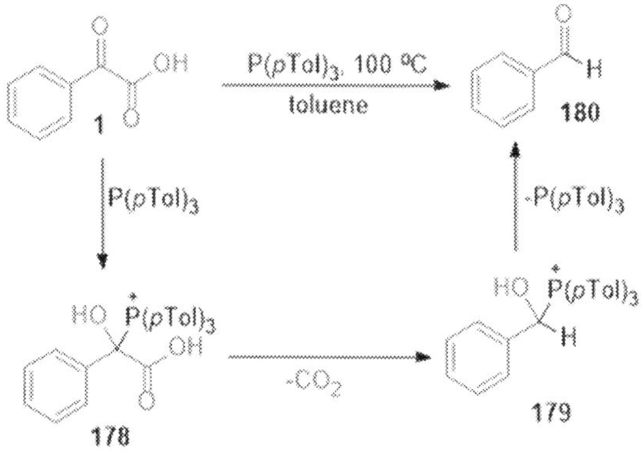

FIGURE 4.83 Formation of aryl aldehydes.

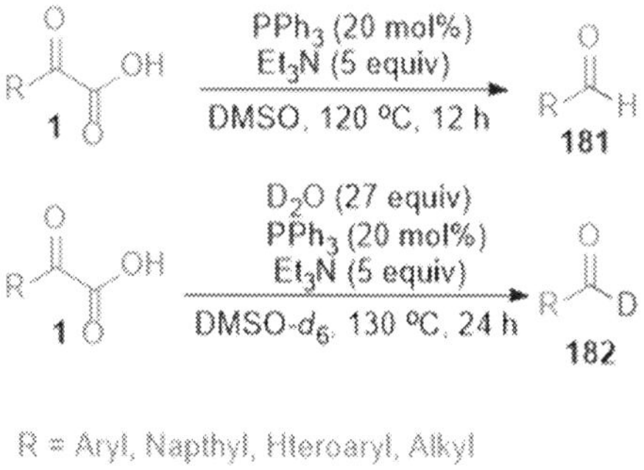

R = Aryl, Napthyl, Hteroaryl, Alkyl

FIGURE 4.84 Triphenylphosphine promoted formation of aldehydes.

A simple and mild decarboxylative method for the construction of alde-hydes **181** from 2-oxoacids **1** was reported by Gary Jing Chuang and cowor-kers[82] in presence of PPh$_3$ and Et$_3$N. The reaction was carried out by treating 2-oxoacids **1** with 20 mol.% of PPh$_3$ as a catalyst and 5 equiv of Et$_3$N as a base in DMSO solvent at 120°C for 12 h to access corresponding aldehydes **181**. Different 2-oxoacids like substituted aryl, heteroaryl, naphthyl, and alkyl were successfully transformed into the corresponding aldehydes. The deuterated aldehydes **182** were also synthesized by using D$_2$O as a deuterium source under the optimized conditions (Fig. 4.84).

The Et$_3$N abstract the acidic hydrogen of 2-oxoacids **1** to form the 2-oxocarboxylate anion **183**, followed by the nucleophilic addition of PPh$_3$ on the keto group of 2-oxocarboxylate anion to form intermediate **184**. The intermediate **184** abstracts proton from protonated triethylamine to form intermediate **185** followed by decarboxylation to give ylide **186**. Protonation of ylide **186** from protonated triethylamine to give intermediate **187** and finally, the elimination of PPh$_3$ affords benzaldehyde **181** (Fig. 4.85).

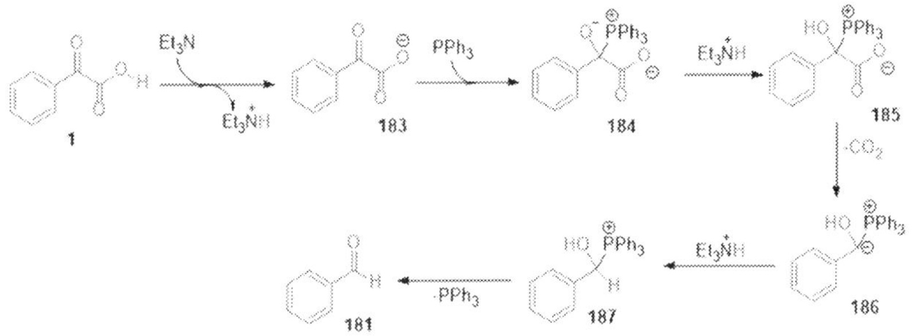

FIGURE 4.85   PPh₃-catalyzed synthesis of benzaldehyde.

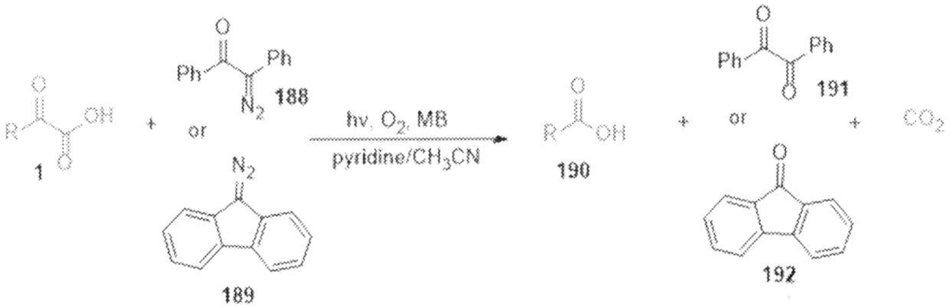

FIGURE 4.86   Visible light promoted formation of benzoic acid and byproducts.

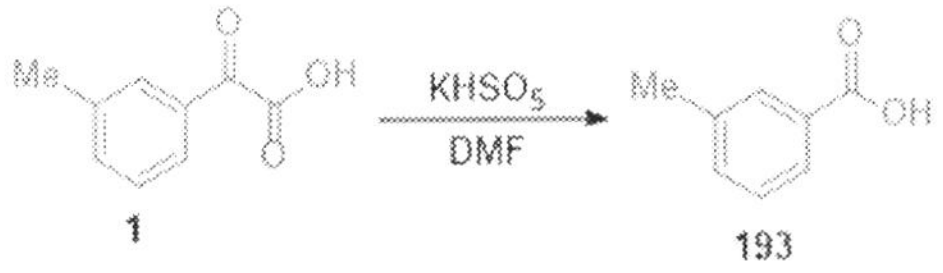

FIGURE 4.87   Formation of benzoic acid derivative.

In 1982, Wataru Ando and coworkers[83] reported an oxidative decarboxylative approach of 2-oxoacids **1** in the presence of diazo compounds (**188** or **189**) and halogen lamp to form acids **190**. A reaction was carried out by taking a solution of phenylglyoxylic acid **1** in acetonitrile followed by the addition of pyridine and diazo compounds (**188** or **189**). The resulting reaction mixture was irradiated by a halogen lamp (500 W) while passing O_2 for 80 min to affords benzoic acid **190** (Fig. 4.86).

One more method for direct transformation of 2-oxoacids **1** into benzoic acids **193** was reported by Anil K. Padala et al. after stirring the reaction mixture with oxone in DMF at 65°C for 4 h (Fig. 4.87).[77]

4.1.9 Miscellaneous reactions

4.1.9.1 Synthesis of oxindoles

In 2013, Guo, Duan, and coworkers[84] reporteda direct and mild approach for the arylacylation of activated alkenes **194** with 2-oxoacids **1** under silver catalysis and $K_2S_2O_8$ as an oxidizing reagent. An equimolar ratio of acetone/H_2O was taken as a solvent and reaction was done at 50°C for 24 h to access the highly functionalized oxindoles **195**. This method shows tolerance against different functional groups of acrylamides **194** and α-oxocarboxylic acids (Fig. 4.88).

As for as the mechanism is concerned, 2-oxoacids **1** in the presence of Ag(I)/$K_2S_2O_8$ undergoes decarboxylation to form the highly reactive acyl radical. The acyl radical behaves as nucleophile and attacks on the double bond of N-arylacrylamides **194** to form another radical intermediate **196**, followed by the intramolecular cyclization to give radical intermediate **197**. Finally, radical intermediate **197** undergoes oxidation in presence of Ag(II) to generate the corresponding carbocation and then loss of H^+ affords the desired oxindole **195** (Fig. 4.89).

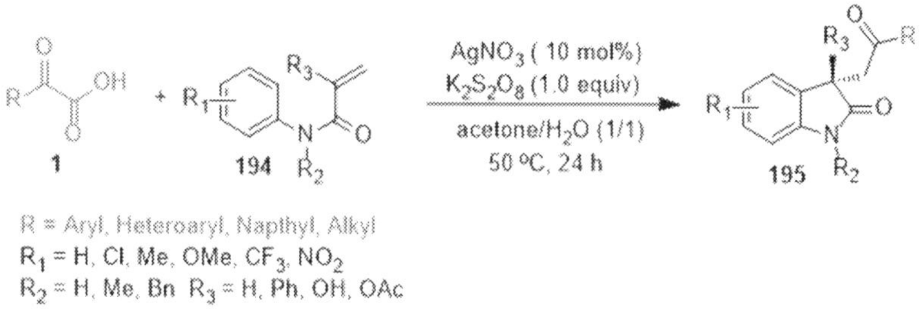
AgNO3 (10 mol%)
K2S2O8 (1.0 equiv)
acetone/H2O (1/1)
50 ºC, 24 h
1
194
195
R = Aryl, Heteroaryl, Napthyl, Alkyl
R1 = H, Cl, Me, OMe, CF3, NO2
R2 = H, Me, Bn R3 = H, Ph, OH, OAc

FIGURE 4.88 Silver catalyzed synthesis of oxindoles.

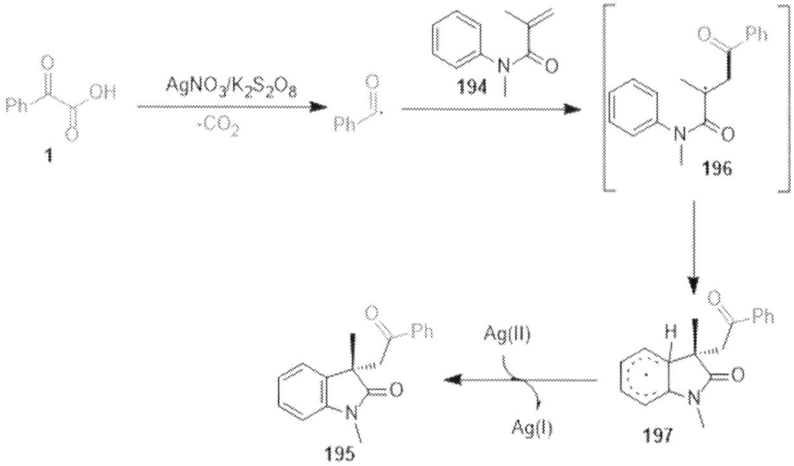
AgNO3/K2S2O8
-CO2
194
196
Ag(II)
Ag(I)
197
195
1

FIGURE 4.89 Proposed mechanism for oxindoles synthesis.

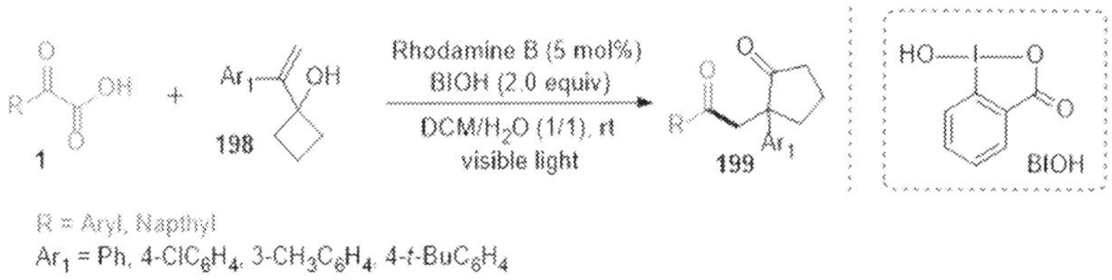

FIGURE 4.90 Visible light induced formation of vinylcyclobutanol.

4.1.9.2 Synthesis of 1,4-dicarbonyl compounds from vinylcyclobutanols

The same group reported the direct acylation and ring expansion in vinylcyclobutanols **198** with 2-oxoacids **1** in presence of rhodamine B as a catalyst and hypervalent iodine reagent (BIOH). Hypervalent iodine reagent (HIR) facilitates the important decarboxylation step of α-keto acid **1** to form the acyl radical. In order to check the substitution effect on the reaction, a library of substituted 1,4-dicarbonyl compounds **199** was prepared and a little effect was observed on the yields of the desired products (Fig. 4.90).[84]

2-oxoacids in the presence of BIOH generates in situ HIR **200** which is oxidized by photoexcited state PC* to generate the acyl radical and the reduced photocatalyst PC⁻ after loss of CO_2. The acyl radical attacks on the C−C double bond of vinyl cyclobutanol **198** to produce the alkyl radical **201** followed by oxidation with BI to give carbocation intermediate **202**. 1,2-migration of alkyl takes place in intermediate **202** to give the desired product **199** along with the loss of H^+. Finally, BI radical oxidizes the reduced photocatalyst PC − to regenerate the catalyst PC (Fig. 4.91).

4.1.9.3 Acylfluorination of styrenes

Furthermore, Xin-Hua Duan and coworkers[85] derived an efficient and mild decarboxylative acylfluorination of styrenes **204** in presence of $AgNO_3$ as a catalyst and selectfluor as a fluorinating reagent. The standard reaction conditions were optimized by stirring a mixture of α-keto acids **1** with styrenes **204** in presence of 20 mol.% of $AgNO_3$, 1.5 equiv of selectfluor, and 1.0 equiv of Na_2SO_4 in a mixture of acetone/H_2O as a solvent at room temperature for 12 h to afford β-fluorinated 3-aryl ketones **205** in good yields. Different aryl, heteroaryl, and napthyl derivatives of 2-oxoacids **1** were treated under the optimized conditions against different aryl and alkyl-substituted styrenes **204** to access the corresponding β-fluorinated 3-aryl ketones **205**. This protocol provides access to β-fluorinated 3-aryl ketones with better selectivity through C−C and C−F bond formation in one step (Fig. 4.92).

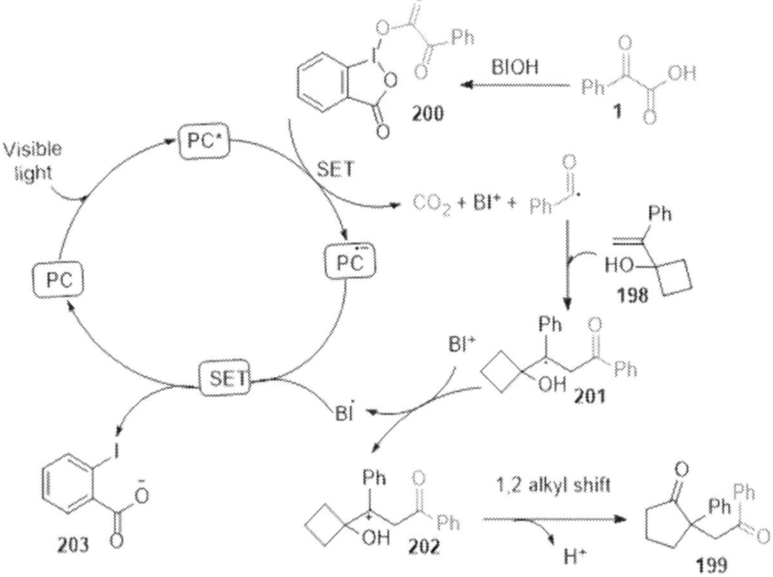

FIGURE 4.91 Proposed mechanism for the synthesis of 1,4-dicarbonyl compounds.

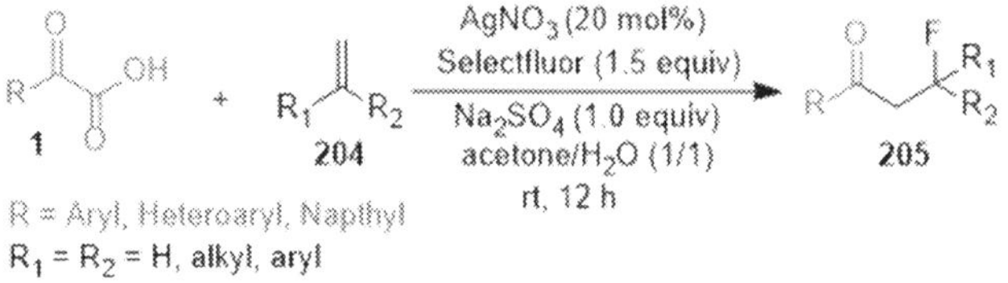

R = Aryl, Heteroaryl, Napthyl
$R_1 = R_2$ = H, alkyl, aryl

FIGURE 4.92 Formation of β-fluorinated ketones.

4.1.9.4 Synthesis of γ,γ-difluoroallylic ketones

A decarboxylative/defluorinated reaction between 2-oxoacids **1** and α-trifluoromethyl alkenes **206** in presence of photocatalyst [Ir $(dFCF_3ppy)_2(dtbbpy)$-PF_6] and LiOH as a base was established by Tiebo Xiao et al.[86] Different γ,γ-difluoroallylic ketones **207** were prepared in DMSO at room temperature under visible-light irradiation for 24 h. The obtained functionalized *gem*-difluoroalkenes behaved as valuable synthetic intermediates for the construction of monofluorinated heterocycles and difluoromethylated compounds in moderate to good yields. In presence of the Grignard reagent or reducing reagent $NaBH_4$ the carbonyl group in γ,γ-difluoroallylic ketone **207a** gives corresponding methyl-substituted γ,γ-difluoro allylic alcohols **209** or γ,γ-difluoro allylic alcohols **208** in 90 to 80% yields respectively. Treatment of γ,γ difluoroallylic ketone with TABF

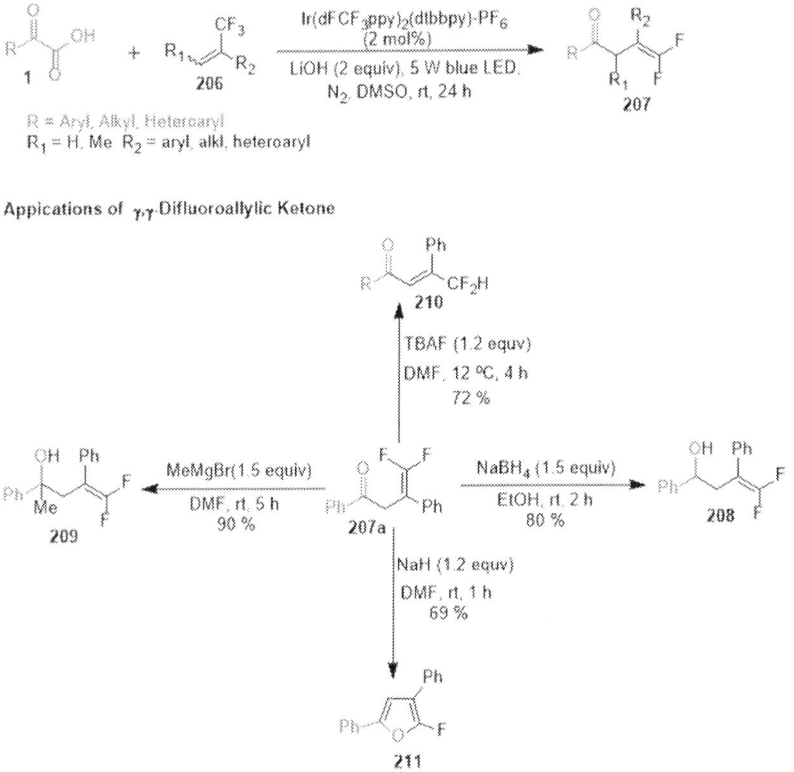

Appications of γ,γ-Difluoroallylic Ketone

FIGURE 4.93   Formation of γ,γ-difluoroallyl ketone.

in DMF affords E-selective α-difluoromethyl α,β-unsaturated ketone **210** in 72% yield and 2-fluorinated furan **211** in 68% yield (Fig. 4.93).

4.1.9.5   Synthesis of γ,γ-difluoroallylic ketones

A condensation reaction between benzothiazoles **212** and 2-oxoacids **1** was successfully achieved by Ze Tan and coworkers[87] for the synthesis of 2-aryl benzothiazoles **213** in presence of $K_2S_2O_8$ as an oxidizing reagent. In one step, a mixture of 1 equiv of 2-oxoacids and 1.5 equiv of benzothiazole in DMSO/H_2O were heated at $100°C$ for 1 h followed by the addition of 1 equiv of $K_2S_2O_8$ and keep the reaction for another 3 h. A variety of 2-aryl benzothiazoles **213** were prepared by using different 2-oxoacids **1** and benzothiazoles **212** in good to moderate yields (Fig. 4.94).

4.1.9.6   1,4-addition of 2-oxoacids with Michael acceptors

A decarboxylative cross-coupling reaction between 2-oxoacids **1** and different Michael acceptors **214** in presence of photocatalyst was demonstrated by

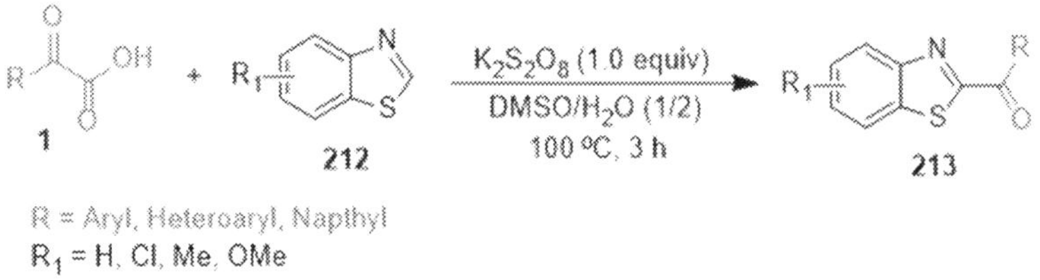

R = Aryl, Heteroaryl, Napthyl
R_1 = H, Cl, Me, OMe

FIGURE 4.94 Formation of 2-aryl benzothiazoles.

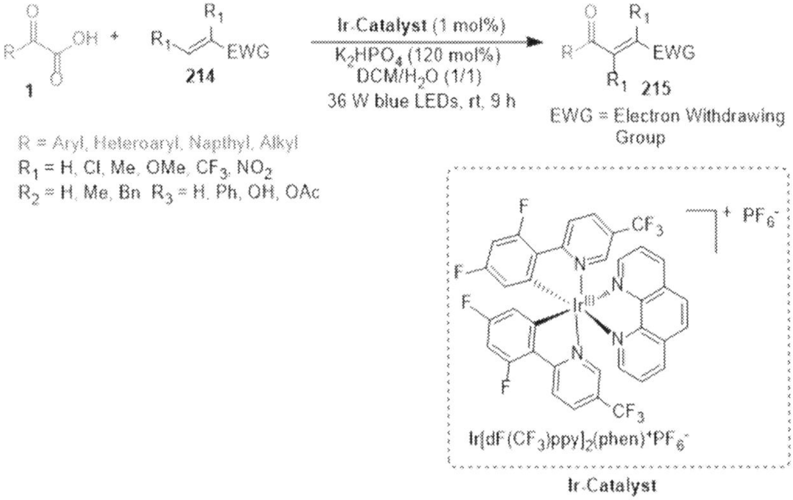

FIGURE 4.95 Iridium catalyzed formation of Michael acceptors.

Shang Fu, and coworkers[88] to access various 1,4-substituted products **215** in good yields. This approach proceeds well with different Michael acceptors like α,β-unsaturated ester, nitrile, amide, sulfone, ketone and aldehyde to favour the formation of acyl addition on Michael acceptors. The reaction isa novel acyl Michael addition by using stable 2-oxoacids which forms acyl anion during photoredox-catalysis (Fig. 4.95).

4.2 Ketoamide and ketoester synthesis

4.2.1 Ketoamide

α-Ketoamides are an important class of organic compounds found in many natural products like caspsase-3-inhibitor, Pin-1, and medicinally important compounds, for example, Orotomide, indibulin, Boceprevir, and polymers or materials.[89] Apart from this, they also serve as key substrates in synthetic organic chemistry.[90] Some of the biologically active ketoamides are depicted in Fig. 4.96. Therefore, due to their crucial role, various developments have taken place for α-ketoamide synthesis.[91]

Ketoamides **217** are prepared by decarboxylative acylation of formamides, on treating α-keto acids **1** with substituted formamides **216** under the catalytic amount of CuBr, di-tert-butyl peroxide (DTBP) as an oxidant and pivalic acid as an additive at 110°C for 18 h. The presence of 4-substituted strong electron-withdrawing and electron-donating groups decreases the yield of the desired product. However, moderate electron-withdrawing and electron-donating groups do not affect the reaction. When R = H, CH$_3$, 4-t-Bu, Cl, Br, the overall yield of the product is 81%, 75%, 69%, 85%, 82% respectively and when R = 4-NO$_2$, 31% of the product is obtained (Fig. 4.97).[92]

The predicted mechanism supposes the formation of Cu(II) carboxylate **I** followed by the formation of organocopper intermediate **II** after the loss of

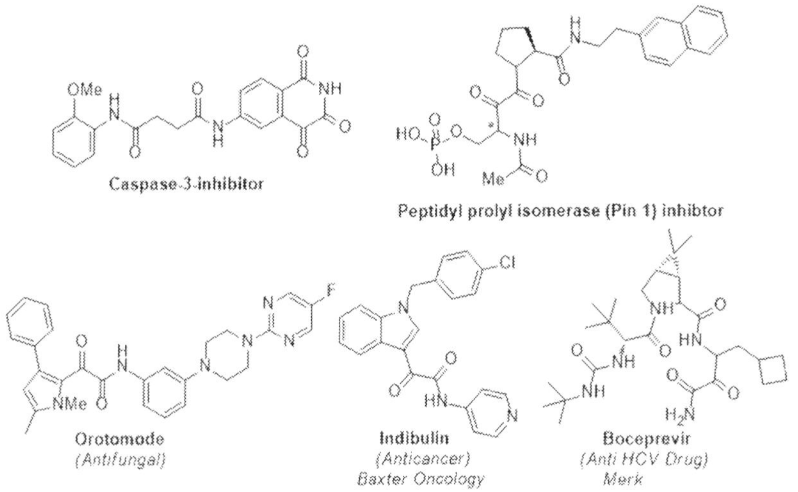

FIGURE 4.96 Examples of biologically active ketoamides.

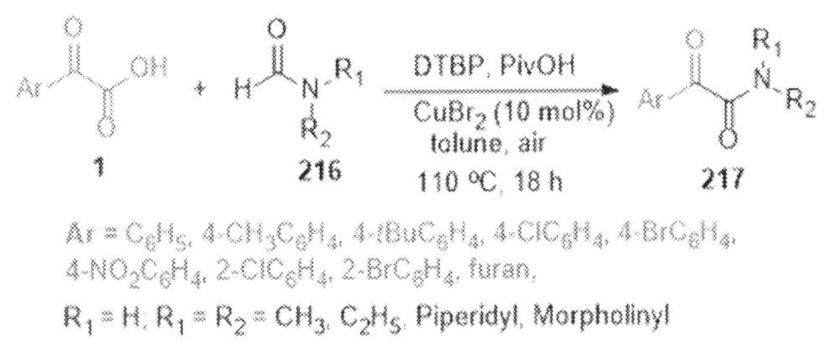

Ar = C$_6$H$_5$, 4-CH$_3$C$_6$H$_4$, 4-tBuC$_6$H$_4$, 4-ClC$_6$H$_4$, 4-BrC$_6$H$_4$, 4-NO$_2$C$_6$H$_4$, 2-ClC$_6$H$_4$, 2-BrC$_6$H$_4$, furan,

R$_1$ = H, R$_1$ = R$_2$ = CH$_3$, C$_2$H$_5$, Piperidyl, Morpholinyl

FIGURE 4.97 Formation of ketoamides.

carbon dioxide. Meanwhile, t-BuO radical captures the carbonyl H from dimethylformamide leading to intermediate **III** which reacts with organocopper intermediate **II** to form intermediate **IV**. The final product is formed after the reductive elimination of **IV** along with Cu(I) which is again oxidized to Cu(II) thereby, starting a new catalytic cycle (Fig. 4.98).

An amidation reaction occurs between arylglyoxylates **1**, 2,3-dibromo pyridine **218** and N, N-alkyl amides **219** in presence of palladium catalyst and ligand to form arylglyoxylic amides **217** in 45%−88% yield.[93] Although, other additives were screened the reaction is, unprecedentedly, activated by 1, 2-dibromopyridine while as N, N-alkyl amides act as amine surrogates. Mechanistically, arylglyoxylate reacts with dihalopyridine and generates traceless activated arylglyoxylic acid pyridyl ester **220**, which further reacts with amine surrogate, N, N-dialkyl amides to form arylglyoxylic amides (Fig. 4.99).

A simple and practical way for the preparation of aliphatic α-keto amides **222** is the reaction of 2-keto acids **1** with formamide **221**. The reaction works under Cu(II) catalysis along with DTBP and tolerates wide range of substrates like heterocycles (furan, thiophene) and naphthyls besides other

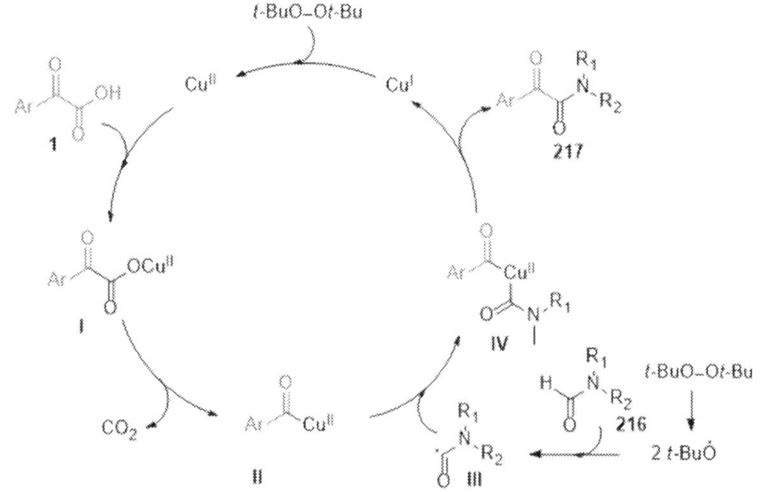

FIGURE 4.98 Reaction mechanism for the Cu(II)-catalyzed decarboxylative acylation.

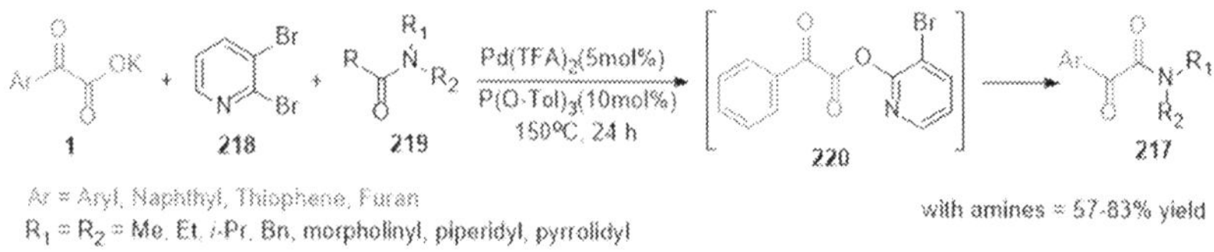

FIGURE 4.99 Pd catalyzed formation of arylglyoxylic amides.

common aryl groups. The reaction proceeds through a free radical pathway and the amide carbon in the product is drawn from **1** which was demonstrated by caring out ^{13}C-labeling studies (Fig. 4.100).[94]

The substituted α-keto amides **223** are obtained from α-ketoacids by using nucleophilic fluorinating reagents, Deoxofluor (bis(2 methoxyethyl) amino sulfur trifluoride) or DAST diethyl amino sulfur trifluoride **224**. The product formation depends on the time of the reaction and the type of ketoacid when R = Ph, α-keto amides are formed in 1 h in 81%-94% yields and α, α-difluoroamides are obtained when the reaction is allowed to run for 36 h to furnish the desired product in 60%−92% yield. When R = Me, Et and thienyl, α-keto amides are formed in 33%−48% yields while as α, α-difluoroamides are obtained in 52%−67% yield after 36 h (Fig. 4.101).[95]

α-Keto amide **226** was obtained by the reaction of 2-ketoacids **1** and amines **225** at ambient temperature for 10 min. in upto 99% yield in a catalyst-free method. The 2-ketoacid is activated in situ by adding of 2, 4, 6-trichloro-1, 3, 5-trizaine (TCT) in stoichiometric ratios. Besides using various substituted and heterocyclic aryl groups like furan, thiophene, naphthalene, the reaction has scope over primary and secondary amines like morpholine, piperidine, pyrrolidine, mono, and dialkyl amines, benzylamine, aniline, pyrrole, piperidine, and

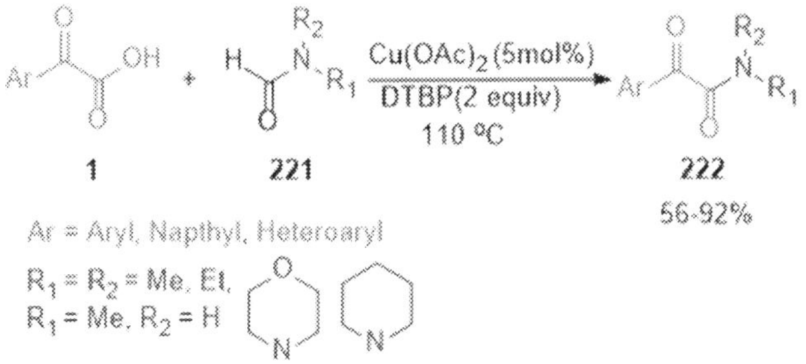

FIGURE 4.100   Cu catalyzed formation of aliphatic α-keto amides.

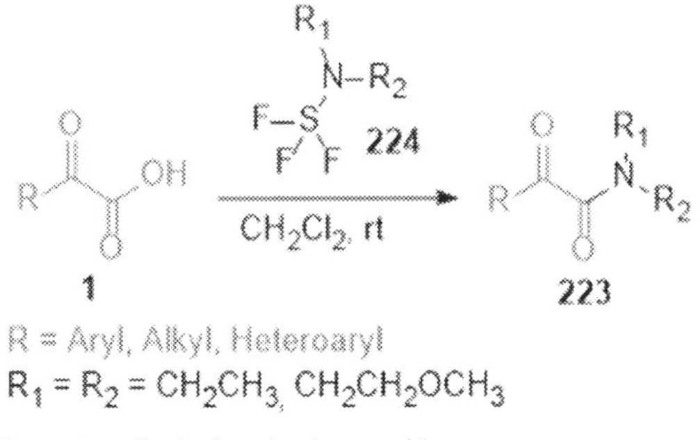

FIGURE 4.101   Formation of substituted α-keto amides.

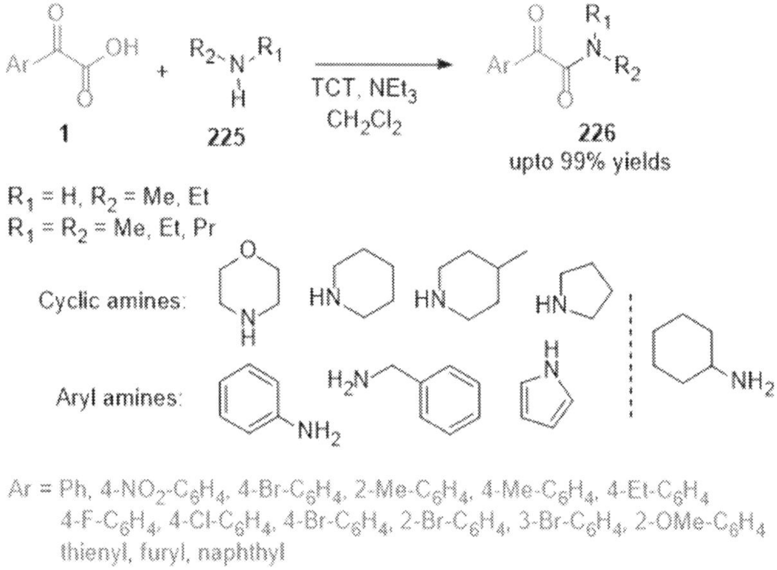

FIGURE 4.102 Formation of aliphatic α-keto amides.

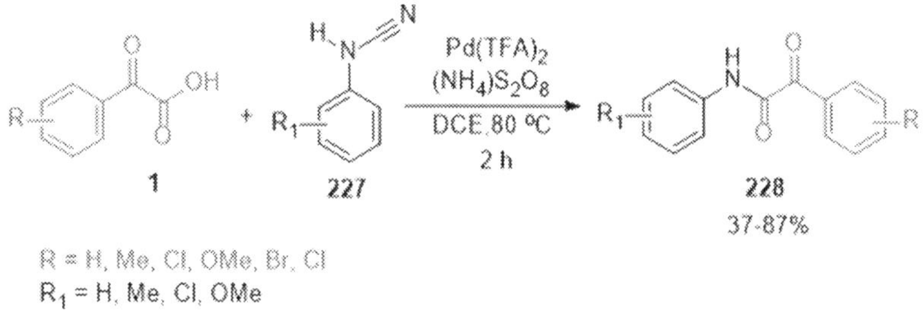

FIGURE 4.103 Formation of aliphatic α-keto amides from phenylcyanamides.

cyclohexylamine. Other reactions for the synthesis of α-keto amides use coupling of amides and α-ketoacids (Fig. 4.102).[96]

N-substituted α-keto amides are obtained by chemoselective insertion of acyl group into the N-monosubstituted cyanamide catalyzed by palladium. Arylglyoxalicacids **1** react with phenylcyanamide **227** in the presence of additive $(NH_4)_2S_2O_8$ which is supposed to initiate the generation of acyl radicals via the decarboxylation of previously formed acid radical. After decarboxylation, 1, 2 carbopalladation from N to C occurs to form corresponding α-keto amides **228** due to in situ hydrolysis caused by H_2O present in commercial DCE. The reaction has wide substrate tolerance with aryl substituents and operates under milder conditions. However, when aliphatic derivatives of amines (cyclohexylamine or benzylamine) and α-glyoxylic acids ($CH_3COCOOH$) are used, the yield of the reaction falls drastically (Fig. 4.103).[97]

A green approach toward the synthesis of α-ketoamides **230** is the reaction of 2-ketoacids, water, and isocyanides **229** promoted by visible light and rose bengal without the use of any external metal. The reaction occurs in very mild aerobic conditions with 56%−85% product formation. Range of electron-donating groups in 2-ketoacids like Me, OMe including electron-withdrawing substituents such as halogens, CF_3, ester, and CN at the *para*-position and substrates with Cl, CF_3 or NO_2 at the *meta*-position work very well for the reaction. Mechanistic studies reveal that decarboxylation occurs followed by radical addition and hydration processes (Fig. 4.104).[98]

4.2.2   Ketoesters

α-Ketoesters are present in many natural products like cephalotexus esters, (S)-camptothecin, and drug molecules, for example, Oxybutynin and so on. In addition, α-ketoesters are used as synthons for the synthesis of drugs,[99] pesticides,[100] naturally occurring[101] and other synthetic molecules (Fig. 4.105).[102]

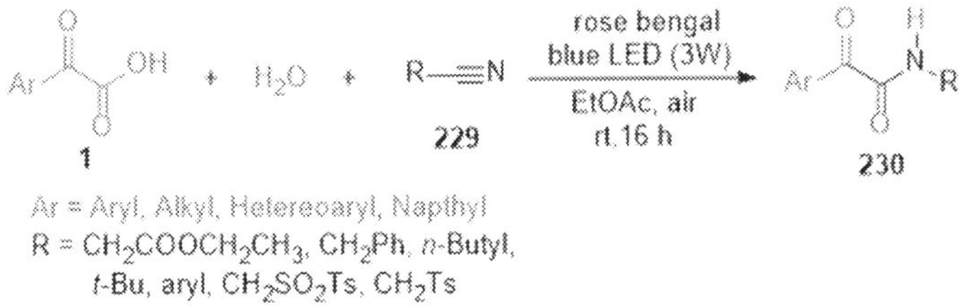

FIGURE 4.104   Light induced formation of α-ketoamides.

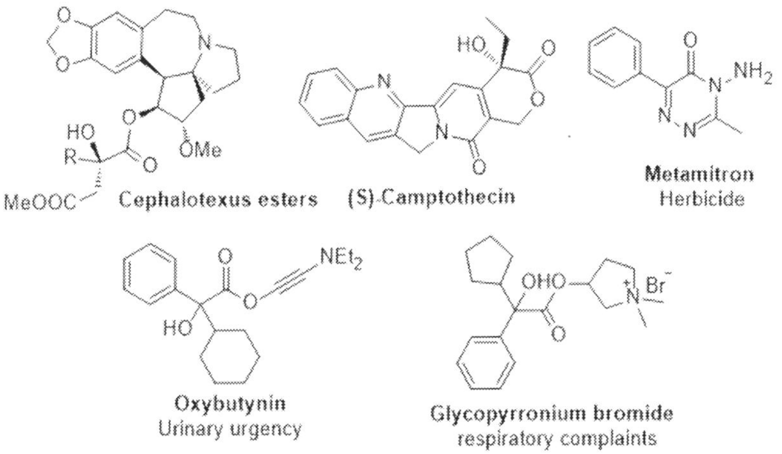

FIGURE 4.105   α-Ketoesters as part of some natural products and drug molecules.

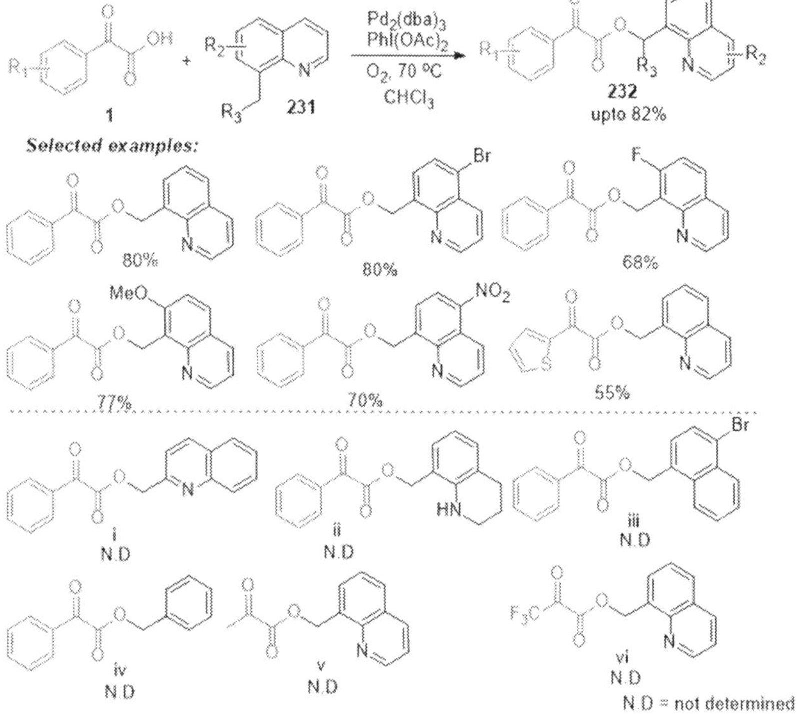

FIGURE 4.106 Pd catalyzed formation of quinoline ketoesters from phenyl-2-oxoacids.

2-ketoacids react with methyl quinoline derivative **231** to form α-ketoester derivative **232**. This is a regiospecific transformation in which 2- ketoacid acts as an esterifying agent and reacts with methyl quinoline derivative in presence of PhI(OAc)$_2$ and Pd$_2$(dba)$_3$ at 70°C in chloroform for 12 h duration in free air. The reaction works well in many simple substrates covering both counterparts and is able to generate the product in gram scale with 1 mol.% of Pd-catalyst. However, the reaction fails to form the desired product 2-methyl quinoline (**i**) or 8-methyl-1,2,3,4- tetrahydroquinoline (**ii**) as the cyclopalladated intermediate is not fully generated. Similarly, due to the same reason, the reaction of alkyl-substituted α-ketoacids with benzoic acid (**v**, **vi**) including phenyl glyoxylic acids with methyl-substituents (**iii**, **iv**) and aromatic carboxylic acids did not lead to any product formation (Fig. 4.106).[103]

Esterification of alkylbenzenes catalyzed by Fe(III) is also an important reaction for the synthesis of ketoesters **234**. Various 2- ketoacid derivatives **1** react with simple alkyl benzenes **233** like methyl, ethyl or propylbenzenes at a higher temperature in presence of catalytic amounts of FeF$_3$ and di-tert-butyl peroxide (DTBP) under argon atmosphere.[103b] This is an atom

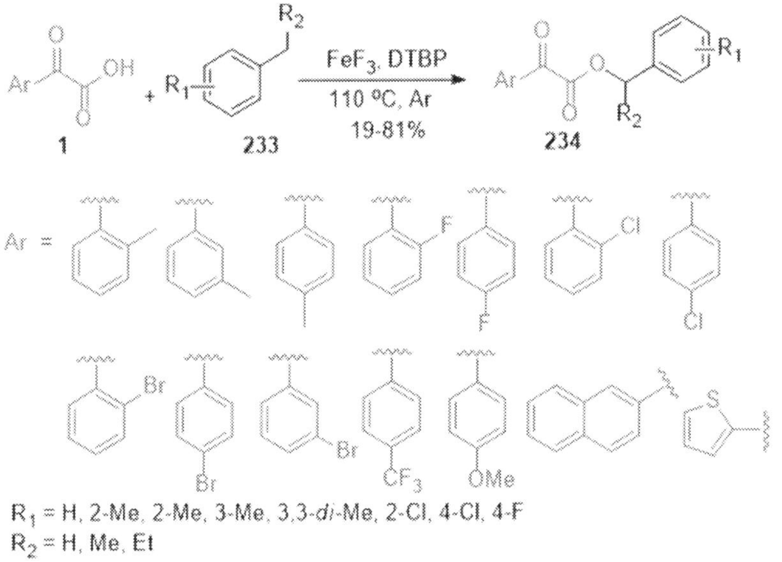

R_1 = H, 2-Me, 2-Me, 3-Me, 3,3-di-Me, 2-Cl, 4-Cl, 4-F
R_2 = H, Me, Et

FIGURE 4.107   Formation of benzyl containing ketoesters.

economical process leading to the formation of linear ketoesters **234** which can play an important role in procuring catalytic designs. Methyl-substituents on 2, 3, or 4 positions did not exhibit any drastic changes in the product formation. 2-ketoacids work well and furnish good yields while lower yields (19%−42%) were obtained with the alkylbenzenes bearing F and Cl atoms (Fig. 4.107).

The plausible mechanism shows that the reaction proceeds through a radical pathway. Tert-butoxyl radical **I** is generated from DTBP, which adds to the active benzylic C-H bond in alkylbenzene to form benzyl radical **II**, which undergoes Fe(III) catalyzed oxidation to generate intermediate carbocation **III**. Product formation takes place after the nucleophilic attack of ketoacid acid on **III**. Fe(II) released after the reduction of Fe(III) during the formation of intermediate **III** is reoxidized to Fe(III) catalyst by **I** to initiate the next catalytic cycle and tertiary alcohol **IV** as a byproduct (Fig. 4.108).

A new kind of method for the protection of saccharides has been devised that includes the reaction of 2- ketoacid with an unprotected saccharide(s). The protection of saccharide **235** is carried in presence of *N*, *N*-diisopropyl carbodiimide (DIC) and 4-dimethyl aminopyridine (DMAP) in DCM at ambient temperature to form protected saccharide **236** in 3 h upto 93% yield. The protocol has characteristic orthogonal stability against various protecting groups and deprotection of the oxophenylacetyl ester (OPAc) can easily be carried in the presence of $KHSO_5$ or timely monitoring by the use of AcCl in methanol (Fig. 4.109).[104]

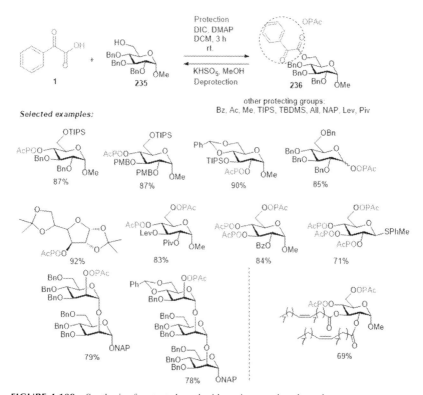

FIGURE 4.108 A plausible mechanism for Fe(III)-catalyzed synthesis of a-keto benzyl ester through C-H functionalization.

FIGURE 4.109 Synthesis of protected saccharides using oxophenylacetyl ester.

The enhanced electrophilic character of OPAc as compared to Bz makes this method more reliable. Oxophenylacetyl (OPAc) are surrogates of benzoate and acetate; hence, they are supposed to work better than OBz and OAc protecting groups. This method has been applied for the preparation of

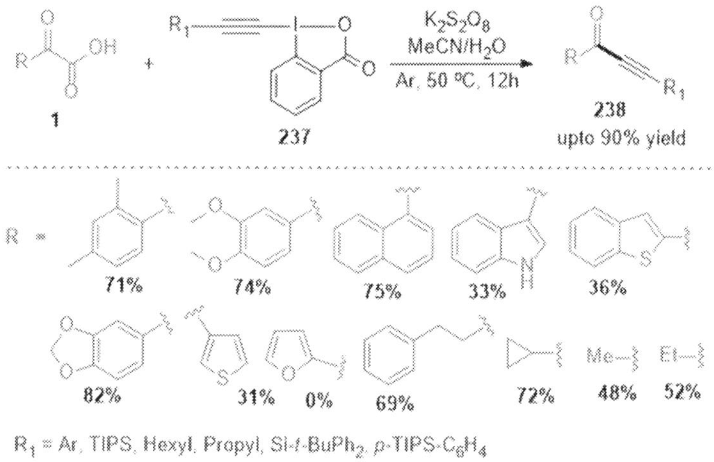

FIGURE 4.110   Formation of ynones from 2-oxoacids.

various mono-, di-, tri-, tetra-, and penta-OPAc protected saccharides including the synthesis of OPAc protected di- and tri-saccharides and lipid model.

4.3   Synthesis of Ynones and Ynedione

Ynones are important Michael species and are very reactive with various nucleophiles in cycloaddition and cyclocondensation reactions. Ynones are used as building blocks for the synthesis of many heterocycles with very beneficial properties.[105]

α-ketoacids react with HIRs 237 to form ynones 238. This reaction is actually alkynylation of ketoacid followed by loss of carbon dioxide promoted by potassium thiosulphate in $MeCN/H_2O$ with upto 92% yield (Fig. 4.110).[106]

The ynonilation reaction is led by persulfate radical species to form 2-ketoacid radical II which after decarboxylation affords acyl radical III. The previously formed BI-alkyne species IV combines with III to generate V which forms the desired ynone along with the iodonium radical VI. Theiodonium radical VI after protonation further converts into 2-Iodobenzoic acid (Fig. 4.111).

2-ketoacids react chemoselectively with 239 under mild conditions in presence of hypervalent iodine(III) reagent benziodoxole to form ynones 240 with 85% yield. The reaction is induced by visible light and tolerates substrates bearing sensitive groups towards transition metal catalysis. The reaction goes very well with thiophene, furan, indole, phenyl, and alkyl-substituted ketoacids. While ynamides 241 are formed with primary and secondary carbamoyl ketocids 1b; ynoates 242 were formed with alkoxycarbonyl ketoacids 1c under

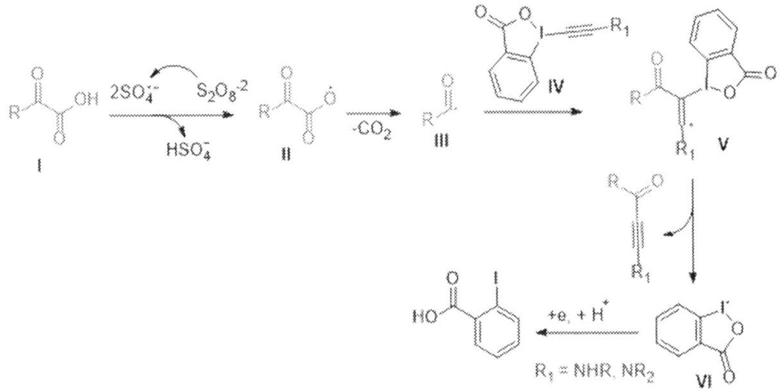

FIGURE 4.111 Decarboxylative alkynylation of α-keto acids.

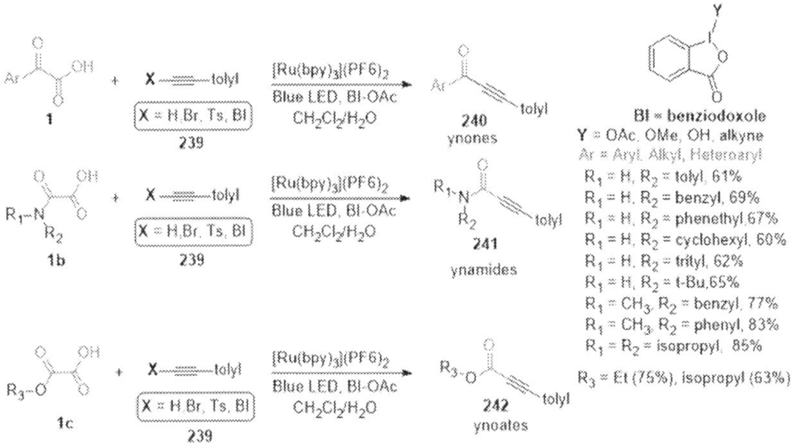

FIGURE 4.112 Formation of tolyl substituted ynones, ynamides and ynoates fro 2-oxoacids.

the same conditions. The acyl radicals generated from α-ketoacid adds to alkyne radical to form the desired product (Fig. 4.112).[107]

On the basis of the proposed mechanism, the in situ formation of benziodoxole-ketoacid **I** undergo decarboxylation and subsequent oxidation by $[Ru(bpy)_3]^{3+}$ to generate an acyl radical **II**. The alkyne then reacts with **II** and gives rise to BI-alkenyl intermediate **III** which releases benziodoxole radical to yield the corresponding ynone (Fig. 4.113).

A rare example of sunlight-driven decarboxylative alkynylation reaction of a-ketoacids **1** with bromoacetylenes **243** gives ynones **244** in 66% yield through a free radical pathway where **245** is supposed to be an intermediate. The reaction is promoted by hypervalent iodine (III) reagent **246**, visible wavelength and ruthenium based photocatalyst (Fig. 4.114). Comparative

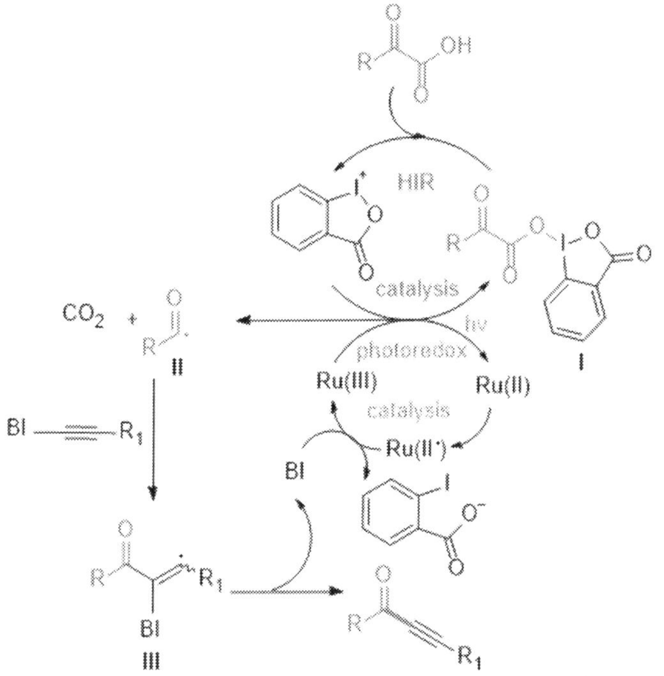

FIGURE 4.113   Mechanistic proposal of the ynonylation reaction.

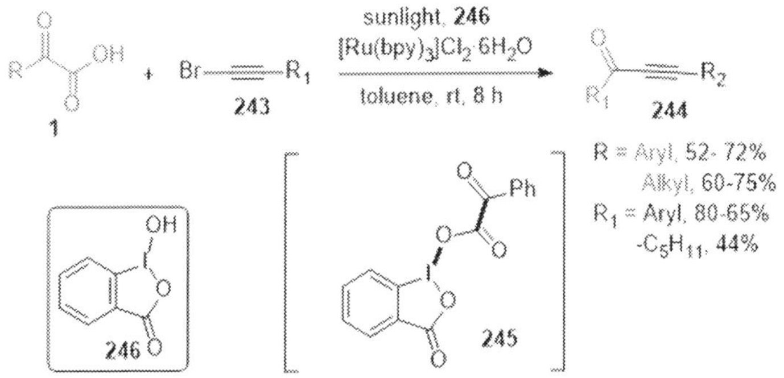

FIGURE 4.114   Visible light promoted formation of ynones.

yields were obtained when blue light (450−455 nm) was used in place of visible wavelength. While as bromoacetylenes with EWGs on benzene ring tend to promote the formation of the desired product, EDGs tend to decrease the product formation.[108]

α-keto acids undergo cycloaddition with 1-iodoalkynes **247** leading to the formation of ynones **248** without the use of HIRs followed by ring-opening pathway. The reaction is a one-pot two-step synthesis, which takes place with catalytic borontrifluoroetherate in toluene to form 4-iodo-5-hydroxyfur-an-2-one intermediate **III** which opens to form ynone **248** after addition of potassium carbonate in DMSO (Fig. 4.115). The carbonyl unit of α- keto acid **1** is retained in the product. The reaction works well over range of substrates including 4-substituted EDG, R = 4-CH$_3$, R = 4-OMe and EWG, R = 4-F with 95%, 85%, and 89% yields respectively. Steric encumbrance in 2-ketoacid did not affect the reaction performance, therefore, α-keto acids bearing substituents at 2 and 3 positions R = 2-CH$_3$ and R = 3-CH$_3$ formed the desired product in 87% and 90% yields respectively.[109]

The proposed mechanism of the reaction shows the generation of carbocation **II** which is accompanied by nucleophilic attack by water molecule to form intermediate **III** which opens into enolate **IV** after the abstraction of a proton. Intermediate **IV** loses carbon dioxide to form the final product in up to 95% yield (Fig. 4.116).

Formation of ynones **250** from α-keto acids and arylpropiolic acids **248** catalyzed by Ag(I) which occurs through double-decarboxylative alkynylation was reported by Cheng and coworkers. In this reaction aryl glyoxylic acid **1** is heated at 50°C for 3 h in presence of catalytic amounts of AgOAc and oxidant $(NH_4)_2S_2O_8$ in DMSO and H$_2$O (1:1 v/v). Since the reaction is not significantly dependent on electronic and steric effects, different substituted ynone derivatives can be prepared. Compounds like ortho-hydroxy phenyl glyoxylic acid **251** and phenyl-4H-chromen-4-one **253** are formed in 71%−93% yields.

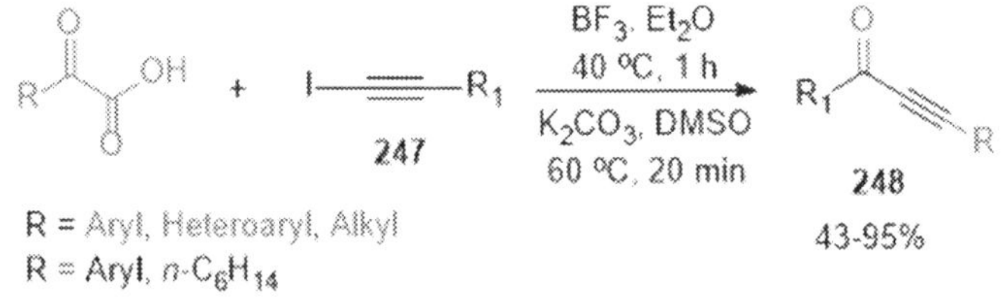

FIGURE 4.115 Lewis acid promoted formation of ynones.

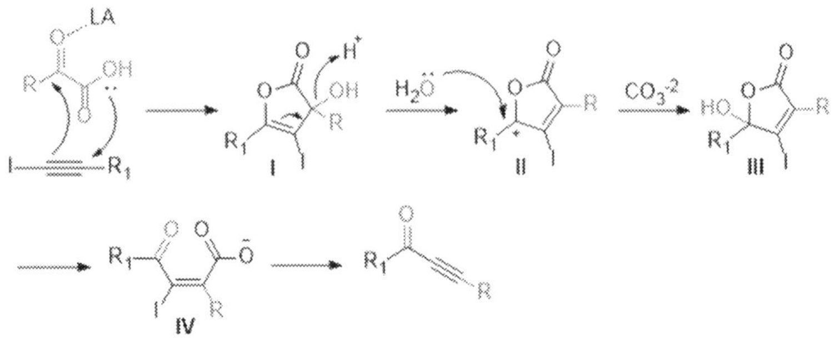

FIGURE 4.116 Proposed mechanism of decarboxylative two-step one-pot synthesis of ynones.

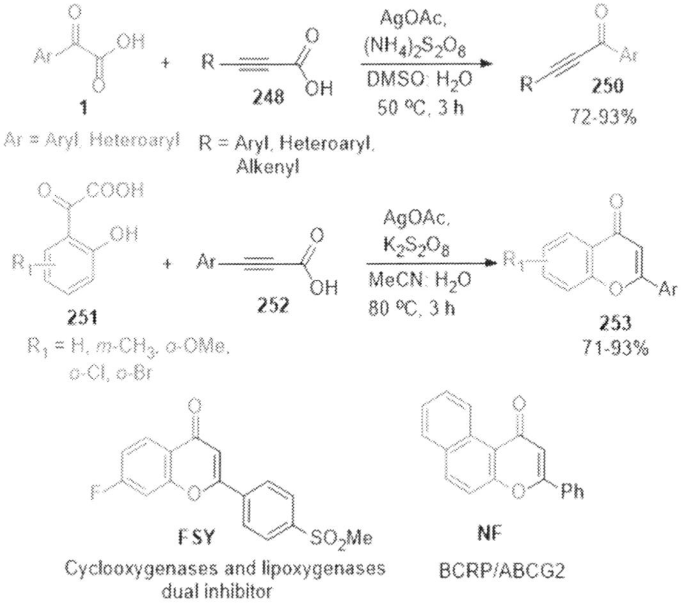

FIGURE 4.117   Silver catalyzed formation of ynones and 4-chromenones.

The reaction is useful in the synthesis of cyclooxygenases and lipoxygenases dual inhibitor (FSY) and breast cancer resistance protein BCRP/ABCG2 (NF) which were formed in 62% and 75% yields (Fig. 4.117).[110]

Ynediones are characterized by 1,2-dione unit and are even more densely functionalized electrophiles than ynones. Formation of ynediones 255 is achieved by alkynylation of a-keto carboxylic acid with oxalylchloride starting from glyoxylation of electron-rich heteroaryl p-nucleophiles with $COCl_2$. The reaction is catalyzed by Cu(I) and in situ generated glyoxylyl chloride 254 which undergoes Stephens-Castro coupling with an alkyne. Synthesis of medicinally important 2-o-tolylaminopyrimidines like necessary building blocks for imatinib and 5-acylpyrazoles, can be obtained by this method (Fig. 4.118).[111]

4.4   Synthesis of heterocyclic compounds

Different N-based heterocycles are synthesized from potassium salts of ketoacid 1a and 2-substituted anilines. The reaction takes place by using $K_2S_2O_8$ as an oxidizing reagent in acetonitrile at 80°C for 8 h. While using 2-aminobenzoic acid 256 and 2-aminobenzamide derivative 258, quinazolin-4-ones 257/259 is formed in 30%−95% yield. On the other hand, 2-amino thiophenol 260, 2-aminophenol 262, and benzene 1, 2-diamines 264 gave benzothiazoles 261, 2-phenylbenzoxazole 263, and benzimidazoles 265

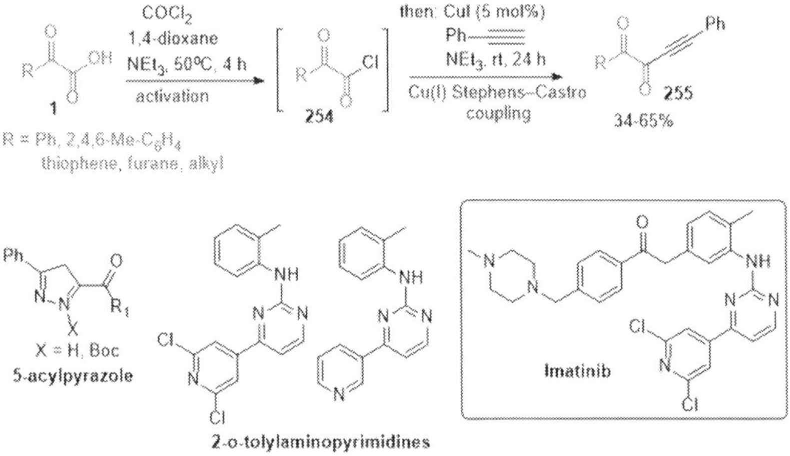

FIGURE 4.118 Formation of ynediones via Stephens-Castro coupling.

respectively in 40%−75% yield. The reaction is not affected much due to electronic effects: when R = OMe and R = F the products are formed in 89% and 87% yield respectively. Heteroaryl substrates were also very successful with 78%−82% product formation. However, lower yields were obtained in the case of 2-aminophenol **262** and benzene 1, 2-diamines **264**. The product formation takes place through imine followed by intramolecular cyclization. This method has found important use in the synthesis of blockbuster drug sildenafil (Fig. 4.119).[112]

The reaction is initiated by the formation of SO_2^{-}. anion from $K_2S_2O_8$ that leads to decarboxylation of the oxocarboxylate ion generating an acyl radical **I** along with CO_2 and SO_4^{2-}. Acyl radical further reacts with SO_2^{-}. to form acyl sulfate intermediate **II**, which is electrophilic in nature. Both intermediates generate HSO_4^{-} anion by either radical or ionic pathway. Radical pathway carries N-acylation at NH_2 group while as ionic pathway caries out the nucleophilic attack over the NH_2 group of the o-substituted aniline to form intermediate **IV** after the intermediate **I** is captured by a sulfate radical-anion to **II**. Intermediate **IV** undergoes HSO_4^{-} and SO_4^{2-} promoted cyclization to form the corresponding products (Fig. 4.120).

α-keto acids **1** can react with tosylmethyl isocyanide (TosMIC) **266** to produce 5-aryloxazoles **267** catalyzed by Cu(I) by using 1,10-Phen as a chelating ligand and potassium hydroxide as a base. The reaction takes place in DMF without using any oxidant within 45% − 94% yield. The method works well over range of substituted aryl ketoacids. However, steric factors dominate the reaction when substituents in the phenyl ring are present at *ortho* position and needs more time to complete the reaction (Fig. 4.121).[113]

The reaction mechanism does not follow the radical path and is initiated by the formation of Cu(I) coordinated oxocarboxylate ion to form

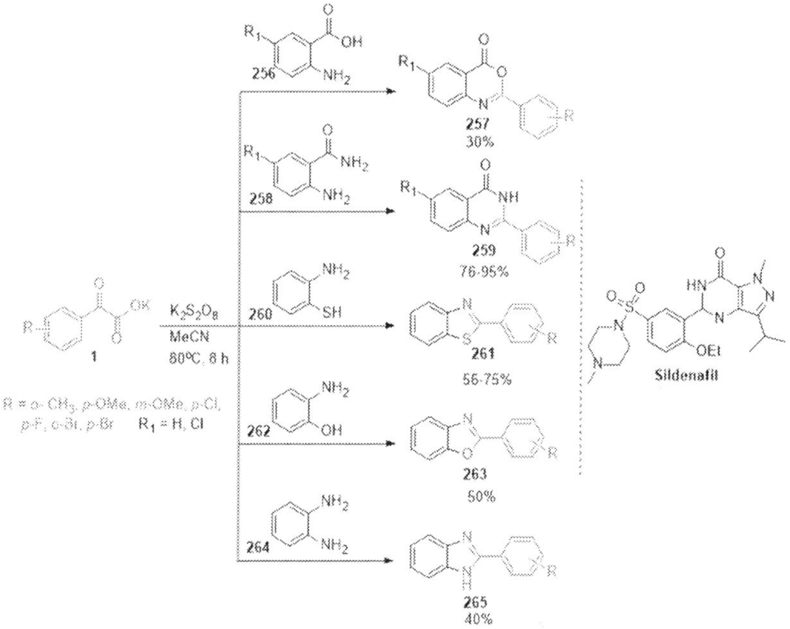

FIGURE 4.119   Oxone mediated formation of heterocycles.

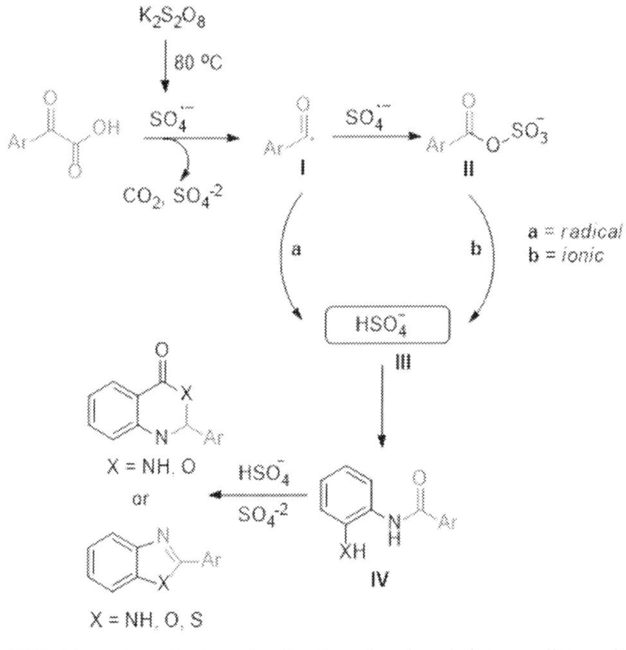

FIGURE 4.120   General mechanism for the free decarboxylative coupling and annulation between 2-ketoacids and amines.

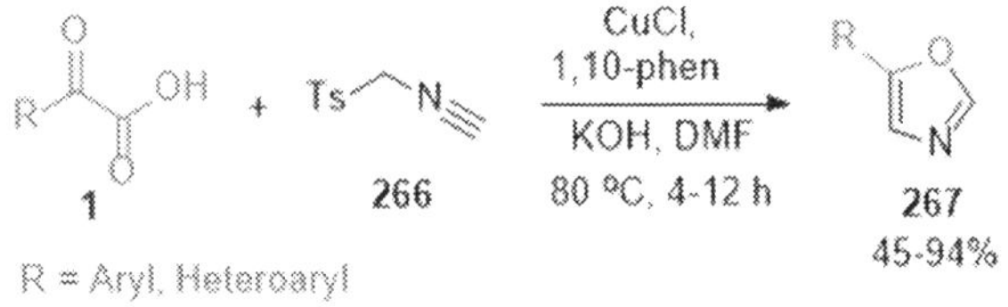

FIGURE 4.121 Formation of oxazoles from 2-oxoacids.

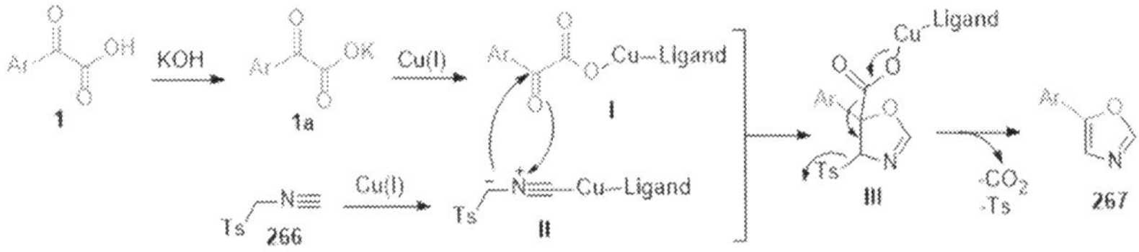

FIGURE 4.122 Reaction mechanism of the Cu(I)-catalyzed decarboxylative cyclization.

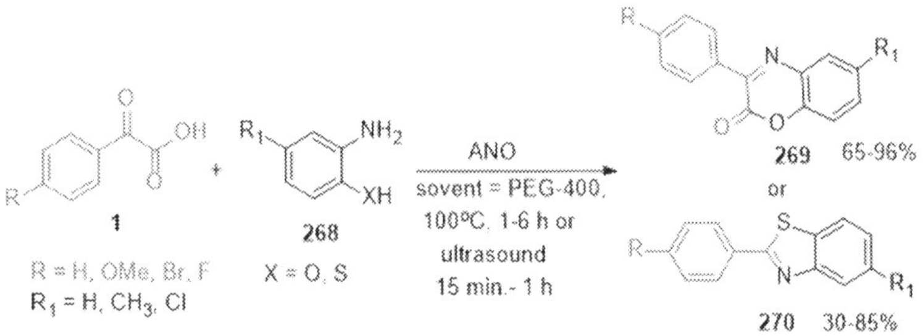

FIGURE 4.123 Formation of aminophenols, benzothiazole from 2-oxoacid.

intermediate **I**. [3 +2] cycloaddition takes place between **I** and cuprioisocya-nide **II** to form the intermediate **III**. Finally, the aromatization occurs with the loss of CO_2 and tosyl group to form the final product (Fig. 4.122).

2- ketoacids react with 2-aminothiophenols and 2- aminophenols **268** to form benzothiazole **270** and benzoxazine-2-one derivatives **269** in 30%−85% and 65%−96% yield, respectively. The reaction is performed by employing catalytic amounts of $NH_4[NbO(C_2O_4)_2(H_2O)x] \cdot nH_2O$ (ANO) in PEG-400 as a solvent. The formation of benzothiazoles takes place by heat-ing the reaction mixture at $100°C$ for 1−6 h whereas benzoxazine-2-ones **269** are formed after 2.5 h of heating at $100°C$. Reaction yields were improved and the time of the reaction was reduced from 15 min to 1 h when the reaction was subjected to ultrasound waves (Fig. 4.123).[114]

2-keto acids undergo decarboxylative condensation with S and Se based bis-ortho-amino dichalcogenides **271** and **273** which leads to the formation of 2-substituted benzothiazoles **272** and benzoselenazoles **274** in 41%−91%

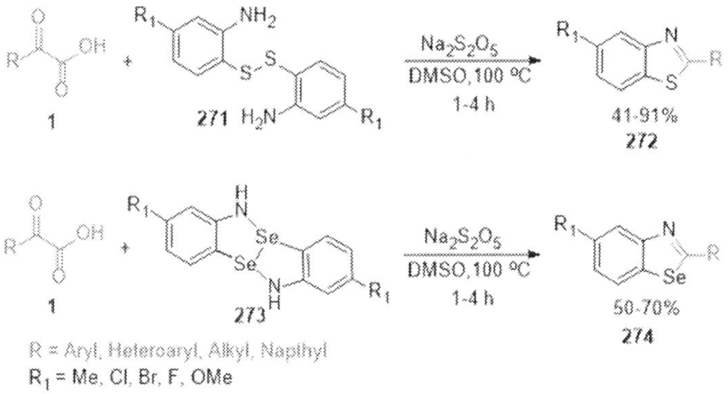

FIGURE 4.124 Formation of benzoselenazoles and benzothiazoles from 2-oxoacids.

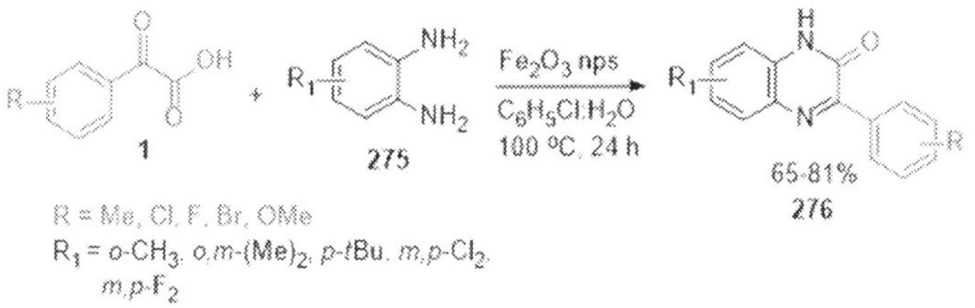

FIGURE 4.125 Formation of quinoxalinones from 2-oxoacids.

and 50%−70% yields respectively. The reaction occurs in the presence of $Na_2S_2O_5$ in DMSO at 100°C after heating for 1−4 h. The disulfide substrates containing electron-releasing substituents like 5-Cl, 4-OMe were obtained in lower yields than unsubstituted derivatives with higher reaction times. However, benzothiazoles **272** were obtained in 87% and 85% yields when the substrates involved possessed at least electron-withdrawing substituents (R = 4-F and 2, 4-Cl) on arylglyoxylic acids **1**. Electronic effects were not prominent in diaryl diselenides **273** and both electron-releasing as well as electron-withdrawing substrates furnished lower but similar yields due to intramolecular Se · · · N bonding (Fig. 4.124).[115]

Superparamagnetic Fe_2O_3 nanoparticle catalyzed reaction of α-keto acids with 2-phenylenediamine **275** derivatives to form quinoxalinones **276** was reported by Le, Phen, and coworkers. The cyclization is achieved in chlorobenzene/water mixture (1.5: 0.5 v/v) after heating at 100°/ for 24 h (Fig. 4.125).[116]

α-ketoacids react in two different conditions to produce benzo[b][1,4] benzoxazine-2-one derivatives **278**. In one of the condition, the reaction is promoted by oxone where 2-ketoacid reacts with benzoxazoles **277** in DMSO/diglyme (5%) at 120°C after 12 h. This method is very successful for

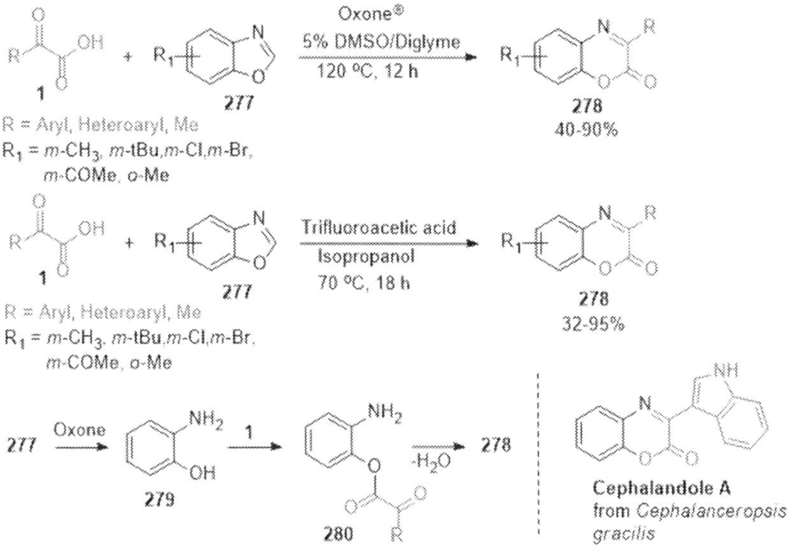

FIGURE 4.126 Formation of benzoxazoles from 2-oxoacids.

the synthesis of benzoxazoles **278** which bear both electron-releasing and electron-withdrawing substituents which were obtained in 40%−90% yields. The second method involves the addition of trifluoroacetic acid to a mixture of 2-ketoacid and benzoxazole **277** while using isopropanol as a solvent. The reaction occurs within 18 h when heated at 70°C with 32%−95% yields. Unsubstituted and electron-deficient arylglyoxylic acids work very well as compared to electron-rich substrates. For instance, when R_1 = H and R_1 = F: the desired product is formed in 94% and 93% yields respectively. This method has found application in the synthesis of natural product cephalandole A. In both cases, benzoxazole ring-opening occurs to form o-aminophenol **279** followed by esterification and subsequent loss of water molecule to form the final product (Fig. 4.126).[117,118]

Butenolides are four-membered lactone heterocyclic compound that is found in many natural and synthetic molecules which are medicinally important.[119] 2-furanone is the simplest butanolide, which is constituted in many naturally occurring molecules like fissohamion, goniobutanolide-A; while ascorbic acid is biologically vital butenolide, Patulin is a mycotoxin from fungal sources and digitixigenin is a steroid natural product containing butenolide core (Fig. 4.127). Butenolides are used as building blocks for the synthesis of various molecules.[120] Therefore, in organic and medicinal chemistry many protocols have been developed for the synthesis of butenolides.[121]

2-ketoacids have been applied for the synthesis of pharmacologically potent γ-butenolides **282**−**284**. Three methods, in which 2-ketoacids react with tertiary alcohols **281/285** and tert-butyl mercaptan **283** have been

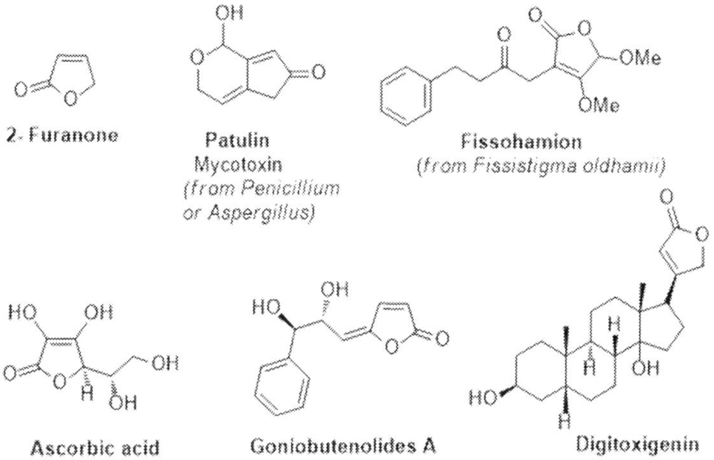

FIGURE 4.127 Some biologically important molecules containing butenolide core.

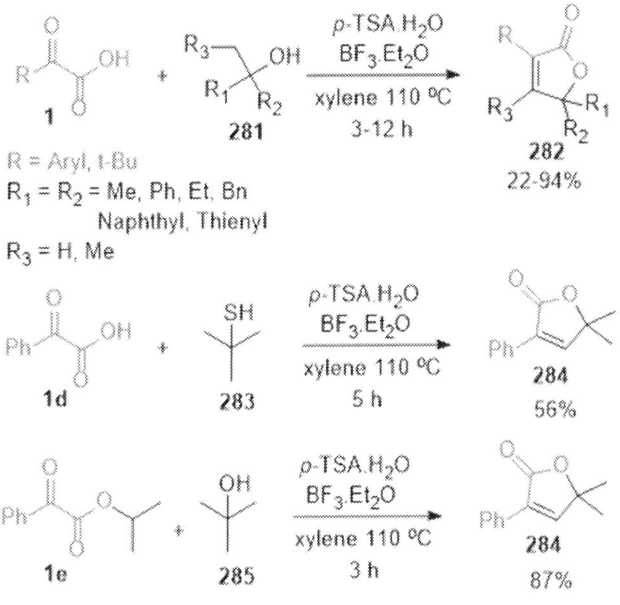

FIGURE 4.128 Lewis acid promoted formation of γ-butenolides from 2-oxoacids.

developed. The common reaction condition for these methods is heating the reaction mixture in the presence of *p*-toluene sulfonic acid and BF_3. OEt_2 in xylene at 110°C for several hours. Various tertiary alcohols were used for the annulation with α-ketoacids (Fig. 4.128).[122,123]

Similarly, different glyoxylic acid substrates were also used to give the desired product in 94% yield after 3 h of reaction. While there are no apparent electronic effects governing the reaction, lower yields were isolated when phenyl glyoxylic acid incorporated any o-substituents due to steric crowding.

In a similar manner, alkynes are also employed in the synthesis of bute-nolides. 2-ketoacids react with both internal alkynes **286** and terminal alkynes **289** to form γ- hydroxybutenolides **287** instead of previously sup-posed α-hydroxybutenolides **288**. The reaction is led by adding a Lewis acid BF₃. OEt₂ in fluorobenzene at 70°C for about 24 h excluding the need of p-TSA. Aliphatic 2-keto acids of aliphatic nature were not successful in the reaction. Electron-donating or electron-rich para-substituted and unsubsti-tuted phenylglyoxylic acid substrates showed less and slow reactivity that lead to the lower amount of product formation within the given time than the electron-deficient substrates which exhibited more reactivity. For example, when the substituent (R) on arylglyoxalic acid (R = H, 4-CH₃ and 4-OMe), the corresponding γ-hydroxybutenolides were formed in 92%, 88%, and 73% yields, respectively, after heating for 5 and 24 h. While as when R = 4-CF₃ and 4-F: 91% and 94% of the desired product was formed respectively only after heating for 4 h. However, depending on the steric factors the reac-tion with ortho-methyl and ortho-bromo-phenylglyoxylic acid provided γ-hydroxybutenolides in 42%−70% yields after 24 h (Fig. 4.129).[124]

The plausible mechanistic pathway is initiated by the addition of alkyne to 2-keto acid **1** to form the cyclized intermediate, α-hydroxybutenolide **I**, which after acid-promoted dehydration gives carbocation **II**. Nucleophilic addition over **II** by a water molecule furnishes γ-hydroxybutenolide (Fig. 4.130).

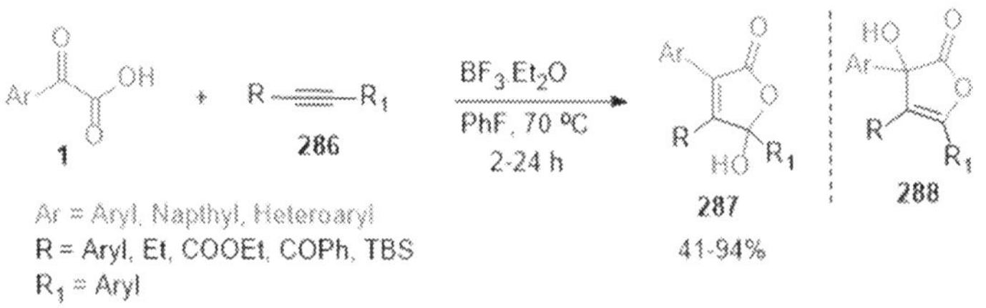

FIGURE 4.129 Formation of γ-hydroxybutenolide from 2-oxoacids.

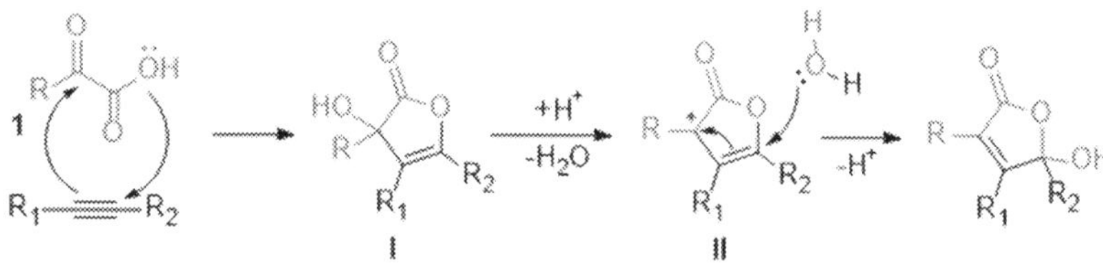

FIGURE 4.130 General mechanism for the Lewis-acid catalyzed cyclization toward γ-hydroxybutenolides.

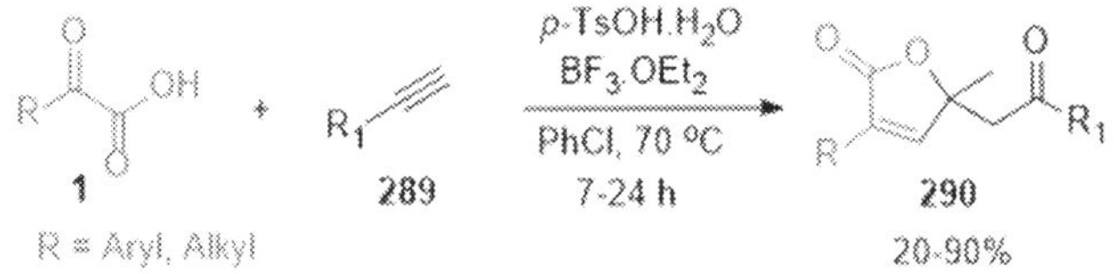

FIGURE 4.131 Formation of substituted butenolide from 2-oxoacids.

When 2-keto acids react with terminal alkynes **289**, more substituted buteno-lides **290** are formed when the reaction mixture is heated in the presence of *para*-toluene sulfonic acid and BF$_3$. OEt$_2$ in chlorobenzene (Fig. 4.131). Unlike the utilization of only one alkyne molecule in the case of internal alkynes; two molecules of terminal alkynes were consumed during the annulation of terminal alkynes with 2-ketoacids **1**. Due to this reason, owing to a new reaction path-way, substitution on the butenolide derivatives was enhanced with the emer-gence of keto functionality that is generated after the hydration at the second alkyne molecule. In contrast to the above reaction electron-withdrawing substi-tuted glyoxylic acids exhibited lesser reactivity than electron-donating glyoxylic acid substrates and in case of aliphatic glyoxalic acid substrates the reaction required longer time periods to complete the reaction.[123]

Another method that leads to the formation of butenolides is the reaction between 2-ketoacid and internal alkynes of aliphatic nature **291** in which γ-ylidenebutenolides **292** are formed in good yields. The reaction is cata-lyzed by using (10 mol.%) of [Cu(MeCN)$_4$]BF$_4$ at 130°C for 20 h in toluene. The reaction has a very wide scope over aryl, carbocyclic alkynes and ali-phatic glyoxylic acids and can also withstand halogen and silyl containing groups in the alkyne part with overall 14%−90% yield. The presence of *p*-substituted electron-donating substituents in arylglyoxylic acid part improves the product formation as compared to electron-deficient substituents. As part of the alkyne, the percentage of Z-isomer is reduced when no substituent is present at the β-position of the butenolide. γ-ylidenebutenolides can further be conveniently transformed into γ hydroxybutenolide, pyrrolone derivatives, γ-keto esters, pyradazinone and cyclopentenone. Another application is the synthesis of bovolide **296** from bromopyruvic acid **1f** and 2-octyne **294** formed after the reduction of butenolide intermediate with zinc and aq. ammo-nium chloride (Fig. 4.132).[125]

Mechanistically, alkenyl − Cu-complex **293** is supposed to be an interme-diate after the reaction of 2-ketoacid and alkyne with copper salt which undergoes intramolecular nucleophilic attack followed by loss of water mole-cule to furnish the γ- ylidenebutenolide **292** in 74% yield.

Arylglyoxalic acids **1** and *N*-aryl-substituted cyclic amines **297** react together following a cascade reaction promoted by Fe(III) to form piperidine fused fura-none derivative **298**. The reaction takes place by using FeCl$_3 \cdot$ H$_2$O, DMAP, and

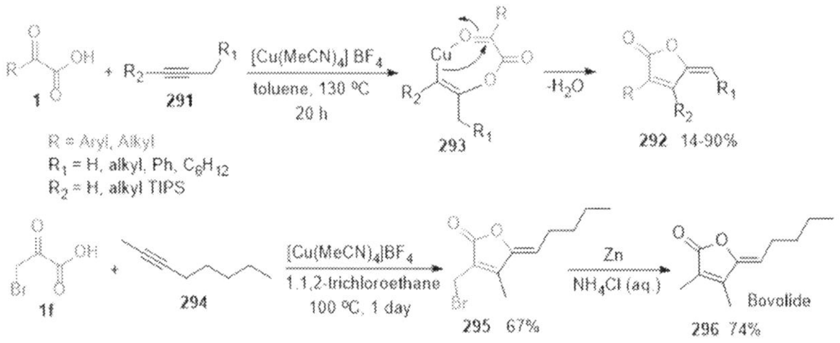

FIGURE 4.132   Cu catalyzed formation of bovolide from 2-oxoacids.

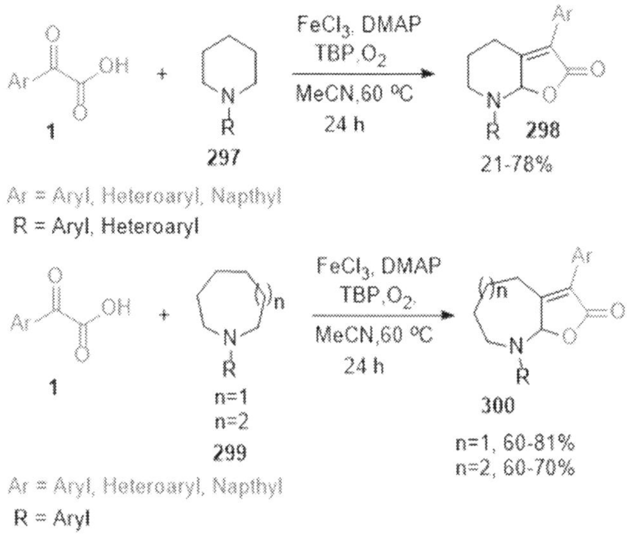

FIGURE 4.133   Fe promoted formation of cyclic amine fused derivatives of butenolides from 2-oxoacids.

di-tert-butyl peroxide (TBP) as an oxidant in acetonitrile after heating at 60°C for 24 h. Electron-releasing groups present at the phenyl group of the glyoxalic acid are known to promote the product formation while the electron pulling groups tend to reduce the yield of the reaction. The same trend is observed with N-aryl attached groups in cyclic amines **299**. The scope of the reaction is further elaborated by the synthesis of N-substituted azopanes and azocanes **300** formed after the reaction of 2-aryl ketoacids with N-aryl cyclic amines of higher molecular weights (Fig. 4.133).[126]

Amino acids also give rise to the formation of butenolides after reaction with 2-ketoacids. In a metal-free process excess of aliphatic

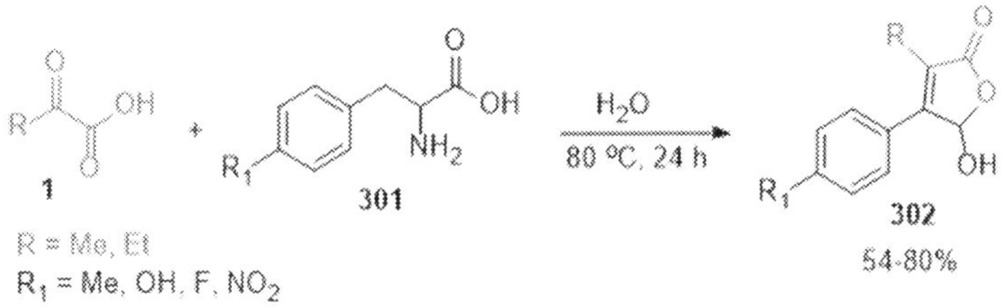

FIGURE 4.134 Formation of γ-hydroxybutenolide from 2-oxoacids.

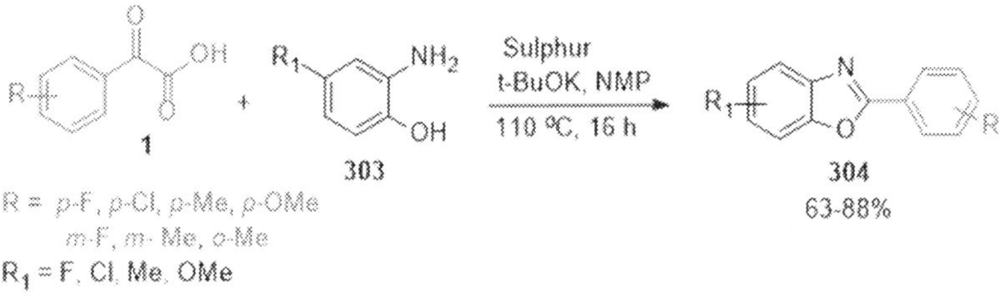

FIGURE 4.135 Formation of benzoxazole derivatives from 2-oxoacids.

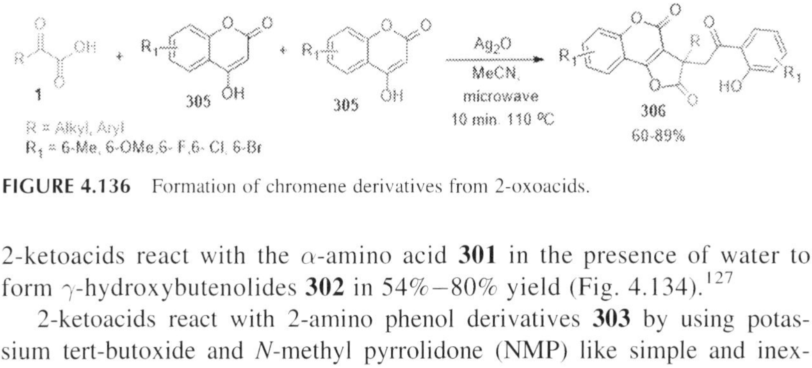

FIGURE 4.136 Formation of chromene derivatives from 2-oxoacids.

2-ketoacids react with the α-amino acid **301** in the presence of water to form γ-hydroxybutenolides **302** in 54%−80% yield (Fig. 4.134).[127]

2-ketoacids react with 2-amino phenol derivatives **303** by using potassium tert-butoxide and N-methyl pyrrolidone (NMP) like simple and inexpensive materials catalyzed by elemental sulfur to give substituted benzoxazole analogs **304**. 2-substituted and 6-substituted 2-keto acids were not successful to form the product. The reaction is supposed to follow a free radical process in which sulfur promotes decarboxylation followed by cyclization which leads toproduct formation after heating at 110°C for 16 h (Fig. 4.135).[128]

An important and simple way for the preparation of chromenes **306** of chiral nature is the reaction between arylglyoxalic acids **1** and 4-hydroxycoumarins **305**. The reaction is promoted by Ag(I) under microwave irradiation at 110°C in a very short time (Fig. 4.136). The reaction works over a wide range of substrates like aryl and heteroaryl and has displayed tolerance against different functional groups with 60%−89% yields. Many

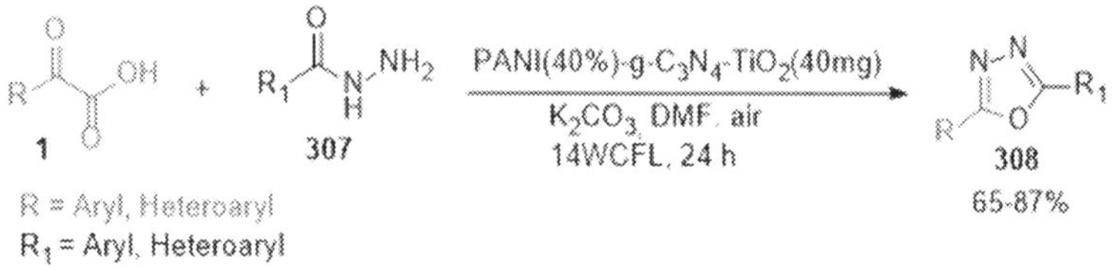

FIGURE 4.137 Formation of oxadiazole derrivatives from 2-oxoacids.

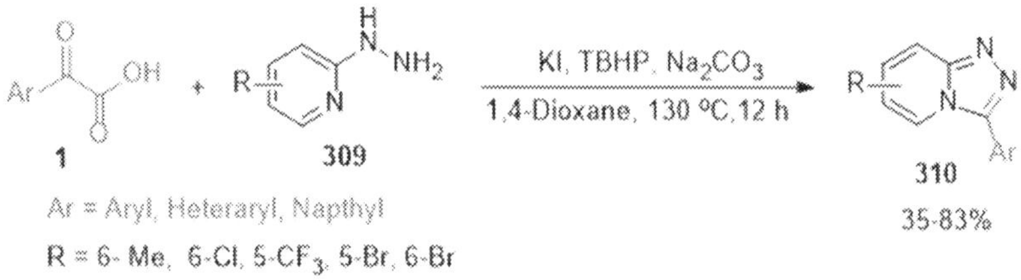

FIGURE 4.138 Formation of triazolopyridine derivatives from 2-oxoacids.

substrates can further be used and easily transformed into other synthetic molecules, for example, iodine-containing products can be used in metal-catalyzed cross-coupling reactions.[129]

In a greener approach, 2-ketoacids react with acyl hydrazines **307** to form 1,3,4-oxadiazole analogs **308** in presence of a photocatalyst in DMF for 24 h at ambient temperature (Fig. 4.137). The reaction passes through imine formation and after subsequent decarboxylation annulation occurs to form the oxadiazoles in 65%−87% yield. A wide range of oxadiazole derivatives is synthesized by using this method. Electron-releasing substituents attached to the phenyl ring of ketoacid work better as compared to the electron-withdrawing groups in increasing the yield of the product and decreasing the reaction times during the product formation. The same principle was followed in case of acyl hydrazines without any apparent role of steric factors.[130]

α-keto acids react with 2-hydrazinopyridines **309** to form triazolo pyridine analogs **310** in the presence of potassium iodide as catalyst, tert-butyl hydroperoxide (TBHP) as oxidant and Na_2CO_3 in 1,4-dioxane at 130°C for 12 h (Fig. 4.138). The reaction covers substrates like aryl, heteroaryl and also tolerates many functional groups in both the participating substrates. Pyruvic acid, however, failed to form any desired product. The general trend in the reaction follows that arylglyoxylic acids possessing electron-releasing groups react faster than those bearing electron attracting groups.[131]

In some reactions, potassium thiosulphate acts as an important reagent or catalyst in promoting several reactions of 2-ketoacids with other coupling partners. In this context, 2-ketoacids react with aryl alkynoates **313** in silver-

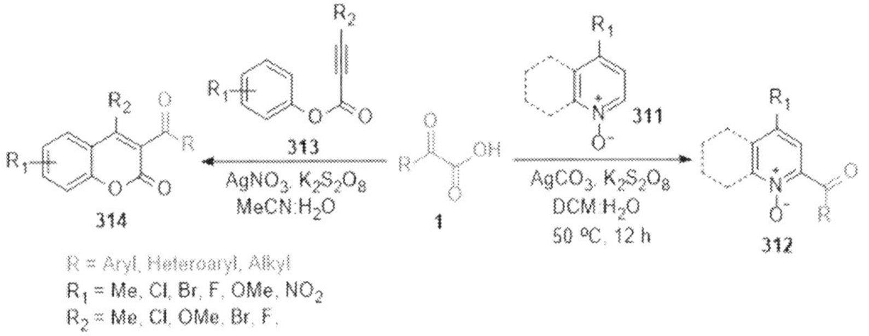

FIGURE 4.139 Formation of pyridine-*N*-oxides from 2-oxoacids.

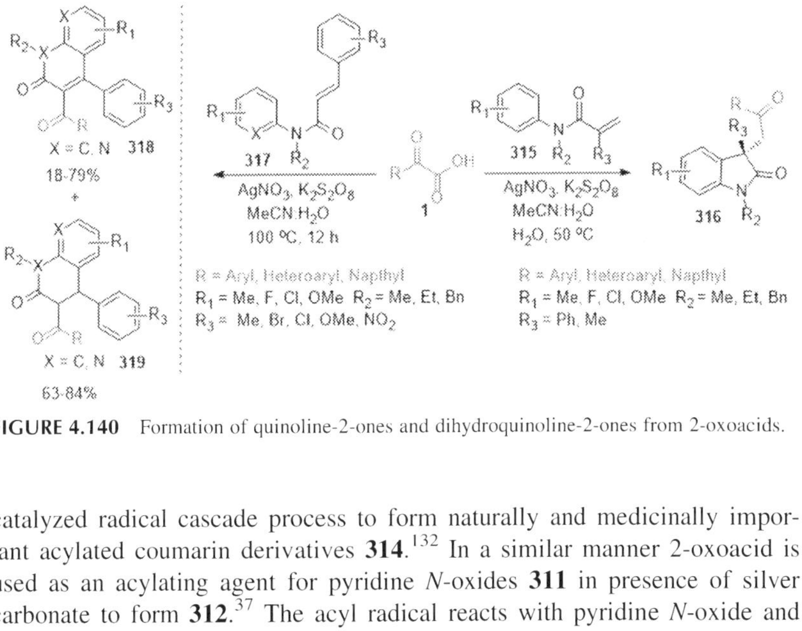

FIGURE 4.140 Formation of quinoline-2-ones and dihydroquinoline-2-ones from 2-oxoacids.

catalyzed radical cascade process to form naturally and medicinally important acylated coumarin derivatives **314**.[132] In a similar manner 2-oxoacid is used as an acylating agent for pyridine *N*-oxides **311** in presence of silver carbonate to form **312**.[37] The acyl radical reacts with pyridine *N*-oxide and sulfate radical generated from $K_2S_2O_8$ captures the hydrogen to form the desired product (Fig. 4.139).

Other reactions favored by the combination of potassium thiosulphate and silver nitrate include the reaction of 2-ketoacids **1** with acrylamides **317** in acetonitrile and water to form quinoline-2-ones **318** and dihydroquinoline-2-ones **319** by using 4 and 2 equivalent of $K_2S_2O_8$ respectively. The reaction once again uses 2-ketoacid as an acylating agent with scope over both N-alkyl and N-aryl acrylamides. Alkyl carboxylic acids can be easily converted into α-ketoacids through this method which can, later on, launch the "acyl" group in the product through 6-endo-trig manner (Fig. 4.140).[133,134]

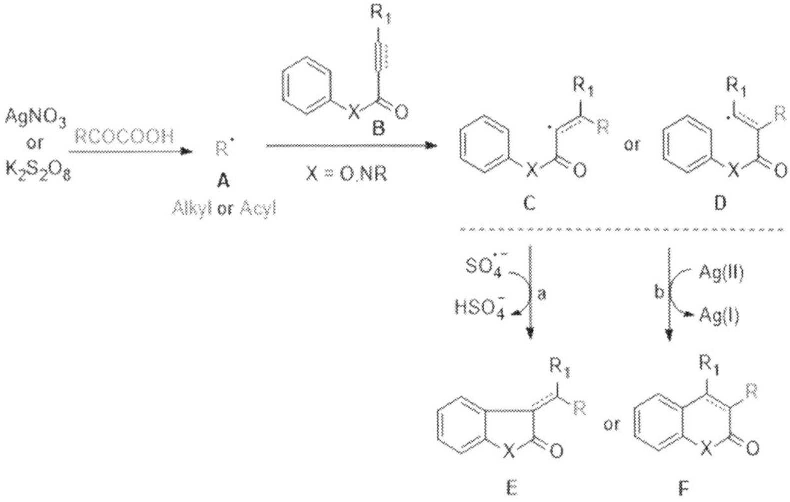

FIGURE 4.141 Ag-catalyzed radical cyclization processes of alkynoates and acrylamides.

Preparation of oxindoles **316** is led by similar processes in which acrylamide derivatives **315** and 2-ketoacids **1** react in presence of silver nitrate and $K_2S_2O_8$ (Eq. 3.41). The proposed mechanism of the Ag-catalyzed radical cyclization processes of alkynoates and acrylamides promoted by potassium thiosulphate is given (Fig. 4.141). The first step is regioselective alkyl or acyl radical addition to alkynoates **B** to form alkene radical intermediate **C** and **D**. The five- or six-membered rings, thus formed, follow 5-exo-trig cyclization rule. Aromatization process could trigger in two different pathways, a hydrogen abstraction could occur by SO4•$^-$ through (Path **a**) to formoxindoles **E** and deprotonation after Ag(II) promoted the formation of the carbocation (Path **b**) leading to dihydroquinoline-2-ones **F**.

However, when $(NH_4)_2S_2O_8$ was employed instead of potassium thiosulphate in acetone/water the reaction of 2-ketoacids with o-cyanoarylacrylamides **320** results in the formation of carbonyl containing quinoline derivatives **321**. This reaction is also accompanied by decarboxylation followed by cyclization cascade through a carbon radical which adds intramolecularly to the nitrile group (Fig. 4.142).[135]

4.5 Multicomponent reactions

A multicomponent reaction of 2-keto acids that leads, stereoselectively, to the formation of biologically important 2, 5-diketopiperazines (DKP) derivatives catalyzed by monoamine oxidase N(MAO-N) and Ugi-Pictet−Spengler process is a good example regarding the divergent reactivity of 2-ketoacid.[136] In this reaction 2-ketoacid reacts with optically pure pyrrolidine **323** and

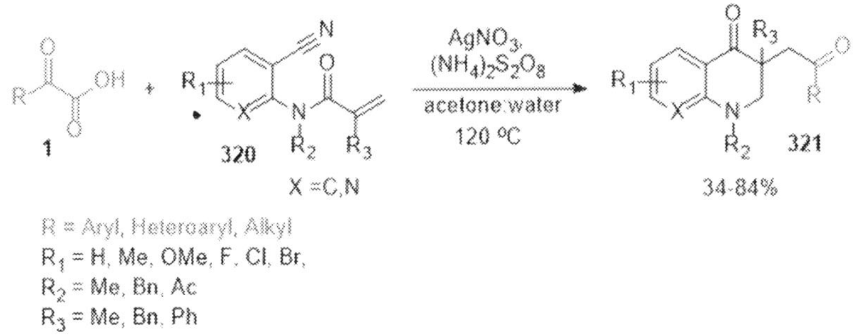

FIGURE 4.142 Formation of quinoline derivatives from 2-oxoacids.

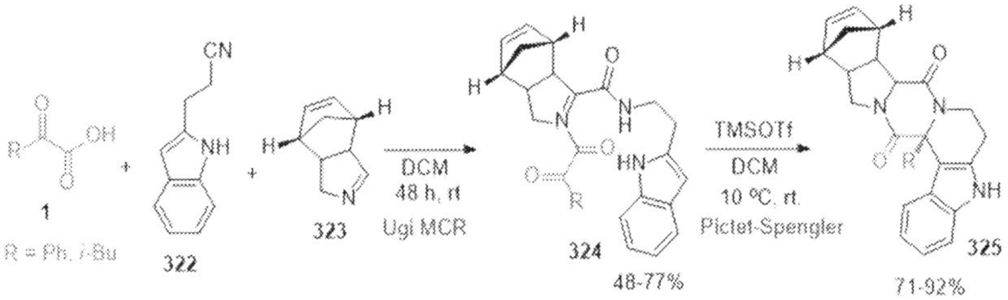

FIGURE 4.143 Formation of 2,5-diketopiperazines from 2-oxoacids.

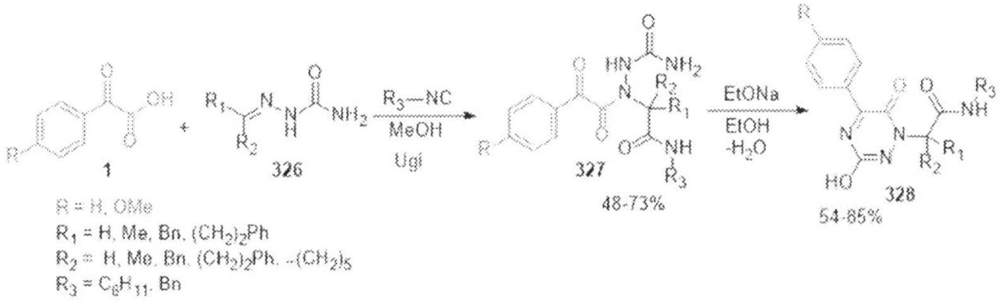

FIGURE 4.144 Formation of triazin-alaninamide from 2-oxoacids.

isocyanide **322** to give substituted prolyl peptides **324** which later on treatment with TMSOTf yields complex 2,5-diketopiperazines (DKPs) **325** which can serve as valuable targets in medicinal chemistry (Fig. 4.143).[137]

Among other important Ugi type reaction is the reaction of 2-ketoacids with **326** or isocyanides or 4-methoxy benzoyl formic acid and semicarbazones to form triazin-alaninamide analogs **328**. The reaction operates in a very simple and ecofriendly conditions to form the Ugi adduct which cyclizes to form the desired product in 54%−85% yield (Fig. 4.144). These products can further be transformed into *O-methyl-* or hydroxy-pseudopeptidic molecules containing different amino acids or other analogs.[138]

The rare four-component Ugi reaction accompanied by aldol condensation has been reported. The reaction takes place between phenylglyoxylic acid **1**, pyruvic aldehydes **329**, 2-fluoro-4-bromoaniline **331** and tert-butyl isocyanide **330** to formpyridoquinoxalinedione **332** after stirring in MeOH followed by the addition of DIPA in DMF and refluxing the reaction mixture at 120°C The reaction proceeds with aldol reaction to form **333** which after tautomerization and nucleophilic substitution gives the desired product in 70% yield (Fig. 4.145).[139]

A new one-pot two-step four-component synthesis of naturally occurring spiropyrrolidinochromanone **338** is reported by exploiting the reactivity of 2- ketoacid **1** with amines **336**, 3-formylchromones **334** and isonitriles **335** in a stereoselective manner (Fig. 4.146). This is a very simple and effective strategy for the

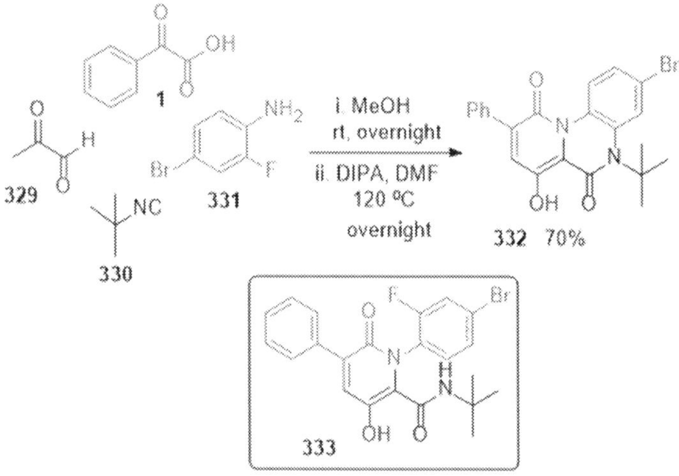
1
329
330
331
i. MeOH
rt, overnight
ii. DIPA, DMF
120 °C
overnight
332 70%
333

FIGURE 4.145 Formation of pyridoquinoxalinedione from 2-oxoacids.

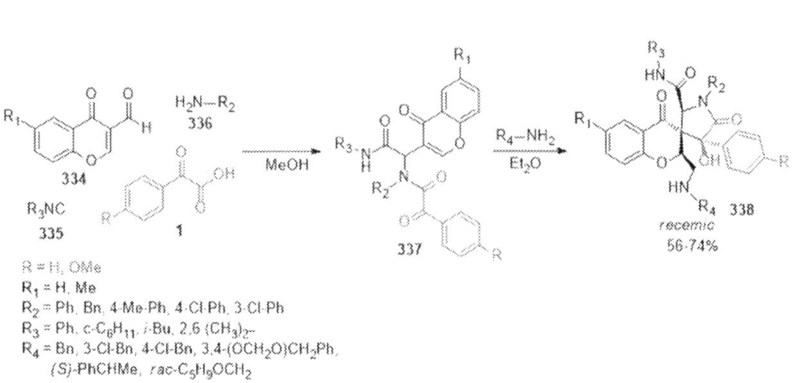
334
336
335
1
MeOH
337
R_4–NH_2
Et_2O
338
racemic
56-74%
R = H, OMe
R_1 = H, Me
R_2 = Ph, Bn, 4-Me-Ph, 4-Cl-Ph, 3-Cl-Ph
R_3 = Ph, c-C_6H_{11}, i-Bu, 2,6 $(CH_3)_2$–
R_4 = Bn, 3-Cl-Bn, 4-Cl-Bn, 3,4-$(OCH_2O)CH_2Ph$, (S)-PhCHMe, rac-$C_5H_9OCH_2$

FIGURE 4.146 Formation of spiropyrrolidinochromanone from 2-oxoacids.

preparation of complex scaffolds which show tolerance to many labile substituents and reveal an important criterion for combinatorial chemistry.[140]

Another Ugi type *pseudo*-Knoevenagel reaction leads to the formation of aziridine derivatives **343** and **344** involving cascade annulation. In this reaction 2-ketoacids **1**, amine **340**, ketoester **339** and isocyanide **341** react together in very mild condition to form **342** in MeOH which is later on transformed into *trans* aziridine **343** in simple basic conditions after heating at 130°C for 10 min by changing the arylamine and isoisocyanide (Fig. 4.147). Aziridine can also be converted into maleimide **345** by using triethanolamine in DMF at 70°C. Many biologically active molecules like **346** which bears potency like Sorafenib[a] is used in the hepatic cancer cell line and could be an asset in cancer drug discovery.[141]

2-ketoacids **1** undergo decarboxylative three-component reaction with amines **348** and alkynes **347** to form propargyl amines **349** catalyzed by Cu (I). The reaction takes place by heating the reaction mixture in toluene at100°C under microwave irradiation (Fig. 4.148). The reaction works well with aliphatic alkynes including hindered or unhindered amines. The plausible mechanism proposes the generation of in situ iminium ion which may lose carbon dioxide molecule under Cu(I) catalysis to give iminium copper ion **a** which reacts with an aliphatic alkyne to form the Cu-complex **b** that finally leads to the formation of the desired product.[142]

In a similar manner, 2-ketoacids **1**, alkynyl carboxylic acids **350**, and cyclic amines **351** furnish propargyl amine **352** (Fig. 4.149).[143] However, when 2-aminopyridine **354** is employed in the presence of a catalytic amount of copper triflate, imidazopyridines **355** are obtained in 55%−81% yield.

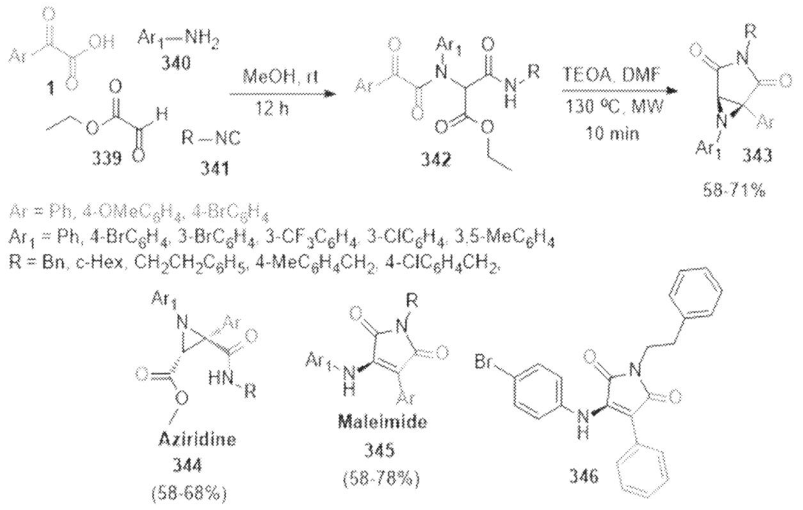

FIGURE 4.147 Formation of aziridine from 2-oxoacids.

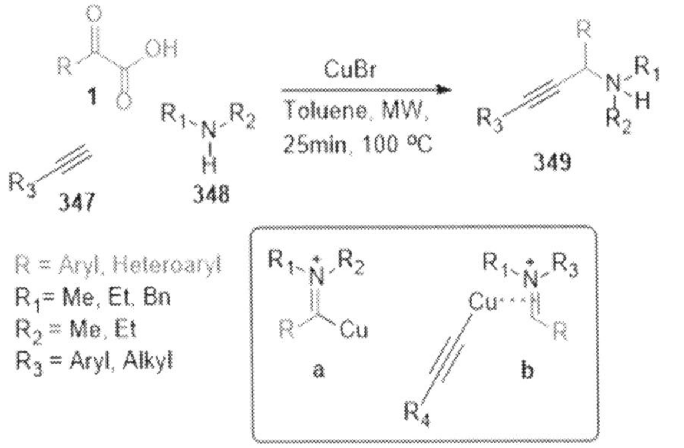

FIGURE 4.148   Formation of propargyl amines from 2-oxoacids.

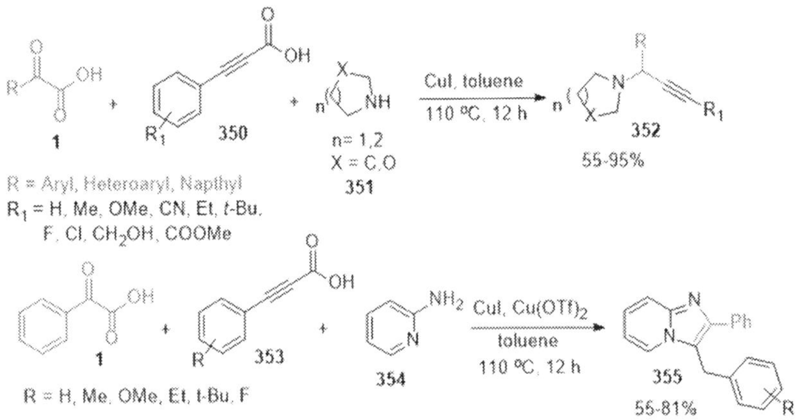

FIGURE 4.149   Formation of propargyl amines and imidazopyridines from 2-oxoacids.

Both alkynyl carboxylic acids as well as glyoxylic acids undergo decarboxylative coupling with amines by using 10 mol.% of CuI.

A different type of decarboxylative coupling to produce azomethines is reported by Lukas J. Gooben and coworkers. This process occurs when the mixture of 2-ketoacids **1**, aryl halides **357**, and amines (aryl or aliphatic) **356** is heated at 100°C in polar aprotic solvent N-methyl pyrrolidone (NMP) in presence of Pd(F$_6$-acac)$_2$, dppf, CuBr as catalyst and 1,10-phenanthroline. The corresponding azomethine **358** was obtained in 92% yield as a mixture of cis/trans isomers (Fig. 4.150). The reaction has scope over electron-donating and electron-withdrawing groups as well as aryl and heteroaryl bromides bearing nitro, thioether, halide or ester groups.[144]

The mechanistic details include the in situ formation of α-iminocarboxylate intermediate **a**, from potassium α-oxocarboxylate and amine to form a Cu-complex **b**. Finally, azomethine is formed as a result of reductive elimination.

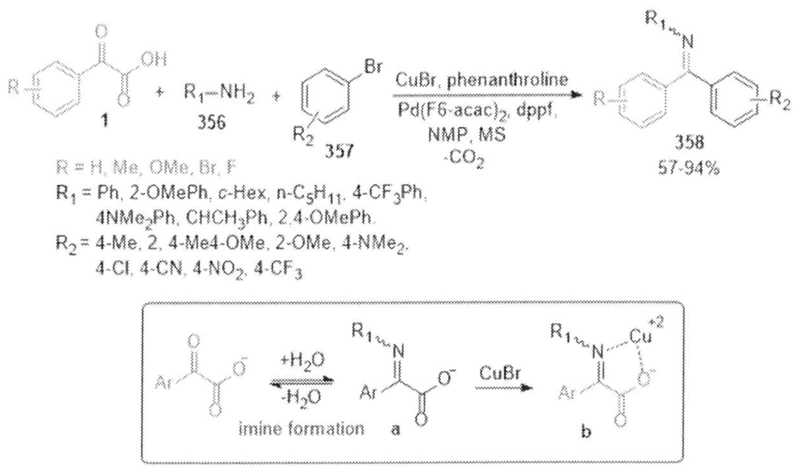

1
R₁—NH₂
356
Br
R₂
357
CuBr, phenanthroline
Pd(F6-acac)₂, dppf,
NMP, MS
-CO₂
R₁
N
R₂
358
57-94%
R = H, Me, OMe, Br, F
R₁ = Ph, 2-OMePh, c-Hex, n-C₅H₁₁, 4-CF₃Ph,
4NMe₂Ph, CHCH₃Ph, 2,4-OMePh.
R₂ = 4-Me, 2, 4-Me4-OMe, 2-OMe, 4-NMe₂,
4-Cl, 4-CN, 4-NO₂, 4-CF₃
Ar
+H₂O
-H₂O
imine formation
a
CuBr
Cu⁺²
b

FIGURE 4.150 Formation of azomethine from 2-oxoacids.

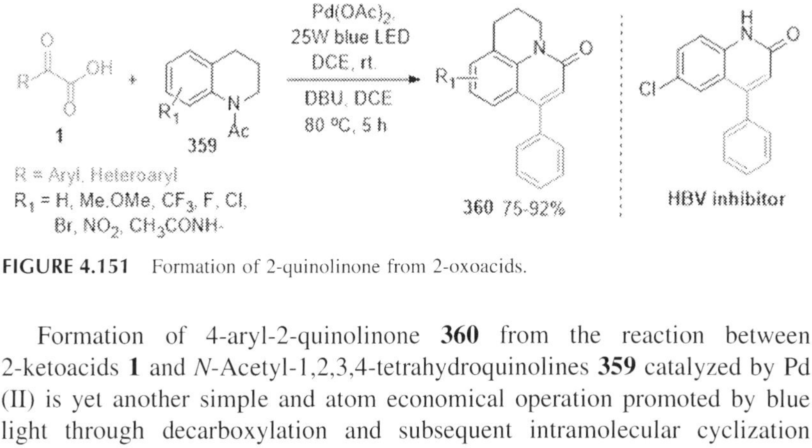

OH
1
R₁
N
Ac
359
Pd(OAc)₂,
25W blue LED
DCE, rt
DBU, DCE
80 °C, 5 h
360 75-92%
Cl
H
N
O
HBV inhibitor
R = Aryl, Heteroaryl
R₁ = H, Me,OMe, CF₃, F, Cl,
Br, NO₂, CH₃CONH-

FIGURE 4.151 Formation of 2-quinolinone from 2-oxoacids.

Formation of 4-aryl-2-quinolinone **360** from the reaction between 2-ketoacids **1** and *N*-Acetyl-1,2,3,4-tetrahydroquinolines **359** catalyzed by Pd (II) is yet another simple and atom economical operation promoted by blue light through decarboxylation and subsequent intramolecular cyclization (Fig. 4.151). When ($R_1 = CF_3$ and NO_2) in tetrahydroquinoline counterpart and (R = thienyl, furyl) in 2-ketoacid, no product formation was observed. The reaction has found application in one-pot synthesis of HBV inhibitor in 85% yield.[145]

A transition metal-catalyzed reaction between phenylglyoxylic acids **1** and benzo[b]thiophenes **361** featuring triple C-H activation and simultaneous C-O/C-C bond formation produces benzothieno benzopyranone derivatives **362** (Fig. 4.152). This is a regioselective oxidative cyclization reaction catalyzed by Ir(III) ligand along with Ag_2O and 1-AdCOOH as an oxidant and additive respectively. The reaction is thought to proceed by a free radical mechanism in which C-H activation steps are reversible in both the reacting partners. Furthermore, phenylglyoxylic acid is supposed to be transformed

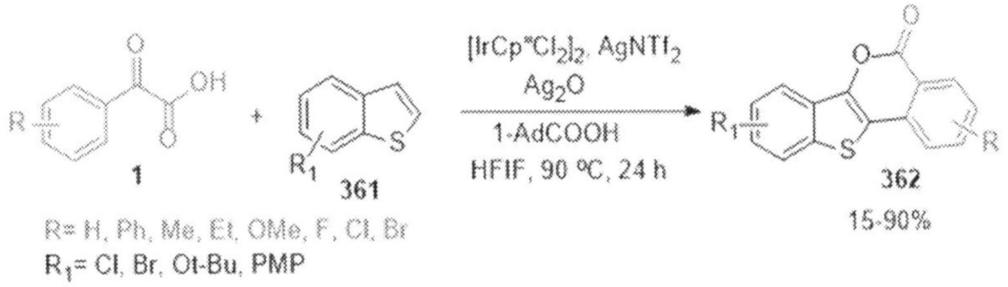

R= H, Ph, Me, Et, OMe, F, Cl, Br
R₁= Cl, Br, Ot-Bu, PMP

FIGURE 4.152   Formation of benzothieno benzopyranone from 2-oxoacids.

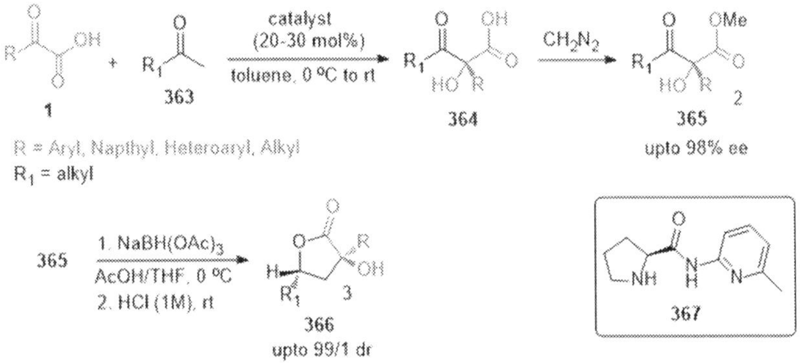

R = Aryl, Napthyl, Heteroaryl, Alkyl
R₁ = alkyl

FIGURE 4.153   Formation of α-hydroxy esters and 2-hydroxy-γ-butyrolactones from 2-oxoacids.

first into benzoic acid followed by decarboxylation promoted by Ag_2O. The reaction works well over many substrates extending to both 2-ketoacid as well as benzthiophene counterparts.[146]

4.6   Reactions at keto group of 2-oxoacids

Apart from the reactions exhibited by the "carboxy" unit, various other reactions selectively occur at the "*keto*" unit of 2-oxoacids. The most common of such reaction is the asymmetric direct aldol reaction of ketones with 2-oxoacids in the presence of various organocatalysts.

In 2007, Xiao-Ying Xu et al. established an efficient asymmetric aldol reaction for the synthesis of β-hydroxyl carboxylic acids with excellent enantioselectivities (up to 98% ee) in presence of catalyst prepared from proline and 6-methyl-2-amino pyridine. A reaction mixture of 2-oxoacids **1** and ketones **363** in toluene was stirred at 0°C to rt in presence of the given catalyst **367** (20−30 mol.%) to access the β-hydroxyl carboxylic acids **364** up to 98% ee followed by the reaction with diazomethane (CH_2N_2) to form β-hydroxyl esters **365** with similar enantioselectivities (Fig. 4.153). The aldol adduct **365** in the presence of $NaBH(OAc)_3$ undergoes a diastereoselective reduction in a

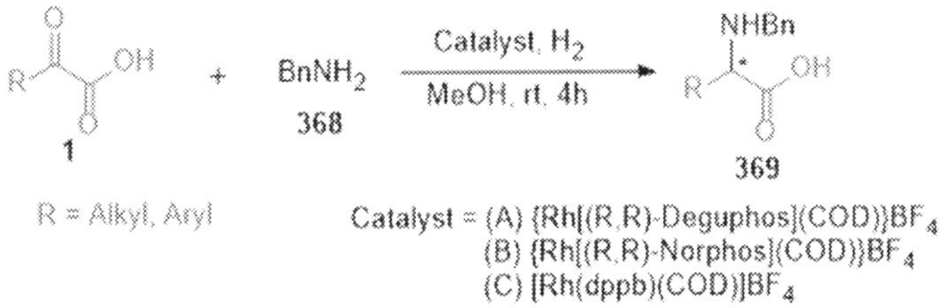

FIGURE 4.154 Formation of chiral benzylamines from 2-oxoacid.

solvent mixture of AcOH and THF at $0°C$ followed by lactonization with HCl (1.0 M) to form 2-hydroxy-γ-butyrolactones **366** in 95/5−99/1 diastereomeric ratios. A wide variety of aryl, heteroaryl, and aliphatic 2-oxoacids were tested against different acyclic ketones under the optimized conditions. Experimental and theoretical studies suggested that two hydrogen-bonding interactions between (1) amide N-H of the catalyst and 'keto' oxygen of 2-oxoacids and (2) N atom of pyridine in organocatalyst and hydroxyl group of 2-oxoacids which are responsible for high reactivity and better enantioselectivity of the reaction.[147]

The highly enantioselective reductive amination of 2-oxoacids **1** in presence of homogeneous catalysts was developed by Renat Kadyrov et al. for the synthesis of α-N-benzyl amino acids **368**. The reaction was performed under homogeneous catalysis between 2-oxoacid and benzylamine **369** in MeOH at room temperature for 4 h. Different phosphorous based ligands were screened to prepare three types of Rh complexes like {Rh[(R, R)-Deguphos](COD)}BF$_4$,{Rh[(R, R) Norphos](COD)}BF$_4$, and [Rh(dppb) (COD)]BF$_4$ which were used against various 2-oxoacid derivatives for better selectivity and yield (Fig. 4.154).[148]

Efficient asymmetric hydrogenation of 2-oxoacids to form α-hydroxycarboxylic acids **370** in the presence of ruthenocenyl phosphino-oxazoline ruthenium complex (RuPHOX-Ru) has been successfully established by Delong Liu and coworkers with up to 97% ee. A reaction was performed with 2-oxoacids, RuPHOX-Ru (0.2 mol.%), AcONa (0.3 equiv), and PPh$_3$ (5 mol.%) in MeOH at $0°C$, followed by 40 bar hydrogen pressure for 24 h (Fig. 4.155). Different substituted aryl, heteroaryl and naphthyl derivatives of 2-oxoacids were effectively transformed into corresponding α-hydroxycarboxylic acids. The gram-scale production of α-hydroxycarboxylic acids was achieved under the same optimized conditions with low catalyst loading.[149]

The α-hydroxy phenyl glyoxylic acid **370** was further transformed into different chiral building blocks that are essential for the synthesis of different biologically active compounds. In one case, **370** is treated with 4-aminophenol **376** in the presence of dicyclohexylcarbodiimide (DCC) to give the corresponding amide derivative **375**. However, etherification of **370** with $(CH_3)_2SO_4$ gives **371** followed by reduction with LiAlH$_4$ to form the

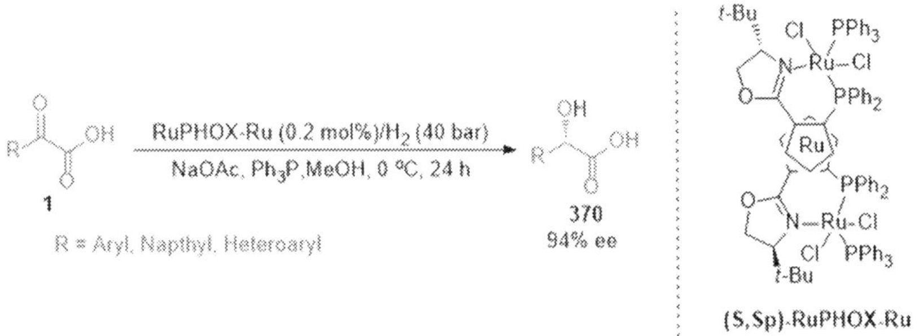

FIGURE 4.155 Formation of chiral α-hydroxy carboxylic acids from 2-oxoacid.

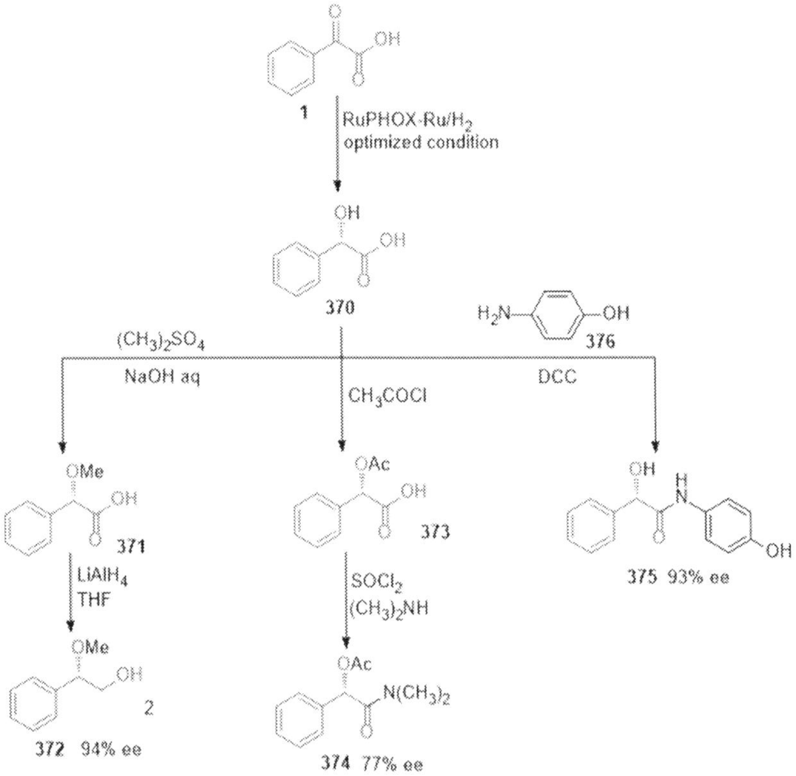

FIGURE 4.156 Formation of chiral carboxylic acids and amides from 2-oxoacids.

alcohol **372**. On the other hand, acetylation of the hydroxyl group by CH₃COCl produces **373**, which after the reaction with SOCl₂ and NH(CH₃)₂ generates the corresponding amide **374** (Fig. 4.156).

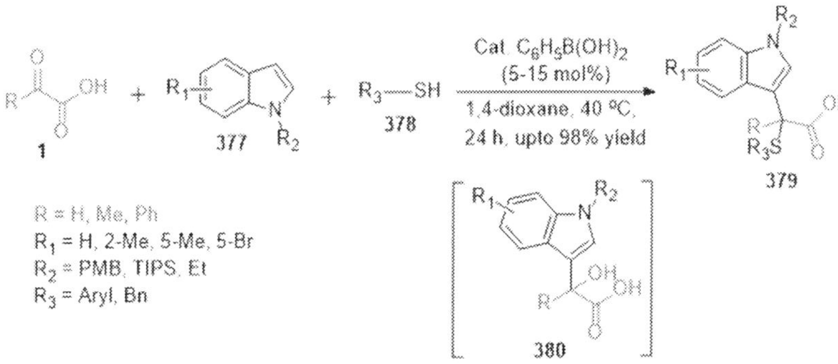

FIGURE 4.157   Formation of indole-3-acetic acid derivatives from 2-oxoacids.

A three-component reaction of glyoxylic acids **1**, indoles **377** and thiols **378** for the synthesis of α-sulfanyl-substituted indole-3-acetic acids **379** was demonstrated by Amrita Das et al. Boronic acid plays an important role in activating the α-hydroxy group of α-hydroxycarboxylic acid intermediate **380** (Fig. 4.157). The reaction was carried out by taking a mixture of glyoxylic acids, indoles, and thiol in 1,4-dioxane followed by the addition of 5−10 mol.% boronic acid as a catalyst at 0°C for 24 h. A library of products was prepared in up to 98% yields by taking different substituted glyoxylic acids, indoles and thiols.[150]

4.7   Synthesis of diketones

The 1,2-dicarbonyl motif is an important structure that is present in natural products and biologically active compounds. Some of the biologically active natural products which possesses 1, 2-dicarbonyl motif are Licoagrodione, isolated from a *Chinese herb* that was found to exhibit antimicrobial activity. Tanshinone IIA is a transcription factor inhibitor and was isolated from Salvia *miltiorrhiza* BUNGE. Mansonone C was isolated from *Mansonia altissima* and displays antifungal activity against P. *parasitica*. Sophoradione was isolated from the roots of *S. flavescens* and is cytotoxic to KB tumor cells (Fig. 4.158).[151] 1,2-dicarbonyl compounds are popular and attractive synthetic scaffolds given their dense number of reactive centers that can be exploited in consecutive reactions or in cascade sequences and could be used in asymmetric organocatalytic transformations.[152] Different starting materials and reagents were used for the construction of different substituted diones.

An efficient and mild silver-catalyzed decarboxylative radical coupling of 2-oxoacids **1** (X = O) and anhydrides **382** was demonstrated by Hua-Xu Zou et al. for the synthesis of 1,2-diketones **381**. The acyl or alkyl radicals are formed in the reaction as a result of the decarboxylation of corresponding 2-oxoacids and anhydrides to form a new C-C bond. The reaction occurs between 2-oxoacids or anhydrides in the presence of $AgNO_3$ (20 mol.%) as a

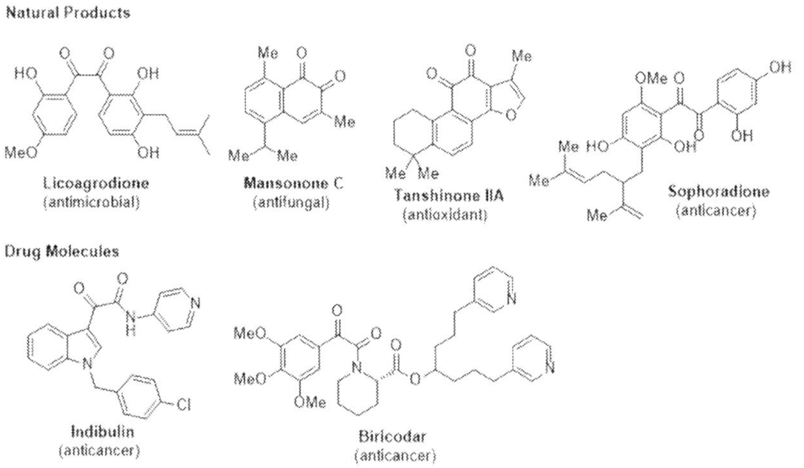
Natural Products
Licoagrodione
(antimicrobial)
Mansonone C
(antifungal)
Tanshinone IIA
(antioxidant)
Sophoradione
(anticancer)
Drug Molecules
Indibulin
(anticancer)
Biricodar
(anticancer)

FIGURE 4.158 Biologically active 1,2-dicarbonyl compounds.

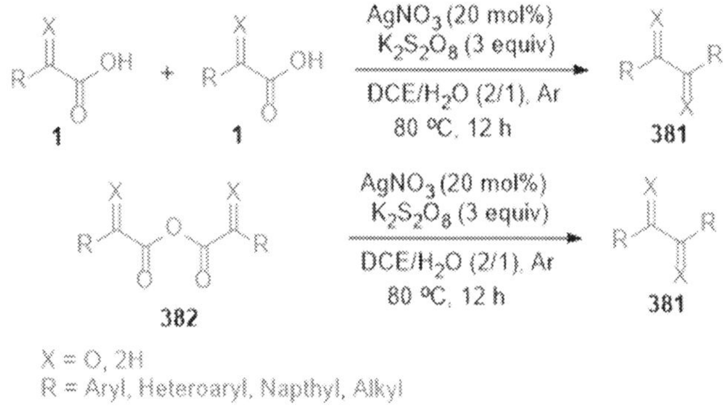
AgNO3 (20 mol%)
K2S2O8 (3 equiv)
DCE/H2O (2/1), Ar
80 °C, 12 h
1
1
381
382
X = O, 2H
R = Aryl, Heteroaryl, Napthyl, Alkyl

FIGURE 4.159 Formation of 1,2-diketones from 2-oxoacids.

catalyst and $K_2S_2O_8$ (3 equiv.) as an oxidizing reagent in a mixture of CH_2ClCH_2Cl (DCE) and H_2O (2:1) as a solvent at 80°C to afforded the corresponding 1,2-diketones **381** (Fig. 4.159).[153]

References

1. Johnathan, C., et al. Organic Chemistry. Oxford University Press. ISBN 978–0–19–850346-0.
2. (a) John, K.; Weitzman, P. D. J. *Kreb's Citric Acid Cycle: Half a Century and Still Turning*, Vol. 57. Biochemical Society: London, 1987.
   (b) Lowenstein, J. M. *Methods in Enzymology, Vol. 13: Citric Acid Cycle;* Academic Press: Boston, 1969.
   (c) Xiao, X.; Liu, M.; Rong, C.; Xue, F.; Li, S.; Xie, Y.; Shi, Y. *Org. Lett.* **2012,** *14*, 5270.

(d) Liu, Y. E.; Lu, Z.; Li, B.; Tian, J.; Liu, F.; Zhao, J.; Hou, C.; Li, Y.; Niu, L.; Zhao, B. *J. Am. Chem. Soc.* **2016**, *138*, 10730.

(e) Lan, X.; Tao, C.; Liu, X.; Zhang, A.; Zhao, B. *Org. Lett.* **2016**, *18*, 3658.

3. (a) Pusterla, I.; Bode, J. W. *Angew Chem. Int. Ed.* **2012**, *51*, 513.

(b) Doll, M. K. H. *J. Org. Chem.* **1999**, *64*, 1372.

(c) Fang, P.; Li, M.; Ge, H. *J. Am. Chem. Soc.* **2010**, *132*, 11898.

(d) He, Z.; Qi, X.; Li, S.; Zhao, Y.; Gao, G.; Lan, Y.; Wu, Y.; Lan, J.; You, J. *Angew Chem. Int. Ed.* **2015**, *54*, 855.

4. (a) Cooper, A. J. L.; Ginos, J. Z.; Meister, A. Synthesis and Properties of the a-Keto Acids. *Chem. Rev.* **1983**, *83*, 321−358.

(b) Carey, J. S.; Laffan, D.; Thomson, C.; Williams, M. T. Analysis of the Reactions Used for the Preparation of Drug Candidate Molecules. *Org. Biomol. Chem.* **2006**, *4*, 2337−2347.

(c) Blanco-Canosa, J. B.; Dawson, P. E. An Efficient Fmoc-SPPS Approach for the Generation of Thioester Peptide Precursors for Use in Native Chemical Ligation. *Angew Chem.* **2008**, *120*, 6957−6961.

5. (a) Berzelius, J. J. Ueber die Destilllationsproducte der Traubensaure. *Ann. Phys.* **1835**, *112*, 1−29.

(b) Berzelius, J. J., et al. On Dry Distilled Racemic Acid. *Acta Chem. Scand.* **1960**, *14*, 1677−1680.

6. Kay, J.; Weitzman, P. D. J. *Biochem. Soc. Symp.*, Vol. 54. The BiochemicalSociety: London, 1987.

7. Voet, D.; Voet, J. G. *Biochemistry*, 3rd ed.; John Wiley & Sons, Inc: New York, 2004.

8. Guo, L.-N.; Wang, H.; Duan, X.-H. *Org. Biomol. Chem.* **2016**, *14*, 7380−7391.

9. Penteado, F.; Lopes, E. F.; Alves, D.; Perin, G.; Jacob, R. G.; Lenardao, E. *J. Chem. Rev.* **2019**, *119*, 7113−7278.

10. (a) Fang, P.; Li, M.; Ge, H. *J. Am. Chem. Soc.* **2010**, *132*, 11898.

(b) Li, M.; Ge, H. *Org. Lett.* **2010**, *12*, 3464.

(c) Li, M.; Wang, C.; Ge, H. *Org. Lett.* **2011**, *13*, 2062.

(d) Wang, H.; Guo, L.-N.; Duan, X.-H. *Org. Lett.* **2012**, *14*, 4358.

(e) Yang, Z.; Chen, X.; Liu, J.; Gui, Q.; Xie, K.; Li, M.; Tan, Z. *Chem. Commun.* **2013**, *49*, 1560.

11. (a) Nicewicz, D. A.; Nguyen, T. M. *ACS Catal.* **2014**, *4*, 355.

(b) Zhang, X. Z. G.; Xiao, W.-J. *Angew Chem. Int. Ed.* **2015**, *54*, 15632.

(c) Hopkinson, M. N.; Sahoo, B.; Li, J.-L.; Glorius, F. *Chem−Eur. J.* **2014**, *20*, 3874.

12. Chaubey, N. R.; Singh, K. N. *Tetrahedron Lett.* **2017**, *58*, 2347−2350.

13. Yao, J.-P.; Wang, G.-W. *Tetrahedron Lett.* **2016**, *57*, 1687−1690.

14. Jing, K.; Yao, J.-P.; Li, Z.-Y.; Li, Q.-L.; Lin, H.-S.; Wang, G.-W. *J. Org. Chem.* **2017**, *82*, 12715−12725.

15. Li, Q.-L.; Li, Z.-Y.; Wang, G.-W. *ACS Omega* **2018**, *3*, 4187−4198.

16. Wang, Q.; Zhang, X.; Fan, X. *Org. Biomol. Chem.* **2018**, *16*, 7737−7747.

17. Lee, P.-Y.; Liang, P.; Yu, W.-Y. *Org. Lett.* **2017**, *19*, 2082−2085.

18. Kim, M.; Park, J.; Sharma, S.; Kim, A.; Park, E.; Kwak, J. H.; Jung, Y. H.; Kim, I. S. *Chem. Commun.* **2013**, *49*, 925−927.

19. Park, J.; Kim, M.; Sharma, S.; Park, E.; Kim, A.; Lee, S. H.; Kwak, J. H.; Jung, Y. H.; Kim, I. S. *Chem. Commun.* **2013**, *49*, 1654−1656.

20. Yang, Z.; Chen, X.; Liu, J.; Gui, Q.; Xie, K.; Li, M.; Tan, Z. *Chem. Commun.* **2013**, *49*, 1560−1562.

21. Li, M.; Ge, H. *Org. Lett.* **2010**, *12*, 3464−3467.
22. Li, H.; Li, P.; Tan, H.; Wang, L. *Chem. Eur. J.* **2013**, *19*, 14432−14436.
23. Miao, J.; Ge, H. *Org. Lett.* **2013**, *15*, 2930−2933.
24. Wang, H.; Guo, L.-N.; Duan, X.-H. *Org. Lett.* **2012**, *14*, 4358−4361.
25. Majhi, B.; Kundu, D.; Ghosh, T.; Ranu, B. C. *Adv. Synth. Catal.* **2016**, *358*, 283−295.
26. Ma, X.; Huang, H.; Yang, J.; Feng, X.; Xie, K. *Synthesis* **2018**, *50*, 2567−2576.
27. Gong, W.-J.; Liu, D.-X.; Li, F.-L.; Gao, J.; Li, H.-X.; Lang, J.-P. *Tetrahedron* **2015**, *71*, 1269−1275.
28. Pan, C.; Jin, H.; L., X.; Chenga, Y.; Zhu, C. *Chem. Commun.* **2013**, *49*, 2933−2935.
29. Ma, Y.-N.; Tian, Q.-P.; Zhang, H.-Y.; Zhoua, A.-X.; Yang, S.-D. *Org. Chem. Front.* **2014**, *1*, 284−288.
30. Yu, L.; Lia, P.; Wang, L. *Chem. Commun.* **2013**, *49*, 2368−2370.
31. Kittikool, T.; Thupyai, A.; Phomphrai, K.; Yotphan, S. *Adv. Synth. Catal.* **2018**, *360*, 3345−3355.
32. Yang, K.; Chen, X.; Wang, Y.; Li, W.; Kadi, A. A.; Fun, H.-K.; Sun, H.; Zhang, Y.; Li, G.; Lu, H. *J. Org. Chem.* **2015**, *80*, 11065−11072.
33. (a) Akai, S.; Peat, A. J.; Buchwald, S. L. *J. Am. Chem. Soc.* **1998**, *120*, 9119−9125.
    (b) Saidi, O.; Marafie, J.; Ledger, A. E. W.; Liu, P. M.; Mahon, M. F.; Kociok-Koohn, G.; Whittlesey, M. K.; Frost, C. G. *J. Am. Chem. Soc.* **2011**, *133*, 19298−19301.
34. Jing, K.; Li, Z.-Y.; Wang, G.-W. *ACS Catal.* **2018**, *8*, 11875−11881.
35. Fontana, F.; Minisci, F.; Barbosa, M. C. N.; Vismara, E. *J. Org. Chem.* **1991**, *56*, 2866−2869.
36. Wang, Q.-Q.; Xu, K.; Jiang, Y.-Y.; Liu, Y.-G.; Sun, B.-G.; Zeng, C.-C. *Org. Lett.* **2017**, *19*, 5517−5520.
37. Suresh, R.; Kumaran, R. S.; Senthilkumar, V.; Muthusubramanian, S. *RSC Adv.* **2014**, *4*, 31685−34688.
38. Li, Y.; Lai, M.; Wu, Z.; Zhao, M.; Zhang, M. *Chem. Sel.* **2018**, *3*, 5588−5592.
39. Bogonda, G.; Kim, H. Y.; Oh, K. *Org. Lett.* **2018**, *20*, 2711−2715.
40. Zeng, X.; Liu, C.; Wang, X.; Zhang, J.; Wang, X.; Hu, Y. *Org. Biomol. Chem.* **2017**, *15*, 8929−8935.
41. Wang, H.; Zhou, S.-L.; Guo, L.-N.; Duan, X.-H. *Tetrahedron* **2015**, *71*, 630−636.
42. Gooßen, L. J.; Rudolphi, F.; Oppel, C.; Rodrguez, N. *Angew Chem. Int. Ed.* **2008**, *47*, 3043−3045.
43. Goossen, L. J.; Linder, C.; Rodriguez, N.; Lange, P. P. *Chem. Eur. J.* **2009**, *15*, 9336−9349.
44. Ji, Y.; Yang, X.; Mao, W. *Appl. Organomet. Chem.* **2014**, *28*, 678−680.
45. Chu, L.; Lipshultz, J. M.; MacMillan, D. W. C. *Angew Chem. Int. Ed.* **2015**, *54*, 7929−7933.
46. Cheng, W.-M.; Shang, R.; Yu, H.-Z.; Fu, Y. *Chem.-Eur. J.* **2015**, *21*, 13191−13195.
47. Li, M.; Wang, C.; Ge, H. *Org. Lett.* **2011**, *13*, 2062−2064.
48. Chang, S.; Wang, J. F.; Dong, L. L.; Wang, D.; Feng, B.; Shi, Y. T. *RSC Adv.* **2017**, *7*, 51928−51934.
49. Panja, S.; Maity, P.; Ranu, B. C. *J. Org. Chem.* **2018**, *83*, 12609−12618.
50. Kakino, R.; Narahashi, H.; Shimizu, I.; Yamamoto, A. *Bull. Chem. Soc. Jpn.* **2002**, *75*, 1333−1345.
51. Cheng, K.; Zhao, B.; Qi, C. *RSC Adv.* **2014**, *4*, 48698−48702.
52. (a) Beletskaya, I. P.; Ananikov, V. P. *Chem. Rev.* **2011**, *111*, 1596−1636.
    (b) Matsumoto, K.; Sugiyama, H. *ACC Chem. Res.* **2002**, *35*, 915−926.

(c) Shen, C.; Zhang, P.; Sun, Q.; Bai, S.; Andy Hor, T. S.; Liu, X. *Chem. Soc. Rev.* **2015,** *44*, 291−314.

(d) Qiao, Z.; Wei, J.; Jiang, X. *Org. Lett.* **2014,** *16*, 1212−1215.

53. (a) Mende, F.; Seitz, O. *Angew Chem. Int. Ed.* **2011,** *50*, 1232−1240.

(b) Keating, T. A.; Walsh, C. T. *Curr. Opin. Chem. Biol.* **1999,** *3*, 598−606.

(c) Khosla, C.; Tang, Y.; Chen, A. Y.; Schnarr, N. A.; Cane, D. E. *Annu. Rev. Biochem.* **2007,** *76*, 195−221.

54. Rong, G.; Mao, J.; Liu, D.; Yan, H.; Zhenga, Y.; Chen, J. *RSC Adv.* **2015,** *5*, 26461−26464.

55. Yan, K.; Yang, D.; Wei, W.; Zhao, J.; Shuai, Y.; Tian, L.; Wang, H. *Org. Biomol. Chem.* **2015,** *13*, 7323−7330.

56. (a) Sahu, N. K.; Balbhadra, S. S.; Choudhary, J.; Kohli, D. V. *Curr. Med. Chem.* **2012,** *19*, 209.

(b) Kumar, D.; Kumar, N. M.; Akamatsu, K.; Kusaka, E.; Harada, H.; Ito, T. *Bioorg. Med. Chem. Lett.* **2010,** *20*, 3916.

(c) Nowakowska, Z. *Eur. J. Med. Chem.* **2007,** *42*, 125.

(d) Tu, H.-Y.; Huang, A.; Hour, T.; Yang, S.; Pu, Y.; Lin, C. *Bioorg. Med. Chem.* **2010,** *18*, 2089.

(e) Damazio, R. G.; Zanatta, A. P.; Cazarolli, L. H.; Chiaradia, L. D.; Mascarello, A.; Nunes, R. J.; Yunes, R. A.; Silva, F. R. M. B. *Eur. J. Med. Chem.* **2010,** *45*, 1332.

57. (a) Nepali, K.; Singh, G.; Turan, A.; Agarwal, A.; Sapra, S.; Kumar, R.; Banerjee, U. C.; Verma, P. K.; Satti, N. K.; Gupta, M. K., et al. *Bio. Med. Chem.* **2011,** *19*, 1950.

(b) Kumar, A.; Rout, S.; Panda, C. S.; Raju, M. B. V.; Ravikumar, B. V. V. *J. Adv. Pharm. Res.* **2011,** *2*, 94.

(c) Song, Q.-B.; Li, X.-N.; Shen, T.-H.; Yang, S.-D.; Qiang, G.-R.; Wu, X.-L.; Ma, Y.-X. *Synth. Commun.* **2003,** *33*, 3935.

(d) Melzer, C.; Barzoukas, M.; Fort, A.; Mery, S.; Nicoud, J.-C. *Appl. Phys. Lett.* **1997,** *71*, 2248.

58. (a) Wei, Y.; Tang, J.; Cong, X.; Zeng, X. *Green Chem.* **2013,** *15*, 3165.

(b) Artok, L.; Kus, M.; Aksin-Artok, O.; Dege, F. N.; Ozkihnc, F. Y. *Tetrahedron* **2009,** *65*, 9125.

(c) Unoh, Y.; Hirano, K.; Satoh, T.; Miura, M. *J. Org. Chem.* **2013,** *78*, 5096.

(d) Schramm, O. G.; Mueller, T. J. *J. Adv. Synth. Catal.* **2006,** *348*, 2565.

(e) Krishnakumar, B.; Swaminathan, M. *J. Mol. Catal. A: Chem.* **2011,** *350*, 16.

(f) Müller, T. J. J.; Ansorge, M.; Aktah, D. *Angew Chem. Int. Ed.* **2000,** *39*, 1253.

(g) Braun, R. U.; Ansorge, M.; Müller, T. J. J. *Chem-Eur. J.* **2006,** *12*, 9081.

59. Wu, S.; Yu, H.; Hu, Q.; Yang, Q.; Xu, S.; Liu, T. *Tetrahedron Lett.* **2017,** *58*, 4763−4765.

60. Zhang, M.; Xi, J.; Ruzi, R.; Li, N.; Wu, Z.; Li, W.; Zhu, C. *J. Org. Chem.* **2017,** *82*, 9305−9311.

61. Zhang, N.; Yang, D.; Wei, W.; Yuan, L.; Nie, F.; Tian, L.; Wang, H. *J. Org. Chem.* **2015,** *80* (6), 3258−3263.

62. Jiang, Q.; Jia, J.; Xu, B.; Zhao, A.; Guo, C.-C. *J. Org. Chem.* **2015,** *80*, 3586−3596.

63. Rodrguez, N.; Manjolinho, F.; Grnberg, M. F.; Gooßen, L. J. *Chem. Eur. J.* **2011,** *17*, 13688−13691.

64. Grünberg, M. F.; Gooßen, L. J. *J. Organomet. Chem.* **2013,** *744*, 140−143.

65. Manjolinho, F.; Grnberg, M. F.; Rodrguez, N.; Gooßen, L. J. *Eur. J. Org. Chem.* **2012,** *25*, 4680−4683.

66. Zhu, Z.; Tang, X.; Li, J.; Li, X.; Wu, W.; Deng, G.; Jiang, H. *Chem. Commun.* **2017,** *53*, 3228−3231.

67. Humphrey, J. M.; Chamberlin, A. R. *Chem. Rev.* **1997,** *97,* 2243−2266.
68. (a) Yoo, W.-J.; Li, C.-J. *J. Am. Chem. Soc.* **2006,** *128,* 13064−13065.
    (b) Dumas, A. M.; Molander, G. A.; Bode, J. W. *Angew Chem. Int. Ed.* **2012,** *51,* 5683−5686.
    (c) Shang, R.; Liu, L. *Sci. China: Chem.*, 54. ; 20111670.
    (d) Wang, H.; Guo, L.-N.; Duan, X.-H. *Org. Lett.* **2012,** *14,* 4358−4361.
69. (a) Baburao, K.; Costello, A. M.; Petterson, R. C.; Sander, G. E. *J. Chem. Soc. C.* **1968,** 2779.
    (b) Klinge, M.; Cheng, H.; Zabriskie, T. M.; Vederas, J. C. *J. Chem. Soc. Chem. Commun.* **1994,** 1379.
    (c) Ke, D.; Zhan, C.; Li, X.; Li, A. D. Q.; Yao, J. *Synlett* **2009,** 1506.
70. Liu, J.; Liu, Q.; Yi, H.; Qin, C.; Bai, R.; Qi, X.; Lan, Y.; Lei, A. *Angew Chem. Int. Ed.* **2014,** *53,* 502−506.
71. Xu, X.-L.; Xu, W.-T.; Wu, J.-W.; He, J.-B.; Xu, H.-J. *Org. Biomol. Chem.* **2016,** *14,* 9970−9973.
72. Xu, N.; Liu, J.; Lia, D.; Wang, L. *Org. Biomol. Chem.* **2016,** *14,* 4749−4757.
73. Xu, W.-T.; Huang, B.; Dai, J.-J.; Xu, J.; Xu, H.-J. *Org. Lett.* **2016,** *18,* 3114−3117.
74. Zhang, S.; Guo, L.-N.; Wang, H.; Duan, X.-H. *Org. Biomol. Chem.* **2013,** *11,* 4308−4311.
75. Chang, L.-M.; Yuan, G.-Q. *Tetrahedron* **2016,** *72,* 7003−7007.
76. Nanjo, T.; Kato, N.; Takemoto, Y.
77. Padala, A. K.; Saikam, V.; Ali, A.; Ahmed, Q. N. *Tetrahedron* **2015,** *71,* 9388−9395.
78. Pimpasri, C.; Sumunnee, L.; Yotphan, S. *Org. Biomol. Chem.* **2017,** *15,* 4320−4327.
79. (a) Ferguson, L. N. *Chem. Rev.* **1946,** *38,* 227.
    (b) March, J. *Advanced Organic Chemistry*, 4th edn; Wiley: New York, 19921183−1184.
80. (a) Nelson, M-A. M.; Baba, S. P.; and Andersonc, E. J.; *Curr. Opin. Pharmacol.* **2017,** *33,* 56−63.
    (b) Sinharoy, P.; McAllister, S. L.; Vasu, M.; and Gross, E. R.; *Adv. Exp. Med .Biol.* **2019,** *1193,* 35−52.
    (c) Ballatore, C.; Huryn, D. M.; Smith, A. B.; *Chem. Med. Chem,* **2013,** *8(3),* 385−395.
    (d) LoPachin, R. M.; Gavin, T.; *Chem. Res. Toxicol.* **2014,** *27,* 1081−1091.
    (e) Hubert, J.; Munzbergova, Z.; Santino, A.; *Pest Manag Sci.* **2008,** *64,* 57−64.
81. Cohen, T.; Song, H. *J. Am. Chem. Soc.* **1965,** *87,* 3780−3781.
82. Niu, G.-H.; Huang, P.-R.; Chuang, G. J. *Asian J. Org. Chem.* **2016,** *5,* 57−61.
83. Ando, W.; Miyazaki, H.; Akasaka, T. *Tetrahedron Lett.* **1982,** *23,* 1197−1200.
84. Zhang, J.-J.; Cheng, Y.-B.; Duan, X.-H. *Chin. J. Chem.* **2017,** *35,* 311−315.
85. Wang, H.; Guo, L.-N.; Duan, X.-H. *Chem. Commun.* **2014,** *50,* 7382−7384.
86. Xiao, T.; Li, L.; Zhou, L. *J. Org. Chem.* **2016,** *81,* 7908−7916.
87. Yang, Z.; Chen, X.; Wang, S.; Liu, J.; Xie, K.; Wang, A.; Tan, Z. *J. Org. Chem.* **2012,** *77,* 7086−7091.
88. Wang, G.-Z.; Shang, R.; Cheng, W.-M.; Fu, Y. *Org. Lett.* **2015,** *17* (19), 4830−4833.
89. (a) Chen, Y. H.; Zhang, Y. H.; Zhang, H.-J.; Liu, D.-Z.; Gu, M.; Li, J.-Y.; Wu, F.; Zhu, X.-Z.; Li, J.; Nan, F.-J. Design, Synthesis, and Biological Evaluation of Isoquinoline-1,3,4-trione. Derivatives as Potent Caspase-3 Inhibitors. *J. Med. Chem.* **2006,** *49,* 1613−1623.
    (b) Xu, G. G.; Etzkorn, F. A. Convergent Synthesis of α-Ketoamide Inhibitors of Pin1. *Org. Lett.* **2010,** *12,* 696−699.
    (c) Hartley, T. F., et al. Efficacy and Tolerance of Fluocortin Butyl Administered Twice Daily in Adult Patients with Perennial Rhinitis. *J. Allergy. Clin. Immunol.* **1985,** *75,* 501−507.

90. (a) Jesuraj, J. L.; Sivaguru, J. *Chem. Commun.* **2010**, *46*, 4791.

    (b) Yang, L.; Wang, D. X.; Huang, Z. T.; Wang, M. X. *J. Am. Chem. Soc.* **2009**, *131*, 10390.

    (c) Tomita, D.; Yamatsugu, K.; Kanai, M.; Shibasaki, M. *J. Am. Chem. Soc.* **2009**, *131*, 6946.

    (d) Yang, L.; Lei, C. H.; Wang, D. X.; Huang, Z. T.; Wang, M. X. *Org. Lett.* **2010**, *12*, 3918.

91. (a) Zhang, C.; Xu, Z. J.; Zhang, L. R.; Jiao, N. *Angew Chem.* **2011**, *123*, 11284.

    (b) Mupparapu, N.; Khan, S.; Battula, S.; Kushwaha, M.; Gupta, A. P.; Ahmed, Q. N.; Vishwakarma, R. A. *Org. Lett.* **2014**, *16*, 1152.

    (c) Zhang, C.; Zong, X. L.; Zhang, L. R.; Jiao, N. *Org. Lett.* **2012**, *14*, 3280.

    (d) Kotha, S. S.; Chandrasekar, S.; Sahu, S.; Sekar, G. *Eur. J. Org. Chem.* **2014**, *2014*, 7451.

    (e) Kotha, S. S.; Sekar, G. *Tetrahedron Lett.* **2015**, *56*, 6323.

    (f) Liu, C. K.; Fang, Z.; Yang, Z.; Li, Q. W.; Guo, S. Y.; Guo, K. *RSC Adv.* **2016**, *6*, 25167.

    (g) Sharma, N.; Kotha, S. S.; Lahiri, N.; Sekar, G. *Synthesis* **2015**, *47*, 726.

92. Li, D.; Wang, M.; Liu, J.; Zhao, Q.; Wang, L. Cu(II)- Catalyzed Decarboxylative Acylation of Acyl C-H of Formamides with α-Oxocarboxylic Acids Leading to α-Ketoamides. *Chem. Commun.* **2013**, *49*, 3640−3642.

93. Laha, J. K.; Patel, K. V.; Tummalapalli, K. S. S.; Hunjan, M. K. Palladium-Catalyzed Serendipitous Synthesis of Arylglyoxylic Amides from Arylglyoxylates and N, NDialkylamides in the Presence of Halopyridines. *ACS Omega* **2018**, *3*, 8787−8793.

94. Wang, H.; Guo, L. N.; Duan, X. H. Copper-catalyzed oxidative condensation of α-oxocarboxylic acids with formamides: synthesis of α-ketoamides. *Org. Biomol. Chem.* **2013**, *11*, 4573.

95. Singh, R. P.; Shreeve, J. M. One-Pot Route to New, -Difluoroamides and -Ketoamides. *J. Org. Chem.* **2003**, *68*, 6063−6065.

96. (a) Zhang, X. B.; Yang, W. C.; Wang, L. *Org. Biomol. Chem.* **2013**, *11*, 3649.

    (b) Liu, F.; Liu, Y.; Chen, Y.; Sun, Z.; Wang, B. Catalyst-Free and One-Pot Procedure for Fast Formation of a-Ketoamides Using a-Oxocarboxylic Acids and Amines at Room Temperature. *Chem. Select* **2017**, *2* 4638−464.

    (c) Li, D. K.; Wang, M.; Liu, J.; Zhao, Q.; Wang, L. *Chem. Commun.* **2013**, *49*, 3640.

97. Guin, S.; Rout, S. K.; Gogoi, A.; Ali, W.; Patel, B. K. A Palladium (II)-Catalyzed Synthesis of a-Ketoamides via Chemoselective Aroyl Addition to Cyanamides. *Adv. Synth. Catal.* **2014**, *11−12*, 2559−2565.

98. Lv, Y.; Bao, P.; Yue, H.; Li, J. S.; Wei, W. Visible-Light-Mediated Metal-Free Decarboxylative Acylations of Isocyanides with α-Oxocarboxylic Acids and Water Leading to α-Ketoamides. *Green Chem.* **2019**, *21*, 6051−6055.

99. Tokuda, O.; Kano, T.; Gao, W. G.; Ikemoto, T.; Maruoka, K. *Org. Lett.* **2005**, *7*, 5103−5105.

100. Bao, J.; Bao, M. CN106187929 (A), 2016.

101. Comins, D. L.; Nolan, J. M. *Org. Lett.* **2001**, *3*, 4255−4257.

102. (a) Ma, G. Z.; Li, P. F.; Liu, L.; Li, W. D. Z.; Chen, L. *Org. Lett.* **2017**, *19*, 2250−2253.

    (b) Kong, J. R.; Krische, M. J. *J. Am. Chem. Soc.* **2006**, *128*, 16040−16041.

103. (a) Li, L.; Zhang, F.; Deng, G.-J.; Gong, H. Palladium-Catalyzed direct αKetoesterification of 8Methylquinoline Derivatives with αKetoacids via Dehydrogenation Coupling Reaction. *Org. Lett.* **2018**, *20* (22), 7321−7325.

    (b) He, Y.; Mao, J.; Rong, G.; Yan, H.; Zhang, G. Iron-Catalyzed Esterification of Benzyl C-H Bonds to Form α-Keto Benzyl Esters. *Adv. Synth. Catal.* **2015**, *357*, 2125−2131.

104. (a) Kumar, A.; Gannedi, V.; Rather, S. A.; Vishwakarma, R. A.; Ahmed, Q. N. Introducing Oxo-Phenylacetyl (OPAc) as a Protecting Group for Carbohydrates. *J. Org. Chem.* **2019**, *84*, 4131−4148.

105. (a) Kirkham, J. D.; Edeson, S. J.; Stokes, S.; Harrity, J. P. *Org. Lett.* **2012**, *14*, 5354−5357.

 (b) Aulakh, V. S.; Ciufolini, M. A. *J. Am. Chem. Soc.* **2011**, *133*, 5900−5903.

 (c) Rooke, D. A.; Ferreira, E. M. *J. Am. Chem. Soc.* **2010**, *132*, 11926−11928.

 (d) Mller, T. J. J. *Top Heterocycl. Chem.* **2010**, *25*, 25−94.

 (e) Bagley, M. C.; Glover, C.; Merritt, E. A. *Synlett* **2007**, 2459−2482.

 (f) Shedvorskaya, R. L. B.; Vereshchagin, L. I. *Russ. Chem. Rev.* **1973**, *42*, 225−240.

106. Wang, P. F.; Feng, Y. S.; Cheng, Z. F.; Wu, Q. M.; Wang, G. Y.; Liu, L. L.; Dai, J. J.; Xu, J.; Xu, H. J. Transition-Metal-Free Synthesis of Ynones via Decarboxylative Alkynylation of αKeto Acids under Mild Conditions. *J. Org. Chem.* **2015**, *80*, 9314−9320.

107. Huang, H.; Zhang, G.; Chen, Y. Dual Hypervalent Iodine(III) Reagents and Photoredox Catalysis Enable Decarboxylative Ynonylation under Mild Conditions. *Angew Chem. Int. (Ed.)* **2015**, *54*, 7872−7876.

108. Tan, H.; Li, H.; Ji, W.; Wang, L. Sunlight-Driven Decarboxylative Alkynylation of a-Keto Acids with Bromoacetylenes by Hypervalent Iodine Reagent Catalysis: A Facile Approach to Ynones. *Angew Chem. Int. (Ed.)* **2015**, *54*, 8374−8377.

109. Zeng, X.; Liu, C.; Yang, W.; Wang, X.; Wang, X.; Hu, Y. A General Two-Step One-Pot Synthesis Process of Ynones from α-Keto Acids and 1-Iodoalkynes. *Chem. Commun.* **2018**, *54*, 9517−9520.

110. Meng, M.; Wang, G.; Yang, L.; Cheng, K.; Qi, C. SilverCatalyzed Double Decarboxylative Radical Alkynylation/Annulation of Arylpropiolic Acids with α-Keot Acids: Access to Ynones and Flavones Under Mild Conditions. *Adv. Synth. Catal.* **2018**, *360*, 1218−1231.

111. Boersch, C.; Merkul, E.; Muller, T. J. J. Catalytic Syntheses of N-Heterocyclic Ynones and Ynediones by In Situ Activation of Carboxylic Acids with Oxalyl Chloride. *Angew Chem. Int. (Ed.)* **2011**, *50*, 10448−10452.

112. Laha, J. K.; Patel, K. V.; Tummalapalli, K. S. S.; Dayal, N. Formation of Amides, Their Intramolecular Reaction for the Synthesis of N-Heterocycles, and Preparation of a Marketed Drug, Sildenafil: A Comprehensive Coverage. *Chem. Commun.* **2016**, *52*, 10245−10248.

113. Zhang, L.-J.; Xu, M.-C.; Liu, J.; Zhang, X.-M. Tandem Cycloaddition-Decarboxylation of α-Keto Acids and Isocyanides Under Oxidant-Free Conditions Towards Monosubstituted Oxazoles. *RSC Adv.* **2016**, *6*, 73450−73453.

114. Penteado, F.; Vieira, M. M.; Perin, G.; Alves, D.; Jacob, R. G.; Santi, C.; Lenardao, E. J. Niobium-Promoted Reaction of α-Phenylglyoxylic Acid with ortho-Functionalized Anilines: Synthesis of 2-Arylbenzothiazoles and 3-Aryl-2H benzo[b][1,4]benzoxazine-2-ones. *Green Chem.* **2016**, *18*, 6675−6680.

115. Lima, D. B.; Penteado, F.; Vieira, M. M.; Alves, D.; Perin, G.; Santi, C.; Lenardão, E. J. α-Keto Acids as Acylating Agents in the Synthesis of 2-Substituted Benzothiazoles and Benzoselenazoles. *Eur. J. Org. Chem.* **2017**, *2017*, 3830−3836.

116. Nguyen, O. T. K.; Phan, A. L. T.; Phan, P. T.; Nguyen, V. D.; Truong, T.; Le, N. T. H.; Le, D. T.; Phan, N. T. S. Ready Access to 3-Substituted Quinoxalin-2-ones under Superparamagnetic Nanoparticle Catalysis. *ChemistrySelect* **2018**, *3*, 879−886.

117. Wang, H.; Yang, H.; Li, Y.; Duan, X.-H. Oxone-Mediated Oxidative Carbon-Heteroatom Bond Cleavage: Synthesis of Benzoxazinones from Benzoxazoles with α-Oxocarboxylic Acids. *RSC Adv.* **2014,** *4,* 8720–8722.

118. Yan, S.; Ye, L.; Liu, M.; Chen, J.; Ding, J.; Gao, W.; Huang, X.; Wu, H. Unexpected TFA-Catalyzed Tandem Reaction of Benzo[d]- oxazoles with 2-Oxo-2-arylacetic Acids: Synthesis of 3-Aryl-2Hbenzo[b][1.4]oxazin-2-ones and Cephalandole A. *RSC Adv.* **2014,** *4,* 16705–16709.

119. Kumar, A.; Singh, V.; Ghosh, S. *Butenolide: A Novel Synthesis and Biological Activities,* 1st ed.; LAP LAMBERT Academic Publishing: Saarbrucken, Germany, 2012.

120. Seitz, M.; Reiser, O. Synthetic Approaches Towards Structurally Diverse γ-Butyrolactone Natural-Product-Like Compounds. *Curr. Opin. Chem. Biol.* **2005,** *9,* 285–292.

121. Ugurchieva, T. M.; Veselovsky, V. V. Advances in the Synthesis of Natural Butano- and Butenolides. *Russ. Chem. Rev.* **2009,** *78,* 337–373.

122. Mao, W.; Zhu, C. Synergistic Acid-Promoted Synthesis of Highly Substituted Butenolides via the Annulation of Keto Acids and Tertiary Alcohols. *Org. Lett.* **2015,** *17,* 5710–5713.

123. Mao, W.; Zhu, C. Efficient Synthesis of Multiply Substituted Butenolides from Keto Acids and Terminal Alkynes Promoted by Combined Acids. *Org. Chem. Front.* **2017,** *4,* 1029–1033.

124. Mao, W.; Zhu, C. Synthesis of Highly Substituted γ-Hydroxybutenolides Through the Annulation of Keto Acids with Alkynes and Subsequent Hydroxyl Transportation. *Chem. Commun.* **2016,** *52,* 5269–5276.

125. Seo, S.; Willis, M. C. A copper(I)-Catalyzed Addition/Annulation Sequence for the Two-Component Synthesis of γ-Ylidenebutenolides. *Org. Lett.* **2017,** *19,* 4556–4559.

126. Shi, X.; He, Y.; Zhang, X.; Fan, X. FeCl3-Catalyzed Cascade Reactions of Cyclic Amines with 2-Oxo-2-arylacetic Acids Toward Furan-2(5H)-one Fused N,O-Bicyclic Compounds. *Adv. Synth. Catal.* **2018,** *360,* 261–266.

127. He, Z.; Fang, F.; Lv, J.; Zhang, J. One-Pot Gram-Scale Synthesis of γ-Hydroxybutenolides Through Catalyst-Free.

128. Yuan, X.; Liu, Y.; Qin, M.; Yang, X.; Chen, B. Elemental Sulfur Participates in the Decarboxylative Coupling of Oxidized 2-Aminophenol and Phenylglyoxylic Acid. *ChemistrySelect* **2018,** *3,* 5541–5543.

129. Zha, D.; Li, H.; Wang, L. Silver-Promoted Cascade Reaction of 4-Hydroxycoumarins With α-Keto Acids Under Microwave Irradiation: One-Step Construction of Quaternary Stereocenters. *Eur. J. Org. Chem.* **2016,** *2016,* 4907–4915.

130. Wang, L.; Wang, Y.; Chen, Q.; He, M. Photocatalyzed Facile Synthesis of 2,5-Diaryl 1,3,4-Oxadiazoles with Polyaniline-g-C3N4-TiO2 Composite Under Visible Light. *Tetrahedron Lett.* **2018,** *59,* 1489–1492.

131. Yang, D.-S.; Wang, J.; Gao, P.; Bai, Z.-J.; Duan, D.-Z.; Fan, M.-J. KI-Catalyzed Oxidative Cyclization of α-Keto Acids and 2- Hydrazinopyridines: Efficient One-Pot Synthesis of 1,2,4-Triazolo- [4,3-a]pyridines. *RSC Adv.* **2018,** *8,* 32597–32600.

132. Yan, K.; Yang, D.; Wei, W.; Wang, F.; Shuai, Y.; Li, Q.; Wang, H. Silver-Mediated Radical Cyclization of Alkynoates and α-Keto Acids Leading to Coumarins via Cascade Double C-C Bond Formation. *J. Org. Chem.* **2015,** *80,* 1550–1556.

133. Mai, W. P.; Sun, G. C.; Wang, J. T.; Song, G.; Mao, P.; Yang, L. R.; Yuan, J. W.; Xiao, Y. M.; Qu, L. B. Silver-Catalyzed Radical Tandem Cyclization: An Approach to Direct Synthesis of 3-Acyl-4-arylquinolin- 2(1H)-ones. *J. Org. Chem.* **2014,** *79,* 8094–8102.

134. (a) Wang, H.; Guo, L. N.; Duan, X. H. Silver-Catalyzed Decarboxylative Acylarylation of Acrylamides with α Oxocarboxylic Acids in Aqueous Media. *Adv. Synth. Catal.* **2013,** *355,* 2222–2226.

(b) Wang, H.; Guo, L. N.; Duan, X. H. Silver-Catalyzed Oxidative Coupling/Cyclization of Acrylamides with 1,3-Dicarbonyl Compounds. *Chem. Chem. Commun.* **2013,** *49,* 10370–10372.

135. Wang, S.; Fu, H.; Shen, Y.; Sun, M.; Li, Y. M. Oxidative Radical Addition/Cyclization Cascade for the Construction of Carbonyl-Containing Quinoline-2,4(1H,3H)Diones. *J. Org. Chem.* **2016,** *81,* 2920–2929.

136. (a) Sinha, S.; Srivastava, R.; De Clercq, E.; Singh, R. K. *Nucleos. Nucleot.* **2004,** *12,* 1815.

    (b) Houston, D. R.; Synstad, B.; Eijsink, V. G. H.; Stark, M. J. R.; Eggleston, I. M.; Van Aalten, D. M. F. *J. Med. Chem.* **2004,** *47,* 5713.

    (c) Kwon, O. S.; Park, S. H.; Yun, B. S.; Pyun, Y. R.; Kim, C. J. *J. Antibiot.* **2000,** *53,* 954.

137. Zonneveld, Z. J.; Janssen, E.; Kanter, F. J. J. D.; Helliwell, M.; Turner, N. J.; Ruijter, E.; Orru, R. V. A. Asymmetric Synthesis of Synthetic Alkaloids by a Tandem Biocatalysis/Ugi/Pictet–Spengler-Type Cyclization Sequence. *Chem. Commun* **2010,** *46,* 7706–7708.

138. Sañudo, M.; Marcaccini, S.; Basurto, S.; Torroba, T. Synthesis of 3-Hydroxy-6-oxo[1,2,4] triazin-1-yl Alaninamides, a New Class of Cyclic Dipeptidyl Ureas. *J. Org. Chem.* **2006,** *71,* 4578–4584.

139. Xu, Z.; Moliner, F. D.; Cappelli, A. P.; Hulme, C. Aldol Reactions in Multicomponent Reaction Based Domino Pathways: A Multipurpose Enabling Tool in Heterocyclic Chemistry. *Org. Lett.* **2013,** *15* (11).

140. Marcaccini, S.; Neo, A. G.; Marcos, C. F. Sequential Five-Component Synthesis of Spiropyrrolidinochromanones. *J. Org. Chem.* **2009,** *74,* 6888–6890.

141. Lei, J.; Ting Song, G. L.; He, J.; Luo, Y. F.; Tang, D. Y.; Lin, H. K.; Frett, B.; Li, H.; Chen, Z. Z.; Xu, Z. G. One-Pot Construction of Functionalized Aziridines and Maleimides via a Novel Pseudo-Knoevenagel Cascade Reaction. *Chem. Commun.* **2020,** *56,* 2194–2197.

142. Feng, H.; Ermolatev, D. S.; Song, G.; Eycken, E. V. V. Microwave-Assisted Decarboxylative Three-Component Coupling of a 2-Oxoacetic Acid, an Amine, and an Alkyne. *J. Org. Chem.* **2011,** *76* 7608–761.

143. Choi, J.; Lim, J.; Irudayanathan, F. M.; Kim, H. S.; Park, J.; Yu, S. B.; Jang, Y.; Raja, G. C. E.; Nam, K. C.; Kim, J., et al. Copper-Catalyzed Double Decarboxylative Coupling Reactions of Alkynyl Carboxylic Acid and Glyoxylic Acid: Synthesis of Propargyl Amines and Imidazopyridines. *Asian J. Org. Chem.* **2016,** *5,* 770–777.

144. Rudolphi, F.; Song, B.; Gooben, L. J. Synthesis of Azomethines from a-Oxocarboxylates, Amines and Aryl Bromides via One-Pot Three-Component Decarboxylative Coupling. *Adv. Synth. Catal.* **2011,** *353,* 337–342.

145. Wang, C.; Qiao, J.; Liu, X.; Song, H.; Sun, Z.; Chu, W. Visible-Light-Induced Decarboxylation Coupling/Intramolecular Cyclization: A One-Pot Synthesis for 4Aryl-2-quinolinone Derivatives. *J. Org. Chem.* **2018,** *83,* 1422–1430.

146. Wang, Z.; Yang, M.; Yang, Y. Ir(III)-Catalyzed Oxidative Annulation of Phenylglyoxylic Acids with Benzo[b]thiophenes. *Org. Lett.* **2018,** *20,* 3001–3005.

147. Xu, X.-Y.; Tang, Z.; Wang, Y.-Z.; Luo, S.-W.; Cun, L.-F.; Gong, L.-Z. *J. Org. Chem.* **2007,** *72,* 9905–9913.

148. Kadyrov, R.; Riermeier, T. H.; Dingerdissen, U.; Tararov, V.; Borner, A. *J. Org. Chem.* **2003,** *68,* 4067–4070.

149. Guo, H.; Li, J.; Liu, D.; Zhang, W. *Adv. Synth. Catal.* **2017,** *359,* 3665–3673.

150. Das, A.; Watanabe, K.; Morimoto, H.; Ohshima, T. *Org. Lett.* **2017,** *19,* 5794−5797.

151. (a) Li, W., Asada, Y. & Yoshikawa, T. Antimicrobial Flavonoids from Glycyrrhiza Glabra Hairy Root Cultures. Planta Med. 64, 746−747.

    (b) ang, S.-I., et al. Tanshinone IIA from *Salvia miltiorrhiza* Inhibits Induciblenitric Oxide Synthase Expression and Production of TNF-α, IL-1β and IL-6 Inactivated RAW 264.7 Cells. *Planta Med.* **2003,** *69,* 1057−1059.

    (c) Bettòlo, G. B. M.; Casinovi, C. G.; Galeffi, C. A New Class of Quinones:sesquiterpenoid Quinones of Mansonia Altissima. *Tetrahedron Lett.* **1965,** *52,* 4857−4864.

    (d) Armistead, M.D., Harding, W.M., Saunders, J.O. & Boger, S.J. Biologically Active Acylated Amino Acid Derivatives. United States Patent 5723459, 1998.

    (e) Bacon, B. R., et al. Boceprevir for Previously Treated Chronic HCV Genotype 1 Infection. *N. Engl. J. Med.* **2011,** *364,* 1207−1217.

152. (a) Ashton, T. D.; Jolliffe, K. A.; Pfeffer, F. M. Luminescent Probes for the Bioimaging of Small Anionic Species in vitro and in vivo. *Chem. Soc. Rev.* **2015,** *44,* 4547−4595.

    (b) Wang, L., et al. Palladium-Catalyzed Oxidative Cycloaddition Through C-H/N-H Activation: Access to Benzazepines. *Angew Chem. Int. Ed.* **2013,** *52,* 1768−1772.

    (c) Sessler, J. L.; Cho, D.-G.; Lynch, V. Diindolylquinoxalines: Effective Indolebased Receptors for Phosphate Anion. *J. Am. Chem. Soc.* **2006,** *128,* 16518−16519.

153. Zou, H.-X.; Li, Y.; Yang, Y.; Li, J.-H.; Xian. *J. Adv. Synth. Catal.* **2018,** *360,* 1439−1443.

Chapter 5

Summary

The chemistry of 2-oxoaldehydes and 2-oxoacids has advanced over the years mainly because of the simplicity of accessing their derivatives. Enormous strategies have been developed for the synthesis of various simple to complicated structures by employing 2-oxoaldehydes and 2-oxoacids chemistry which are separately described in this book. The development of new methods in this area has also allowed us to understand the mechanistic pathway followed during the reaction, which is helpful in designing new biologically important molecules. Moreover, different patterns of reactivity, as well as the double-sited reactivity, can be simultaneously exploited for the efficient preparation of many target molecules in single or several steps. This book aimed to cover various applications of 2-oxoaldehydes and 2-oxoacid, particularly their aryl counterparts in the synthesis of hetero- and non heterocyclic molecules and other building blocks including their use in medicine and biology.

Primarily, **Chapter 1**, "Synthesis and Physical Properties of 2-Oxoaldehydes and 2-Oxoacids" and **Chapter 2**, "Structure and Spectral Characteristics of 2-Oxoaldehydes and 2-Oxoacids" are the brief introductory part of the content highlights the basic processes which are usually carried out for the synthesis of the 2-oxoaldehydes and 2-oxoacids along with their basic physical properties related to structure, reactivity, stability, and medicinal or biological significance. Since, the reactions, basically, stem from the divergent reactivity of these molecules, this part, therefore, further extends on to clarifying the basic idea behind the comparative difference in the reactivity of (1) simple aldehydes and 2-oxoaldehydes and (2) simple acids and 2-oxoacids. However, more theoretical studies are required in this area to understand the basic nature of their structure and reactivity, which can be very useful in planning novel reactions in the future.

Furthermore, species like imines, iminium ions, oximes, and acetals, which are derived from glyoxal substrates also show inherent differences in reactivity and selectivity. Hence, these species also hold a very unique place in the construction of heterocycles. In addition, spectral characterization (^{1}H and ^{13}C-NMR) of the titled molecules is a necessary part of the content, which is useful in understanding their basic spectral properties. These aspects may help the reader to understand the fundamental applications and create

Chemistry of 2-Oxoaldehydes and 2-Oxoacids. DOI: https://doi.org/10.1016/B978-0-12-824285-8.00003-0

synergy among basic nature, reactivity, and characterization of the given compounds.

The reactions of 2-oxoaldehydes still continue to serve chemists on various platforms; the same value is explored in this book in **Chapter 3**, "Applications of 2-Oxoaldehydes" by showing their applications mainly in heterocyclic synthesis. The division in different reactions of 2-oxoaldehydes is described on three points: First, those reactions that take place only at the aldehyde unit are covered, which include synthesis of molecules containing O, N, S, or more than one heteroatom in the ring. Similarly, the formation of three-, four-, five-, six-, and seven-membered or even fused ring structures are also presented. Second, those reactions where both aldehydes, as well as 2-oxo units, are involved simultaneously are covered, which also constitutes the synthesis of a variety of heterocycles and their functionalized derivatives. Both styles of these reactions meet with necessary mechanistic details. Similarly, the third possibility where a 2-oxo unit alone may take part in the reaction exists. The applications further extend to domino reactions through in situ generations of glyoxal intermediates and multicomponent synthesis in Ugi, Ugi-Wittig, Pictet-Spengler cyclo condensation, cycloaddition, and other condensations such as aldol reaction. All of these reactions arising from arylglyoxals or its analogs can be of great significance in the synthesis of pharmaceuticals, naturally occurring molecules, and advanced materials.

Chapter 4, "Applications of 2-Oxoacids", have also expanded significantly in organic synthesis extending from decarboxylative reactions to multicomponent synthesis and acylating agents. This sequential development has lead to the formation of new molecules and reagents through new methodologies. 2-Oxoacids give rise to carbonylated scaffolds in decarboxylative cyclization reactions and work as an efficient reagent in metal-free coupling reactions. Multicomponent synthesis through Ugi reaction leads to many complicated ring structures with the generation of new chiral centers. Moreover, 2-oxoacid is regarded as an ecofriendly alternative source for acyl transfer reagents over aldehydes, acid chlorides, anhydrides, and so on, and has been validated on substrates like alkenes, alkynes, aryl, and heteroaryl in the presence of other directing groups. This fact is also supported by the development of new catalytic systems and the use of ultrasound and light as the energy sources in 2-oxoacid chemistry. These developments need further amendments in making the catalytic systems more cost-effective and economical instead of employing very expensive metals like Pd, Ru, Au, Ag, and so on under harsh conditions. Based on the divergent reactivity of 2-oxoacids, its reactions also emerge from two functional units present in its structure as mentioned previously. Wide ranges of reactions which are mostly decarboxylative in nature occur at the acid unit and mainly afford acylated products. The reaction at the acid unit delivers important molecules like ketones, thioesters, esters, ynones, and so on, and their derivatives. Annulation reactions results when both the oxo unit, as well as the acid unit,

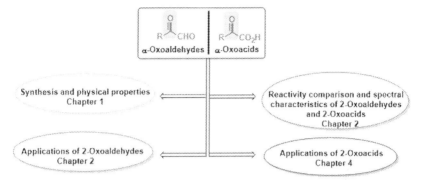

FIGURE 5.1   Flowchart of different chapters in this book.

take part in the reaction. While a few reactions are known to form the β-hydroxyl acids or β-hydroxyl esters and benzylamines at the oxo center without the involvement of the acid part.

Despite recent advancements in 2-oxoacid chemistry, there are some challenges that can be better addressed. More research based on the greener perspective of 2-oxoacid as a reagent system and reduction in its stoichio- metric use will have positive results on the environment. Moreover, looking forward to optimizing milder conditions for the reaction instead of vigorous conditions will reduce the formation of side products and may be beneficial in enhancing the yield of the reaction. In addition, more efforts are required in order to fortify the application of 2-oxoaldehydes and 2-oxoacids in mate- rial chemistry, total synthesis, and natural product chemistry.

We anticipate that this book provides an easy means to decipher funda- mental facts and develop a clear understanding of the chemistry of 2-oxoaldehydes and 2-oxoacids, which can be useful in further expanding and reinforcing the scope of the subject in organic chemistry. This may result in more scope in the area of heterocyclic synthesis of those molecules in the future, which today are either inaccessible or still in imagination. This book is classified into different chapters as depicted in the flowchart in **Fig. 5.1**.

Index

Note: Page numbers followed by "*f*" refer to figures.

Printed in the United States
by Baker & Taylor Publisher Services